T0136990

Urban Computing

Series editors

The Urban Computing book series publishes high-quality research devoted to the study and application of computing technology in urban areas. The main scope is on current scientific developments and innovative techniques in urban computing, bringing to light methods from computer science, social sciences, statistics, urban planning, health care, civil engineering, anthropology, geography, and other fields that directly address urban problems using computer-based strategies. The series offers publications that present the state-of-the-art regarding the problems in question.

Suggested topics for prospective titles for the Urban Computing series include:

Architectures and Protocols for Urban Environments
Case studies in Urban Environments
City Life Improvement through Mobile Services and Big Data
Crowdsourced data acquisition in Urban Environments
Cyber-physical systems
e-Health and m-Health
e-Government
Green Computing in Urban Environments
Human Mobility
Intelligent Transportation Systems
Internet of Things in Urban Areas
Location-based Services in Urban Environments
Metering Infrastructures in Urban Environments
Mobile Cloud Computing
Mobile Sensing
Navigation in the Urban Environment
Recommendation Systems in Urban Spaces
Reliability and Security/Trust in Urban Computing
Semantic Sensing for Urban Information Gathering
Smart Cities
Smart Environment
Smart Grid and Urban Energy Distribution
Social Computing
Standards for Urban Computing
Urban Computing
Urban Economy Based on Big Data
Urban Heterogeneous Data Management
Urban Human-Computer Interaction
Urban Planning using Big Data
User-data interaction in Urban Environments
Using data in heterogeneous environments and Vehicular Sensor Networks

More information about this series at http://www.springer.com/series/15552

Vitor Nazário Coelho • Igor Machado Coelho
Thays A. Oliveira • Luiz Satoru Ochi
Editors

Smart and Digital Cities

From Computational Intelligence to Applied Social Sciences

Editors
Vitor Nazário Coelho
Institute of Computer Science
Universidade Federal Fluminense
Niterói, RJ, Brazil

Igor Machado Coelho
Instituto de Matemática e Estatística
Universidade do Estado do Rio de Janeiro
Rio de Janeiro, RJ, Brazil

Thays A. Oliveira
Department of Engineering and Information and Communication Technologies
Universitat Pompeu Fabra
Barcelona, Spain

Luiz Satoru Ochi
Institute of Computer Science
Universidade Federal Fluminense
Niterói, RJ, Brazil

ISSN 2510-2575 ISSN 2510-2583 (electronic)
Urban Computing
ISBN 978-3-030-12257-7 ISBN 978-3-030-12255-3 (eBook)
https://doi.org/10.1007/978-3-030-12255-3

This Springer imprint is published by the registered company Springer Nature Switzerland AG.
The registered company address is: Gewerbestrasse 11, 6330 Cham, Switzerland

Contents

About the Editors

Vitor Nazário Coelho is a researcher engaged in scientific and social causes. He is a technician in electronic instrumentation, control, and automation engineering and has received his Ph.D. in Electrical Engineering from the Federal University of Minas Gerais. Doctor in Electrical Engineering in the field of Computational Intelligence, by the Pós-Graduação em Engenharia Elétrica (PPGEE) the Universidade Federal de Minas Gerais (UFMG), finishing his Ph.D. in 2 years, with 25 years old, without attending master studies. During his Ph.D., Vitor had an opportunity to study on exchange at the University of Sheffield (England), in partnership with the Rolls Royce Technology Center in Control and Systems Engineering, and at ORT Braude College of Engineering (Israel), as a Marie Curie researcher for the European Commission as part of the Seventh Framework Program (FP7). His thesis was awarded as the best one in the field of engineering at UFMG in the year 2016. Between 2016 and 2018, he was a postdoctoral researcher in partnership with the Universidade Federal Fluminense, supported by the outstanding PDR-10/FAPERJ program. Currently, he acts as an autonomous developer, researcher, investor, and partner of different projects.

Igor Machado Coelho received his B.S. in Computing from the Federal University of Ouro Preto, Brazil, and his M.S. and Ph.D. in Algorithms and Optimization in 2015 at the Computing Institute of Fluminense Federal University, Niterói, Brazil. He is currently Adjunct Professor in Computer Science Department at State University of Rio de Janeiro. He is one of the creators and maintainers of the OptFrame project, a framework for combinatorial optimization, and co-founder of distinct open-source projects and initiatives. His research interests include the resolution of challenging combinatorial optimization problems in fields such as vehicle routing, open-pit mining operational planning, smart grid, and the development of novel algorithms for emerging computing architectures.

Thays A. Oliveira received her B.S. in Business from the Federal University of Ouro Preto, Brazil, and her M.S. in Strategic Management, Marketing, and Innovation in 2016 at the Federal University of Lavras, Brazil. Currently, she is a Ph.D. candidate at the Department of Engineering and Information and Communication Technologies at Universitat Pompeu Fabra, Barcelona, Spain. Her research interests include citizens; works with population integration in urban centers, applying personal and web questionnaires; and studies including consumer and citizens behavior. In the technological side, her Ph.D. research is focused on the relationship between citizens and digital cities, aiming to understand citizens' necessities in the cities.

Luiz Satoru Ochi is full professor at the Universidade Federal Fluminense, Institute of Computer Science in Niteroi, Rio de Janeiro, Brazil. He obtained his B.Sc. in Mathematics and his Ph.D. in Computing and Systems Engineering at COPPE in the Federal University of Rio de Janeiro (UFRJ), Brazil. He was a visiting professor at the University of Colorado at Boulder, USA. He was the General Coordinator of the CAPES-COFECUB Project (Brazil and France) in Graphs and Algorithms, including UFF, UFRJ, and the University of Grenoble, from 1997 to 2000. He was coordinator of the graduate course (master's and Ph.D.) in Computer Science at UFF. His is a Researcher 1C level with a grant from the National Council for Scientific and Technological Development (Conselho Nacional de Desenvolvimento Científico e Tenológico-CNPq), Brazil. His research interests are mixed linear integer programming, operations research, metaheuristics, computational intelligence, and vehicle routing problem. He is the author of more than 290 research articles in chapter-books and specialized journals.

Chapter 1
Introduction

Vitor N. Coelho, Igor M. Coelho, Thays A. Oliveira, and Luiz S. Ochi

Abstract Cities are the core of modern society. Humans civilizations have been transformed for plenty of years and are still under development. From a multi-set of cultures, knowledge has been passed throughout generations and now it is carved in a sea of big data mined by high performance computers. Guided by these technological transformations, this book presents novel insights about the use of computational intelligence tools aligned with a social science perspective. In the faith that data and information are the path for an advanced society, we believe that digital and smart cities summarizes a need of our generation.

This book was designed to be a collection of selected contributions that were managed by the editors and some collaborators during the year of 2017 and 2018:

- Computational Intelligence (CI) for Smart Cities (SC) special session at the XLIX SBPO, Blumenau/SC, Brazil, 2017
- CI for SC special session at the IEEE Symposium Series on Computational Intelligence, Honolulu/Hawaii, USA, 2017
- 1st workshop on SC and CI at the XIII Congresso Brasileiro de Inteligłncia Computacional, Niteri/RJ, Brazil, 2017

V. N. Coelho (✉) · L. S. Ochi
Institute of Computer Science, Universidade Federal Fluminense, Niterói, RJ, Brazil
e-mail: satoru@ic.uff.br; http://vncoelho.github.io/

I. M. Coelho
Instituto de Matemática e Estatística, Universidade do Estado do Rio de Janeiro, Rio de Janeiro, RJ, Brazil
e-mail: igor.machado@ime.uerj.br

T. A. Oliveira
Department of Engineering and Information and Communication Technologies, Universitat Pompeu Fabra, Barcelona, Spain
e-mail: thaysaparecida.deoliveira01@estudiant.upf.edu

V. N. Coelho et al. (eds.), *Smart and Digital Cities*, Urban Computing,
https://doi.org/10.1007/978-3-030-12255-3_1

- 2nd workshop on SC and CI at the International Conference of Neural Network, co-joined with World Conference on Computational Intelligence/IEEE, Rio de Janeiro/RJ, Brazil, 2018

Most of the chapters included here were presented in these aforementioned proceedings, written in Portuguese. While those authors made an effort on improving their ideas and studies while translating their paper to the English version, some other precise contributions were included as additional chapters.

The main focus was to connect CI and the hot topic that is currently surrounding the scope of the SC. On the other hand, we try to give focus on the idea that digital and technological scenarios are becoming reality and much more visible/present in our daily life, a context in which digital cities are emerging.

Cities' constant evolution has been driving human beings footsteps. We mean, society has been moving toward a noble cause that is related to the union and sum of efforts of all of us. It motivates humans to explore the earth, sometimes, unfortunately, in inefficient and unplanned manners, but also makes us dream about the possibility of even reaching other planets and ecosystems. Allied to the advancement of machines, cities are evolving into a new paradigm, being called smart/digital cities. This evolution, closely related to devices equipped with high-performance computational skills, is happening in urban and rural areas. Besides promoting a decentralization of the current system, the new cities open doors for different autonomous agents to optimize their own interests. The adjective autonomous is used in order to highlight the robustness and flexibility of our modern devices. In fact, we do not believe that they will be fully autonomous compared to humans' capabilities and fulfilled with unimaginable artificial intelligence skills. On the other hand, no one can deny the current range of applications that can be performed by our actual systems.

In this context, combinatorial optimization methods play a fundamental role for more precise, efficient, and balanced decision making. As should be noticed, decision making is something done by all of us during our entire life. Furthermore, the decisions reached by others directly affect our trajectory; thus, we are influenced by the decisions made by other entities. In this sense, we highlight the importance of being careful when agreements are being reached, and, as much as possible, with the consideration of a wide range of variables and factors. In summary, we should use the current potential of our devices for the benefit of society, taking hand of high-performance computation tools and methods.

The background of the editors made us select several chapters that involve the resolution of combinatorial optimization problems. While mathematical programming models have been driving society toward the comprehension of several complex problems, it is nowadays known that the best solution(s) for some cases will never be found. Reaching that precise and desired solutions would be like finding a specific atom at the whole universe! For this purpose, computer scientists are moving their efforts in the search for better methods for solving and understanding the nature of NP-Hard problems, which may guide the evolution of cities into efficient problem solvers organisms.

In addition, along this book, the interaction with citizens is not overlooked, new tools are struggling to contribute to society by promoting a more transparent and participatory economy. On the other hand, such advance must take place in a sustainable manner; thus, for this purpose, new ideas and ways of harnessing existing resources are being studied. This universe of possibilities opens doors for researchers to see distinct scenarios of complex and multicriteria decisions.

This book comprises studies and insights of researchers focused on a wide range of topics, such as:

- Computational Intelligence (CI) applied for urban planning and engineering solutions;
- Smart and digital cities solutions with metaheuristics and/or decentralized systems;
- Cities and cryptocurrencies;
- CI and blockchain inspired technologies;
- CI applied for the insertion of renewable energy resources with microgrids;
- CI for smart cities (SC) and smart grids (SG);
- CI applied for SC logistics;
- SC and internet of things;
- CI for smart/green homes;
- CI for decentralized environments;
- CI and citizens;
- Green computing and green operations research;
- High-performance computing and SC;
- City of things and CI testbed for SC;
- Reviews, trends, and state-of-the-art ideas for SC and combinatorial optimization problems; and
- Artificial and computational intelligence, citizens, IoT, SG, renewable resources, among others.

Some of these topics are not handled here in a deep manner; however, readers could have insights about possible applications.

In particular, we think that the collection of chapters and ideas presented along this book can be used to motivate students, engineers, cities' decision makers, and citizens for achieving a better comprehension about the sea of possibilities that are showing themselves to modern society. This book has also the dream of promoting awareness about the potential that researchers could reach together when their efforts are merged and put into reality.

Along this book, contributions are divided into four parts, which comprise 19 chapters including this introduction and one final brainstorming compilation. We decided to start with those chapters that are more dedicated to system development and applications of cutting edge technologies (Parts I and II). Part III possesses a more dedicated focus on social science perspectives, while Part IV presents some insights about emerging technologies.

1.1 Part I: High-Performance Computational Tools Dedicated to Computational Intelligence

This first collection of chapters focuses on some examples of complex problems that are nowadays being dealt within our daily life. Three chapters focus on transportation aspects, while two of them (Chaps. 2 and 3) could fit Part III because of their focus on urban transport and citizens, the other one is mainly focused on challenges for optimizing logistics of urban transportation systems (Chap. 4). Undoubtedly, the application and embrace of the ideas comprised in these three first chapters could lead society to reduce transportation costs, increase efficiency and investors' profit, as well as assisting citizens in the use of cities' services. This second collection of papers also includes considerations about smart allocation of urban space (Chap. 5), efficient use of recyclable materials (Chap. 6), and telecommunication protocols (Chap. 7), all of them considered in the scope of the future digital and smart cities, managed with assistance of the best available computational intelligence inspired tools. The scope of efficient autonomous greenhouses could also be obtained by reading the chapter that deals with a review on forecasting methods for energy consumption (Chap. 8), a component that is intrinsically connected with the future of the digital and interconnected microgrid/houses.

1.2 Part II: Simulating the Possibilities and Getting Ready for Real Applications

While reading the second part of this book, readers will face contributions mostly focused on the simulation of cities' behaviors and applied applications. As expected, innovative models and techniques should be validated before being put into practice. Chapter 9 deals with the concept of an assistive city, in which ubiquitous accessibility is taken into account. Chapter 10 discusses and analyzes the behavior of smart vehicles networks, in which messages flow in a decentralized fashion. Finally, a LoRaWan inspired system is taken into account and analyzed in terms of their contributions in the popularization of technology in Chap. 11.

1.3 Part III: The Social Science Behind the Cities of the Future

The third part contains three chapters. Chapter 12 focuses on the connection between cities' management and citizens. Chapter 13 points out the possibility of accessing non-dominated solutions in order to reach balanced and transparent decisions (Chap. 13). Finally, cryptocurrencies and blockchain applied for territorial development are discussed in Chap. 14.

1.4 Part IV: Emerging Cities' Services and Systems

This last part provides insights regarding the cities' integration and emerging technologies. The contributions of decentralized renewable energy resources is dealt within the scope of microgrids at Chap. 15. The next three chapters mainly focus on green technologies, embedded into promising equipments, such as hybrid/electric vehicles (Chap. 16) and drones (Chap. 17). Finally, Chap. 18 deals with the concept of software certification, an important component for the future IoT devices spread all over our cities.

Society is moving toward the use of efficient and less costly technological devices. We hope that free knowledge exchange will promote transparent competition and the opportunity for developers to dive into complex codes and projects. As a final achievement, industry will probably abstract some of the concepts and possibilities described along this book, and, consequently, several parts of the described tools will become a common mechanism that will be transparent and easy accessible for most part of world's population. In addition, we would like to highlight that these tools could even reach isolated areas and communities, however, its insertion, which is probably undeniable, should be done in a way that citizens and local people are respected and their wishes are really taken into account.

We wish everyone a pleasure journey into the concepts and dreams idealized throughout this insight about cities and communities of the future.

Part I
High-Performance Computational Tools Dedicated to Computational Intelligence

Chapter 2
Urban Mobility in Multi-Modal Networks Using Multi-Objective Algorithms

Júlia Silva, Priscila Berbert Rampazzo, and Akebo Yamakami

Abstract The multi-modal transportation problem is a type of shortest path problem (SPP). Its goal is to find the best path between two points in a network with more than one means of transportation. This network can be modeled using a weighted directed colored graph. Hence, an optimization method can be applied to find a better path between two nodes. There are some digital applications, like the well-known Google Maps, that present solution to this problem. Most of them choose the best path according to one objective function: the minimum travel time. However, this optimization process can consider multiple objective functions if different user's interests are treated in the model such as the cheapest or the most comfortable path. In this work, we implemented and compared two different approaches of algorithms with multiple objectives. One of them is an exact method based on Dijkstra's algorithm added by sum weight method, and the other one is a heuristic approach based on the non-dominated sorting genetic algorithm (NSGA-II). The computational results of the two methods were compared. The comparison shows that the heuristic method is promising due to the low execution time—around 20 s—and the quality of the results. This quality was measured by the closeness of the points found by the two methods in the objective domain. The runtime and the quality of the results can indicate that this modeling is suitable to a real-time problem, for instance, the multi-modal transportation problem.

J. Silva (✉) · A. Yamakami
School of Electrical and Computer Engineering, University of Campinas, Campinas, Brazil
e-mail: julia@dt.fee.unicamp.br; akebo@dt.fee.unicamp.br

P. B. Rampazzo
School of Applied Sciences, University of Campinas, Campinas, Brazil
e-mail: priscila.rampazzo@fca.unicamp.br

V. N. Coelho et al. (eds.), *Smart and Digital Cities*, Urban Computing,
https://doi.org/10.1007/978-3-030-12255-3_2

2.1 Introduction

Transport is one of the primary drivers of production and development economic. Transportation services determine people's access to education, health, work, leisure, etc. Companies also rely on transportation to get input from their suppliers and bring their products to consumers. The necessary changes to new scenarios of urban mobility and urban logistics come from planning and optimized use of transport networks. The more the diversity of means of transportation a region has, the more efficient the flow of traffic can be. A study sponsored by the Rockefeller Foundation https://www.rockefellerfoundation.org/ states that it is necessary to coordinate multi-modal locomotion systems that include the most diverse options. Also, priority actions for multi-modal public transport should be proposed mainly in large cities.

The multi-modal transportation problem can be defined as a shortest path problem (SPP) applied in a network that involves more than one means of transportation. In this work, we consider an application in urban transport networks. However, the model is easily widespread in other types of network transportation problems, like a goods transportation plan around the country.

In urban mobility, the routing condition can include many means of transportation, for instance, car, bicycle, subway, train, etc. Besides that, there are many users with different interests related to alternative paths. Technological applications have become prominent among users indicating faster paths, considering the flow in the pathways and providing alternative routes if necessary. This situation prompts for a multi-objective approach in a multi-modal network. Given a starting and a destination point, the proposal of this work is to find the best route based on two main objectives: the financial cost and a travel time path. We used a heuristic and a deterministic optimization method to solve this problem.

The deterministic method implemented is based on classical Dijkstra's algorithm that uses a heap as a data structure into the optimization process. We consider it as a suited algorithm because this algorithm is widely used in SPP and the implementation with heap structure is indicated when there are no many connections between nodes on such case like multi-modal transportation network.

The heuristic approach implementation is based on one of the most important genetic algorithms in the multi-objective area: the non-dominated sorting genetic algorithm (NSGA-II) [4]. This method uses the concept of dominance defined by the values of the objective functions. This criterion is used to classify the possible solutions and determine which of them will pass to the next iteration. In the event of a tie, the method decides on the solution which has the higher value of crowding distance.

We did some computational experiments, and then we collected and analyzed the results. The results obtained by the heuristic algorithm show that this approach can achieve a great approximation of the Pareto Frontier. It means that the algorithm found some points which belong to the real Pareto frontier. This statement can be proved by the comparison with the solutions obtained by the deterministic approach.

Also, the heuristic algorithm found more different points in the domain of the objective functions than the deterministic. Therefore, we can say that the heuristic is more suitable than the deterministic approach for the multi-modal transportation problem with multiple objectives, providing better support to the user because it presents many alternative paths.

This work is organized as follows: In Sect. 2.2, we present the basic concepts needed to understand the methods and the application better. In Sect. 2.3, it is some related works in the multi-modal transportation problem, and it is reported how the authors have solved it in a literature review. Then, in Sect. 2.4, we describe the implementations based on each approach. After that, in Sect. 2.5, it is detailed how we did the computational experimentations. Finally, in Sect. 2.6, we discuss the analysis of the implementation and the experiments. Then, we conclude the work and the possible future works.

2.2 Basic Concepts

A colored graph can be defined as $G = \{N, E, L\}$, where N is a limited set of nodes, E is a limited set of edges, and L is a limited set of labels. In our case, the nodes represent a geographic position, the edges are the connections between two nodes, and the labels represent the means of transportation. Besides, each one of this triple has two values associated: a real number that characterizes the edge weight, and a binary parameter, component of an incidence matrix, that defines when the edge composes the graph or not.

A path $C = (n_O^-, n_1^{l_1}, \ldots, n_D^{l_D})$ can be defined as a sequence of nodes $\{O, 1, \ldots D\}$ and labels $\{l_1, \ldots, l_D\}$, forming colored edges $(n_{i-1}^{l_{i-1}}, n_i^{l_i})$ that link an origin O to a destination D. The weighted path is defined by Eq. (2.1). It is the sum of the edges weights which belong to the path C. In multi-modal transportation problem, these weights can be a financial cost, travel time, distance, etc.

$$w(C) = \sum_{i=O}^{D} w\left(n_{i-1}^{l_{i-1}}, n_i^{l_i}\right). \tag{2.1}$$

The shortest path problem is defined as a minimum weighted path between two nodes in a graph. It can be formulated by Eq. (2.2).

$$W(O, D) = \begin{cases} \min\{w(C) : O \rightsquigarrow D\}, & \text{if there is a path from } O \text{ to } D \\ \infty, & \text{otherwise.} \end{cases} \tag{2.2}$$

Several methods can be used to solve this type of problem. Overall, they can be classified into two main classes: the label-setting and the label-correcting algorithms [1]. Both of them try to assign the shortest distance between the nodes as a label for each node. The main differences are that the label-correcting methods can be applied in negative arcs weights and the label setting cannot, and this last group of methods assigns a label to a node when the process can guarantee the shortest distance, whereas the other group of methods considers all the labels temporarily until the last step. Dijkstra's algorithm is the most popular label-setting method [1].

2.3 Related Works

Focusing on a dynamic network, [11] developed an exact algorithm with only one objective which it refers to a total time of the path. In a multi-modal network, it is necessary to consider the timetable of public transportation and the delays in commuting points. These factors compose both [9] and [8] proposals with the difference that the last one applies a speed-up technique to accelerate the convergence of the solution.

It is hard to find a work that includes all particularities: multi-objective problem, a multi-modal network including private and public transportation, and dynamic network. The nature of this problem is multi-objective [2]. Transportation problem with the passenger could imply in different and sometimes subjective preferences for a route.

A hybrid approach has been an explored area to solve this kind of problem. In [3] the focus is to solve the vehicle routing problem with multiple time windows proposing a novel hybrid genetic variable neighborhood search. It calculates the optimal routes in terms of transportation cost for goods and service demands. This problem has several constraints.

The transportation problem could be for both passenger and goods. The proposal of using multi-criteria decision making method in this kind of problem is an approach adopted by Dotoli et al. [7].

Related to the transportation of goods, [10] propose a swarm intelligence method by applying the fireworks algorithm to solve the transportation problem. There are two objectives to find a route: minimizing the cost and the time window.

It is a strong tendency to use the heuristic methods in problems with a multi-modal network. In [5], it applies a memetic algorithm mixing two meta-heuristics: a genetic algorithm with variable neighborhood search (VNS). This work models the network separated and then connects them using transfer arcs. This is a critical process of the proposal because it could determine how the algorithm handles the solution in all procedures. It could affect the individual generation, and it could allow considering some constraints or not. Also, it could influence the number of network arcs and directly impact the algorithm performance. Therefore, this process regarding our proposal is presented in the next section.

2.4 Implementations

This work treats the problem considering two objective functions described by f_1 and f_2, in Eqs. (2.3) and (2.4), respectively. The first equation is a math formulation under the financial cost of the path and the second is a formulation under the total travel time. The variable x_{ijl} is a decision variable, and it could be 1 or 0, depending on whether the edge from node i to node j by the mode l belongs to the path or not.

$$\min \quad f_1 = \sum_{(i,j,l)\in E} c_{ijl}\, x_{ijl} \tag{2.3}$$

$$f_2 = \sum_{(i,j,l)\in E} t_{ijl}\, x_{ijl} \tag{2.4}$$

$$\text{subject to} \quad \sum_{l\in L} \sum_{j:(i,j)\in E} x_{ijl} - \sum_{l\in L} \sum_{j:(j,i)\in E} x_{jil} = 0 \,, \quad i \neq o, d \tag{2.5}$$

$$\sum_{l\in E} \sum_{j:(i,j)\in E} x_{ijl} - \sum_{l\in L} \sum_{j:(j,i)\in E} x_{jil} = 1 \,, \quad i = o \tag{2.6}$$

$$\sum_{l\in E} \sum_{j:(i,j)\in E} x_{ijl} - \sum_{l\in L} \sum_{j:(j,i)\in E} x_{jil} = -1 \,, \quad i = d \tag{2.7}$$

$$x_{ijl} \in \{0, 1\} \quad \forall (i, j, l) \in E. \tag{2.8}$$

The constraints are showed in Eqs. (2.5)–(2.7), and they represent the flow balancing of the network. In Eq. (2.5), it means that the intermediate nodes (the path nodes are different than origin and destination nodes) have to be a balancing equal to 0. In Eq. (2.6), it means that the origin node has to be a balancing equal to positive 1. And in Eq. (2.7), it means that the destination node has to be a balancing equal to negative −1.

2.4.1 Deterministic Approach

In the deterministic approach, we adopted the Dijkstra's algorithm to lead with a multi-modal network. We chose this algorithm because it is one of the most important methods for the shortest path problem. Besides, it was relatively easy to include the weight sum method which provides us to obtain a set of solutions. The pseudo-code of the implemented procedure is shown in Algorithm 2.1.

Algorithm 2.1 High-level code of implementation based on Dijkstra's algorithm

for all $u \in N$ **do**
  $C(u) \leftarrow +\infty, \quad color(u) \leftarrow white, \quad P(u) \leftarrow nil.$
  if $u = u_o$ **then**
    $C(u) \leftarrow 0$, insert u in Q, $color(u) \leftarrow gray$
  end if
end for
while $Q \neq \emptyset$ **do**
  Remove u in Q, such that $C(u)$ is minimum and $color(u) = black.$
  for all $v \in A(u)$, such that $color(v) \neq black$ and $v \neq u_d$ **do**
    $tmp \leftarrow C(u) + w(u, v).$
    if $tmp < C(v)$ **then**
      $P(v) \leftarrow u$ and $C(v) \leftarrow temp.$
    end if
    if $color(v) = gray$ **then**
      Update position of v in Q: HeapSort
    else
      Insert v in Q and $color(v) \leftarrow gray$
    end if
  end for
end while

where

- N is the number of nodes in the graph;
- $C(u)$ is the weight from origin to node u;
- u_o is the origin;
- u_d is the destination;
- cor is a vector with length = N, where, in each position, there is a bit that can indicate three states of the node in process:
  - $white$ if the node was not visited yet,
  - $gray$ if the node was already visited, but its weight is not permanent, and
  - $black$ if the node was visited and its weight is permanent;
- $P(u)$ is the predecessor of node u;
- Q is a priority queue as min-heap structure;
- $A(u)$ is the adjacent list of node u; and
- $w(u, v)$ is the weight from u to v.

2.4.2 Heuristic Approach

We decide to use a genetic algorithm (GA) because it is the most used in the heuristic area. One of the most representative GAs in the multi-objective area is the NSGA-II because it has two main instruments to handle the population and support the goals of multi-objective methods in the objective domain. The two main goals are to find points in the objective domain as close as possible of Pareto Frontier and that points have to be well dispersed over the frontier founded.

These mechanisms are controlled by two fields that each individual in the population has. The first is called rank, and it classifies the individuals by the non-dominance concept. If the individual has a rank equals one, it means that there is no individual that dominates it. This rank number is used as a criterion to decide which individual survives in the next generation. If there is a tie, the algorithm calculates a crowding distance and the individuals that have the higher numbers are going to survive. In Figs. 2.1 and 2.2 is shown in a bi-dimensional domain, how these concepts work, respectively.

In Fig. 2.1, the points regarding class $r1$ are not dominated by any other point in the graph. Those regarding class $r2$ are dominated by the points in class $r1$ but they dominate the points in $r3$. The points belonging to this last class are dominated by all other points in the graph.

In Fig. 2.2, it shows some elements used to calculate the crowding distance. For example, the crowding distance of point called i depends on neighboring points $i2$ and $i3$. With their coordinates, this forms a cuboid, and as bigger, as this cuboid will be, it means that this point represents a less density area, so it is preferred.

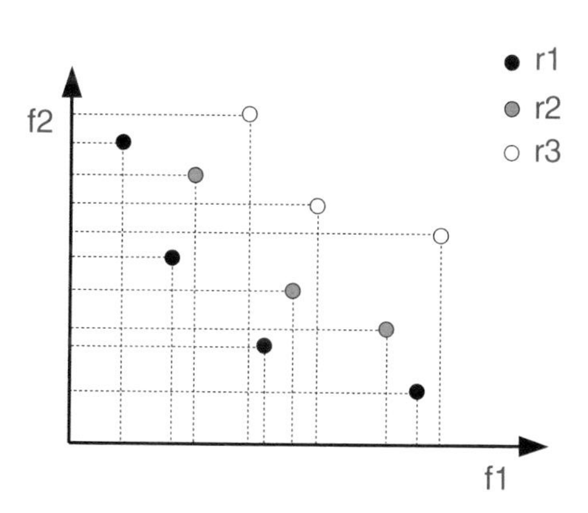

Fig. 2.1 Rank concept in bi-dimensional domain

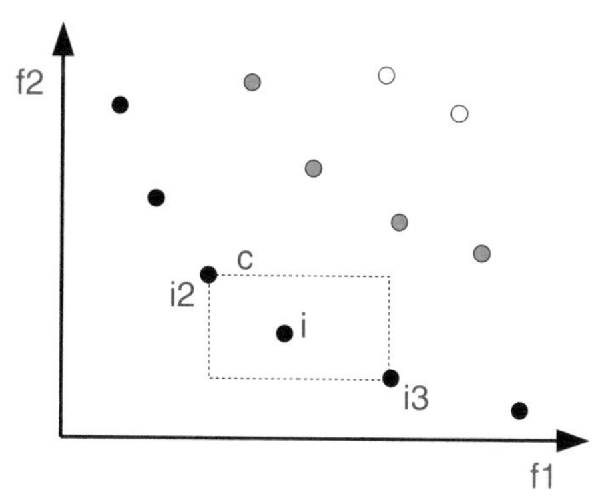

Fig. 2.2 Crowding distance concept in bi-dimensional domain

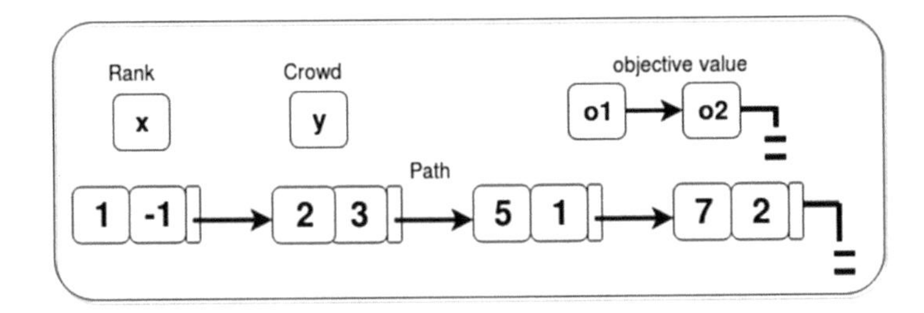

Fig. 2.3 Individual encoding

2.4.3 Encoding

The map of the multi-modal transportation network is represented by a directed graph in which the nodes are stop points, integration terminals between them, or points of origin and destination of the route.

An individual is represented by a structure composed of four fields: (1) a linked list representing the path where each element of the list contains two integers: the first indicates the node and the second indicates the means of transportation, (2) a linked list that stores the values of the objective functions, (3) a field to keep the rank of the individual in population, and (4) the value of crowding distance. These codification is presented in Fig. 2.3.

The paths of the individuals can have different sizes, but necessarily the first node must be the origin and the last one the destination node. All individuals handled by the algorithm are feasible. The domain of the objective functions is defined in real set, that is, $(f_1, f_2) \rightarrow (\Re, \Re)$ and the functions calculations are independent of each other.

2.4.4 Initialization

The initial population is generated by using the Dijkstra's algorithm structure to produce each individual. The original algorithm was adapted for multi-modal networks and it supports the obtainment of only feasible paths. We suppress the sorting part of the algorithm for this initialization phase. This decision was based on two reasons: because of the need to find a feasible path and not a minimum path (in this moment of the algorithm) and second because it is an expensive computational process that serves to guarantee the minimum cost. Therefore, to compose the first population, we assigned weights for each edge randomly and executed the adapted algorithm. This process has, as a result, one individual, and it repeats until the number of individuals in the population had completed.

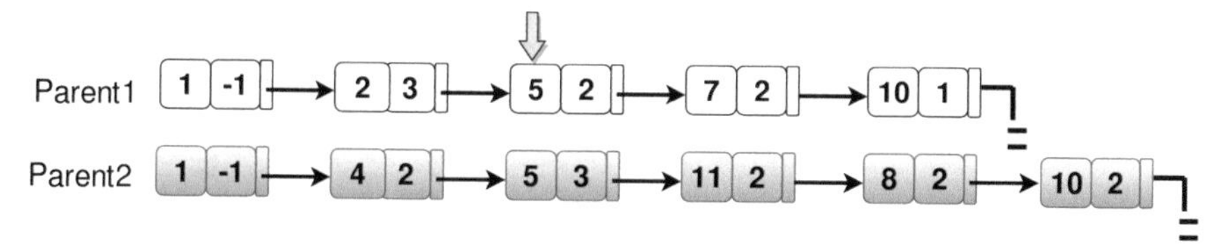

Fig. 2.4 Parents selected by the selection operator

2.4.5 Genetic Operators

Commonly, a basic genetic algorithm has two genetic operators that are responsible for generating the population of offspring: the crossover and the mutation. In our encoding, we use these two operators to gain the diversity of new individuals and try to expand the good genes over the generations. Another operator typically of GA is the selection operator. This one will use a method to choose a number of individuals in the current population and, in this case, will determine which of them will be parents in crossover operator.

In our encoding, the selection operator has half chance to choose a random parent to crossover and half to perform a tournament between two individuals that are randomly selected in population. This operator is applied twice, and the individuals resulting in this process become parents of the crossover operator. In Fig. 2.4, there are two paths of two individuals representing the result of the selection operator. These paths will be submitted to the crossover operator.

To create new individuals in the population, we developed three types of crossover. It was necessary because we handle only feasible solutions (or individuals) and to guarantee this, we cannot do this only in a random way. Then, to keep logical paths, the first trial to combine genes of two individuals is looking for an arc to link them. To generate offspring with the number of arcs as short as possible, the search starts in the first node of the parent 1 and last node of the parent 2. The search ends when the trial arrives in the middle of two parents or find a connect arc.

The second trial is a two-point crossover. It consists of a random selection of two nodes in parent 1 and two nodes in parent 2. We use the adapted Dijkstra's algorithm to find a path between the first point of the parent 1 to the second point of parent 2. This crossover type intends to give some diversity to the population since it is possible to bring new genes by the exact algorithm. This process considers the average between financial cost and time to find a minimum route that connects the two parents.

If all the previous operators generated any offspring, the last crossover type is a one-point crossover for parents that have a common node. It is the easier and computationally cheapest type of crossover, but it compromises the diversity of the population. Figure 2.5 represents two possible offspring generated by this operator based on the parents of Fig. 2.4.

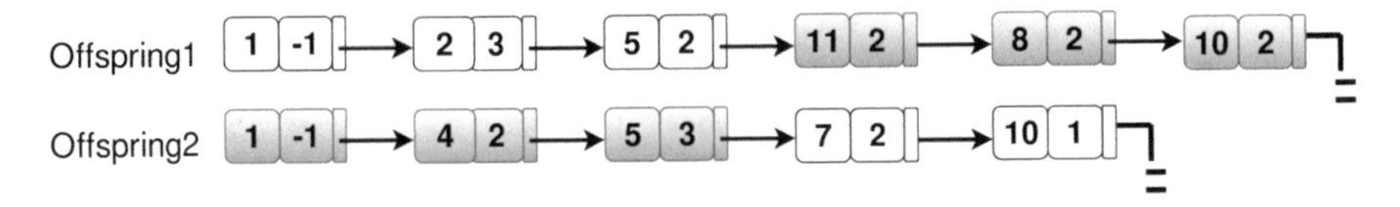

Fig. 2.5 Offspring generated by the one-point crossover type

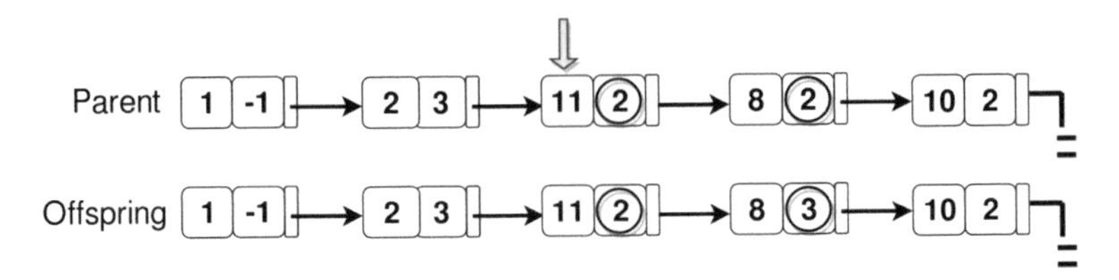

Fig. 2.6 Example of applied mutation operator

In this case, the origin point is the node 1, and the destination point is the node 10. Also, the operator chooses the node number 5 to apply the crossover of genes because this node is common in both paths. Finally, if nonetheless any offspring is generated, the parent is sent to the mutation operator necessarily.

The mutation operator focuses only on the means of transportation in the individual. A random point is selected in a path and the number that represents the means of transportation changes to another mode available. This operator also analyzes which means of transportation is arriving and leaving at that point selected. If there are the same modes, the algorithm tries to change it. Otherwise, it tries to find the same mode to execute the mutation. In Fig. 2.6, it is illustrated this operator.

Given this new population added by the offspring, the algorithm applies the Pareto-dominance concept rating on individuals. This process classifies the individuals by the rank explained above. Then, they are sorted following this rank and because of the elitism approach, the new generation is composed of the individuals with the lowest value of rank which is one in the first iteration.

If the number of the individuals with rank one is less than the size of the population, the algorithm completes the new generation with individuals in rank equals two and so on until the size of the population is met. If the number of the individual selected by their rank is largest than the size of the new population, the tiebreaker criteria are obtained by calculating the crowding distance between the solutions with the same rank on the objective domain as well as the NSGA-II works. This process can be observed in Fig. 2.7.

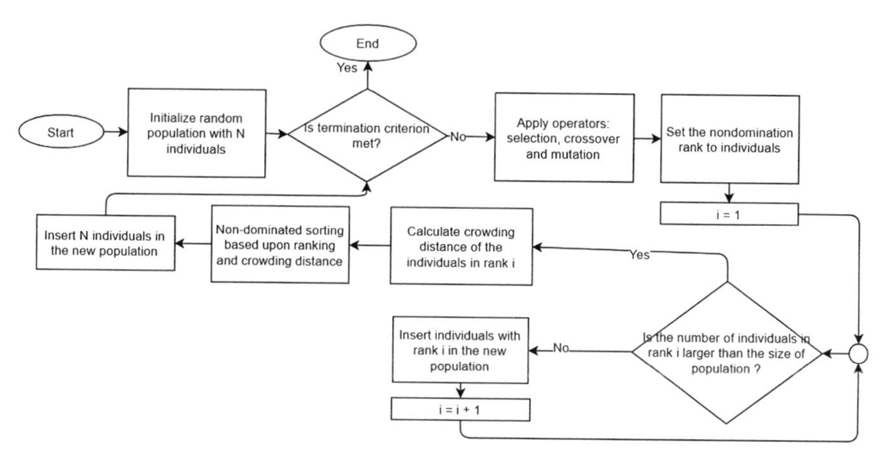

Fig. 2.7 NSGA-II flow chart

2.5 Computational Experiments

To evaluate and compare the methods implemented, we performed some computational experiments. They run on a Linux platform, Intel Core i5 processor with 1.8 GHz and 4 Gb of RAM.

Due to the challenges to have access to the real multi-modal transportation network, we created an artificial to perform as an instance in this paper. The parameters used to define the coverage of each modal and the attributes of financial cost and time for each type of edge are presented in Table 2.1. All the information is based on files of TSPLIB, an online repository of real distances between geographic points, mainly in Europe territory. This database was developed to assist the traveling salesman problem (TSP), but we consider that it could be useful in our case.

Concerning these features, the network has 400 nodes, 7732 arcs, and four means of transportation: pedestrian, subway, car, and bus. The financial costs of the subway and bus are similar. The car is the most expensive type, and the pedestrian is the cheapest one. However, the pedestrian is the slower means of transportation, followed by bus, subway, and car, in that order.

Table 2.1 Parameters used to create the artificial multi-modal network

Id.	Modal	Coverage (%)	Average speed (km/h)	Cost (R$)
0	Subway	20	41.3	4.50
1	Bus	30	11.4	3.50
2	Pedestrian	80	7.02	0.00
3	Car	70	12.15	2.75 + 1.50/km + 0.26/min

Table 2.2 Heuristic approach results

f_1	f_2	Path
0	9.77	$346 \xrightarrow{2} 1 \xrightarrow{2} 78 \xrightarrow{2} 322 \xrightarrow{2} 351 \xrightarrow{2} 374$
3.5	8.34	$346 \xrightarrow{2} 1 \xrightarrow{1} 78 \xrightarrow{2} 322 \xrightarrow{2} 351 \xrightarrow{2} 374$
4.5	7.42	$346 \xrightarrow{2} 1 \xrightarrow{0} 208 \xrightarrow{2} 78 \xrightarrow{2} 322 \xrightarrow{2} 351 \xrightarrow{2} 374$
8	6.79	$346 \xrightarrow{2} 49 \xrightarrow{0} 1 \xrightarrow{1} 78 \xrightarrow{2} 322 \xrightarrow{2} 351 \xrightarrow{2} 374$
9	5.87	$346 \xrightarrow{2} 49 \xrightarrow{0} 1 \xrightarrow{0} 208 \xrightarrow{2} 78 \xrightarrow{2} 322 \xrightarrow{2} 351 \xrightarrow{2} 374$
19.8	5.56	$346 \xrightarrow{2} 49 \xrightarrow{0} 1 \xrightarrow{0} 208 \xrightarrow{3} 78 \xrightarrow{2} 322 \xrightarrow{2} 351 \xrightarrow{2} 374$
26.1	5.31	$346 \xrightarrow{3} 49 \xrightarrow{0} 1 \xrightarrow{0} 208 \xrightarrow{2} 78 \xrightarrow{2} 322 \xrightarrow{2} 351 \xrightarrow{2} 374$
36.9	**5**	$346 \xrightarrow{3} 49 \xrightarrow{0} 1 \xrightarrow{0} 208 \xrightarrow{3} 78 \xrightarrow{2} 322 \xrightarrow{2} 351 \xrightarrow{2} 374$

Table 2.3 Deterministic approach results

f_1	f_2	Path
0	9.77	$346 \xrightarrow{2} 1 \xrightarrow{2} 78 \xrightarrow{2} 322 \xrightarrow{2} 351 \xrightarrow{2} 374$
4.5	7.42	$346 \xrightarrow{2} 1 \xrightarrow{0} 208 \xrightarrow{2} 78 \xrightarrow{2} 322 \xrightarrow{2} 351 \xrightarrow{2} 374$
9	5.87	$346 \xrightarrow{2} 49 \xrightarrow{0} 1 \xrightarrow{0} 208 \xrightarrow{2} 78 \xrightarrow{2} 322 \xrightarrow{2} 351 \xrightarrow{2} 374$
17	5.39	$346 \xrightarrow{1} 342 \xrightarrow{0} 58 \xrightarrow{0} 1 \xrightarrow{0} 208 \xrightarrow{2} 78 \xrightarrow{2} 322 \xrightarrow{2} 351 \xrightarrow{2} 374$
36.9	**5**	$346 \xrightarrow{3} 49 \xrightarrow{0} 1 \xrightarrow{0} 208 \xrightarrow{3} 78 \xrightarrow{2} 322 \xrightarrow{2} 351 \xrightarrow{2} 374$

The experiments with the deterministic algorithm used a discretization of 0.05 for the weight of the objective functions. For example, in the first algorithm running, f_1 starts with minimum weight, zero, and f_2 starts with the maximum weight, equals one. In the second running, the weights were 0.05 and 0.95, respectively. And so on until the f_1 receives the maximum weight and f_2 receives the minimum weight. Therefore, we performed 20 execution with the weighted variety.

In Tables 2.2 and 2.3 are presented the results of the heuristic and deterministic approaches experiments, respectively. The bold values in the first and second columns are the minimum values found for each objective function (f_1 and f_2). We can observe that the heuristic one found more paths with different objective values than the deterministic approach. It is expected because of the weight sum method applied to the algorithm. It becomes the objective function of an approximate function, while the heuristic algorithm treats the multiple objectives explicitly. The results showed in Fig. 2.8 can confirm that this exact method is not a better way to gain the diversity of solutions in the Pareto frontier.

Moreover, we can validate the heuristic results with the exact method. Only one frontier point found by the Dijkstra's algorithm could not be found by the heuristic approach. However, the heuristic algorithm finds the largest number of solutions. It means that, in this case, the customer has more alternatives to choose.

It is worth mentioning that the choice of the weight sum method was due to the easy adaptation to the Dijkstra's algorithm. However, as the main difficulty of the weight sum method is that points of the concave part of a non-convex Pareto boundary are not estimated.

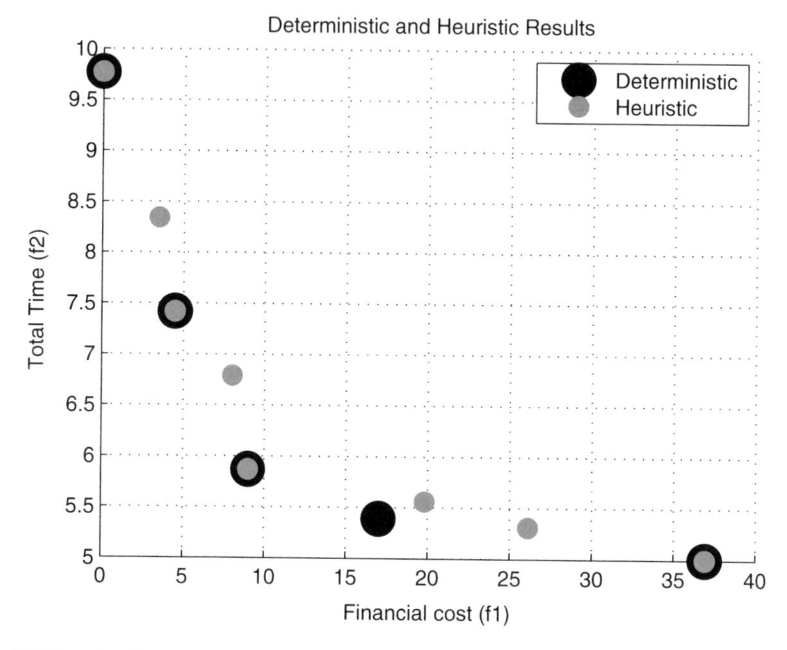

Fig. 2.8 Results of the deterministic and heuristic algorithms

In Tables 2.2 and 2.3, they have presented the final solution values in both objective and variable domains. The first column describes the f_1 value, the second is the f_2 value, and the last column represents the path with these values concern about. The start point is the node 346, and the end point is the node 374 for all the paths because it was the longest distance in the graph. These tables are sorted by the extreme solution in f_1 objective function until the extreme point in objective function f_2. Table 2.2 presents the eight solutions found by the heuristic algorithm and Table 2.3 presents the five results of the deterministic algorithm.

The path is represented by the sequence of nodes which users have to pass through them, and the number above the arc indicates the means of transportation needed to go from a node to another.

2.6 Conclusion

The genetic algorithm seems to be a more appropriate approach for a multi-objective problem because of the nature of the algorithm by handling a set of solutions. Furthermore, this approach needs only one execution to give a set of candidates routes, solutions that are good enough to answer the customer requirements in reasonable computational time.

Due to the weight sum method, the classical algorithm based on Dijkstra's [6] can lose some intermediate point of the frontier. Although to find the extreme points, the exact algorithm is a fitting procedure. There is an alternative way to find a higher number of intermediate points: decrease the gap between the weights of the objective functions used by the sum weight method. However, it requires a higher number of executions, and it will increase the computational time.

There is an advantage to use a population algorithm like a GA, to solve this multi-objective problem. This type of algorithm handles with a set of candidate solutions since its first interaction. It properly fits with the multi-objective character of obtaining a set of solutions as a result instead of a single solution. Therefore, this proposal offers several alternative routes for the customer (besides the extreme points) because of its leads with more than one objective functions simultaneously.

Acknowledgements The authors would like to thank CNPq (National Council for Scientific and Technological Development) for the financial support received.

References

1. Ahuja, R.K., Magnanti, T.L., Orlin, J.B.: Network Flows: Theory, Algorithms, and Applications. Prentice Hall, Upper Saddle River (1993)
2. Ambrosino, D., Sciomachen, A.: A shortest path algorithm in multimodal networks: a case study with time varying costs. In: Proceedings of International Network Optimization Conference, Pisa (2009)
3. Belhaiza, S., M'Hallah, R., Ben Brahim, G.: A new hybrid genetic variable neighborhood search heuristic for the vehicle routing problem with multiple time windows. In: IEEE Congress on Evolutionary Computation (CEC) (2017)
4. Deb, K., Agrawal, S., Pratap, A., et al.: A fast elitist non-dominated sorting genetic algorithm for multi-objective optimization: NSGA-II. In: Schoenauer, M., et al. (eds.) Parallel Problem Solving from Nature PPSN VI. PPSN 2000. Lecture Notes in Computer Science. Springer, Berlin (2000)
5. Dib, O., Manier, M.A., Caminada, A.: Memetic algorithm for computing shortest paths in multimodal transportation networks. Transp. Res. Procedia **10**, 745–755 (2015)
6. Dijkstra, E.W.: A note on two problems in connexion with graphs. Numer. Math. **1**, 269–271 (1959)
7. Dotoli, M., Epicoco, N., Falagario, M.: A technique for efficient multimodal transport planning with conflicting objectives under uncertainty. In: IEEE In Control Conference (ECC) (2016)
8. Kirchler, D.: Efficient routing on multi-modal transportation networks. Ecole Polytechnique X (2013)
9. Liu, L., Mu, H., Yang, J.: Toward algorithms for multi-modal shortest path problem and their extension in urban transit network. J. Intell. Manuf. **28**(3), 767–781 (2017)
10. Mnif, M., Bouamama, S.: Firework algorithm for multi-objective optimization of a multimodal transportation network problem. Procedia Comput. Sci. **112**, 1670–1682 (2017)
11. Ziliaskopoulos, A., Wardell, W.: An intermodal optimum path algorithm for multimodal networks with dynamic arc travel times and switching delays. Eur. J. Oper. Res. **125**(3), 486–502 (2000)

Chapter 3
Urban Transport and Traffic Systems: An Approach to the Shortest Path Problem and Network Flow Through Colored Graphs

Juliana Verga Shirabayashi, Akebo Yamakami, Ricardo Coelho Silva, and Wesley Vagner Inês Shirabayashi

Abstract Urban transport systems generally present complex topologies and constraints, and consequently model them and propose solutions that are not simple. Because of the importance of proposing solutions that improve urban mobility and the people's quality of life, in this work, we propose two algorithms applicable to transport network and traffic systems. The first algorithm approaches the shortest path problem in colored graphs. In this case, the graphs' coloration is used in a different and innovative form: each transport mode is represented by a color (label) and various edges exist between two nodes of the graph (each edge represents a transport mode). The second algorithm, in addition to the coloration used to find the shortest path, considers the multimodal flow network problem by an incremental process. Some examples are presented and the behavior of both algorithms is shown.

J. V. Shirabayashi (✉)
Federal University of Paraná, Jandaia do Sul, PR, Brazil
e-mail: juliana.verga@ufpr.br

A. Yamakami
Department of Systems and Energy, FEEC, University of Campinas, Campinas, SP, Brazil
e-mail: akebo@dt.fee.unicamp.br

R. C. Silva
Department of Statistics and Applied Mathematics, Federal University of Ceará, Fortaleza, CE, Brazil
e-mail: rcoelhos@dema.ufc.br

W. V. I. Shirabayashi
Department of Mathematics, State University of Maringá, Maringá, PR, Brazil
e-mail: wvishirabayashi@uem.br

V. N. Coelho et al. (eds.), *Smart and Digital Cities*, Urban Computing,
https://doi.org/10.1007/978-3-030-12255-3_3

3.1 Introduction

The fast growth of cities has brought with it the need for a better plan and to make urban mobility more efficient, that is, fully influenced by transport systems. Besides, in the last few years, the called Smart City became an object of study in the most diverse areas of knowledge, with the objective of becoming big centers, integrated, better planned, sustainable, and with a better quality of life. In this context, we study how the planning of transport networks is very important and is one of the principal problems of Smart Cities.

Urban transport networks are characterized by traffic jams and their corresponding impact on individual accessibility, air pollution, and the development of economic urban activities [14]. In most cities, people's workplaces are often far away from home, which means that they use more than one mode of transportation to move from their home to the workplace.

Transport networks are considered multimodal when they have various transport modes, such as bus, subway, and train. Thus, passengers can use various transport modes in one journey and are usually not willing to change means of transportation often.

In the literature, there are several works that approach multimodal network transport and related problems. These works seek to obtain solutions for users, and for the administrators of the network transport, because a complicating factor is the poor condition of public transport and consequent growth of the fleet of private vehicles. The methods used in the search for a solution to the multimodal network transport problem are varied, ranging from classic methods of shortest path, optimization, and genetic algorithms, to artificial neural networks and *fuzzy* methods.

The main objective of this work is to present two applicable approaches to network problems. The first addresses the shortest path problem in a multimodal network using colored graphs, where each transport mode is represented by a color and there may be several edges between each pair of nodes in the graph. During the resolution procedure, the method detects if there was a change of mode and a cost of this change is added to the path cost. Finally, a multimodal shortest path is obtained with information about the mode used to move through each edge of the path. The second approach is based on incremental loading of the transport network flow, and uses the first approach to obtain the shortest path in each iteration. In this case, the costs depend on the flow. In the incremental loading of flow, instead of passing all the available flow once, we divide it into increments and distribute it to the shortest paths found in each iteration. The edges costs update is made via the travel time function (Bureau of Public Roads, BPR) [6].

This work is organized as follows: in Sect. 3.2 a brief bibliographic review is presented; in Sect. 3.3 the shortest path problem is shown in colored graphs; in Sect. 3.4 the minimum cost flow problem is demonstrated in colored graphs; and Sect. 3.5 presents the final considerations.

3.2 Brief Bibliographic Review

The multimodal transport network problem has been studied by several authors, who propose different approaches, generally based on the concept of graph theory: hierarchical graphs [4, 26], hypergraphs [15], among others.

Approaches using variants of classical methods are widely used in the resolution of transport network problems, for example, Dijkstra's algorithm [7], Floyd's algorithm [2], and algorithms of k-shortest paths [2].

Modesti and Sciomachen [17] presents an approach based on the classical shortest path problem to find the multi-objective shortest path in multimodal transport networks with the objective of minimizing the total cost of travel and the discomfort of users associated with the paths used.

Some authors [4, 14] focus on minimizing the total cost of travel, taking into account the sequence of modes of transport used, so that the solution set presents the so-called viable paths. A path is viable if the sequence of modes is feasible with respect the restriction set. Lozano and Storchi [15] extend the algorithm to obtain viable hyperpaths.

In the work of Lillo and Schmidt [12], real multimodal transport systems are analyzed experimentally. Real road–rail networks in Denmark, Hungary, Spain, Norway, and New Zealand are built based on a set of map scans obtained from various Geographic Information System (GIS) libraries. These networks are modeled using colored graphs to be used as the main input by an algorithm for multimodal networks that computes a set of optimal paths based on Dijkstra's algorithm. The cardinality of the resulting set is analyzed and it is concluded that the connectivity of the vertices and the shape of the network considerably affect the total number of optimal paths.

The utilization of the GIS is often found in works that seek solutions to the shortest path problem in transport networks [4, 16, 18].

Lam and Srikanthan [11] describe a technique of clustering that improves the performance of the conventional calculus of k-shortest paths in multimodal transport networks. The network is remodeled in acyclic representation, where the cycles are identified and clustered.

Loureiro [13] presents an algorithm to solve the multicommodity multimodal network problem. The algorithm proposed consists of a heuristic procedure of decomposition based on a column-generating technique, which allows the solution of large-scale problems in a reasonable time.

In [28], Ziliaskopoulos and Wardell present an algorithm for the intermodal shortest path problem dependent on time in multimodal transport networks taking into account delays in the modes and in the points of exchange. The convergence and the computational complexity of the algorithm are presented.

Other approaches involving heuristics, evolutionary algorithms, and artificial neural networks have been used to solve the multimodal transport network problem. In [1], Abbaspour and Samadzadegan propose an evolutionary algorithm where the chromosomes have varying sizes to solve the multimodal shortest path problem dependent on time. Qu and Chen, in [21], using a hybrid method for multicriteria decision-making, combining the *fuzzy* analytic hierarchical process (AHP) and artificial neural networks. This approach is used to find the best path from a given origin to a particular destination.

Heid et al. [10] present a hybrid approach to solving the multimodal transport problem dependent on time. The proposed approach is based in Dijkstra's algorithm [7] and ant colony optimization.

Yu and Lu [27] propose a genetic algorithm to solve the multimodal planning routing problem. They use a method of multicriteria evaluation into *fitness* function for the selection of optimal solutions.

Moreover, techniques and methods based on *fuzzy* sets theory have been widely used in recent studies, as can be seen in [5, 9, 20, 23], among others. The work of Verga et al. [24] presents a bibliographic revision of formulations to the transport problem, in addition to methods that have been used to solve them. In addition, a method for multimodal networks taking into consideration uncertainties with regard to the costs and the capacities of edges using *fuzzy* set theory.

As for the approaches to the problem in question using colored graphs, we find a work [25] that addresses the multimodal transport networks problem based on the classic Dijkstra's algorithm [7].

The works cited above, for the most part, deal with the classical shortest path problem in transport networks using different approaches to finding it. Some deal with the *fuzzy* shortest path problem and the distribution of flow in the network.

3.3 Shortest Path Problem in Colored Graphs

A well-known problem in graph theory involves coloring in graphs, the assignment of color to the elements of a graph according to certain constraints. Coloring in nodes is a variant of this problem; in this case, the objective is to color the nodes of a graph using the least number of colors such that two adjacent nodes have different colors. Similarly, the problem of coloring in edges deals with the assignment of colors at the edges of a graph using the least number of colors such that two adjacent edges have different colors.

The coloring of graphs used in this work is significantly different than the types cited above. The coloring in of the edges is not related to determined constraints, but in this case the colors represent the different transport modes in a network and the purpose is to find the optimal structure in the network across shortest paths. As mentioned above, Viedma [25] addresses the problem of multimodal transport networks using coloring in graphs based on Dijkstra's algorithm [7].

Let $G = (N, A, L)$ be a directed multigraph and colored consisting of a set of nodes ($n \in N$), a set of colors (or labels) $l \in L$, and a set of rotulated edges (I, j, l), which are triple in $N \times N \times L$, with $i, \; j \in N, l \in L$. Each color (label), $l \in L$, represents a mode of transport; thus, (i, j, l) represents an edge connecting the node i to node j with mode of transport l.

The shortest path problem in multimodal transport networks can be formulated as follows:

$$\min \; z = \sum_{(i,j)\in A} \sum_{l\in L} c_{ijl} x_{ijl}$$

$$\text{s.t} \sum_{l\in L} \sum_{j:(i,j)\in A} x_{ijl} - \sum_{l\in L} \sum_{j:(j,i)\in A} x_{jil} = \begin{cases} 1, & i = o \\ 0, & i \neq o, d \\ -1, & i = d \end{cases} \tag{3.1}$$

$$x_{ijl} = \{0, 1\}, \;\; \forall (i, j) \in A, \;\; \forall l \in L,$$

where:

- o: origin node.
- d: destination node.
- x_{ijl}: decision variable.

The first constraint guarantees that the graph is balanced, the second constraint guarantees the non-negativity of the decision variables in addition to the values that it can assume: 1 if the arc (i, j, l) belongs to the path or 0 if the arc (i, j, l) does not belong to the path. The objective function minimizes the sum of the costs of the edges that make up the shortest path.

3.3.1 *Algorithm 1*

The proposed algorithm is an adaptation of the classical Bellman–Ford–Moore algorithm [3] for colored graphs to be applied in multimodal transport networks. This algorithm is iterative, having as a criterion stopping the number of iterations, or there is no change in the costs encountered in the previous iteration in relation to the current iteration.

In the generalization made, each node has a set of labels $etq(j, l_j, ant, rot, cust)$ where:

- j: is the node under analysis.
- l_j: is the number of labels of the node j.
- ant: is the node i that precedes the node j by the considered path.
- rot: stores the mode (or label) used to arrive at the node i until j
- $cust$: stores the accumulated cost of origin until the node j (including the cost of the mode change, if any).

Notations for the Algorithm

- N: number of nodes in the graph.
- Γ_j^{-1}: set of nodes that precedes the node j.
- it: counter of iterations.

- d_{ijl}: edge cost (i, j) by mode l.
- $Nmode$: number of modes considered
- ϵ: cost of mode change, which in this case is fixed.

The following are the steps of the algorithm.

- Setp 0: Initialization:
  - $ant = [\ ]$.
  - $rot = [\ ]$.
  - $cust^0(origin) = 0$.
  - $cust^0(j) = \infty$ if $j \neq$ origin.
  - $it \leftarrow 1$.

- Setp 1:

  For all $j \in N, i \in \Gamma^{-1}(j), l_i, l \in Nmode$:

  If $rot(j) \neq rot(i)$: $cust^{it}(j) = cust^{it-1}(j) + d_{ijl} + \epsilon$.

  Otherwise

  - $cust^{it}(j) = cust^{it-1}(j) + d_{ijl}$.
  - $ant(j) = i$.
  - $rot(j) = l$.

  End otherwise

  End if

  End j

- Step 2: Stop criterion:
  1. If $it = it + 1 \geq$ edge number, or if $cust^{it} = cust^{it-1}, \quad \forall j \in N$, go to Step 3;
  2. Otherwise, go to Step 1.

- Step 3: Recompose the paths from the labels constructed in Step 1.

In this algorithm, it is necessary that the number of labels per node is limited, as it tends to increase rapidly as the number of nodes, the number of edges, and the number of transport modes increases, making it expensive to calculate each piece of information. The classic Bellman–Ford–Moore algorithm examines all nodes until no improvements are possible, thereby accepting edges with negative costs. As the proposed algorithm is based on the Bellman–Ford–Moore algorithm, the structure is maintained.

3.3.2 *Examples*

We present the computational tests performed in some instances with the algorithms proposed in the previous subsection. The algorithms were implemented in MATLAB 7.0.1 in a platform Intel I3, 6 Gb of RAM memory, with Windows 7 operating system.

It is noteworthy that owing to the great difficulty in finding, in the literature, instances to test Algorithm 1, all the examples presented in this subsection were created by the authors. This difficulty is because the algorithm is innovative.

3.3.2.1 Instance 1

This example illustrates the model proposed in Sect. 3.3 where we consider two transport modes between each pair of nodes in the graph. This is a small network, for illustrative purposes only, containing four nodes and eight edges, Fig. 3.1.

The edges, the costs, and the transport modes are defined in Table 3.1. The blue edges refer to mode 1 and the red edges refer to mode 2.

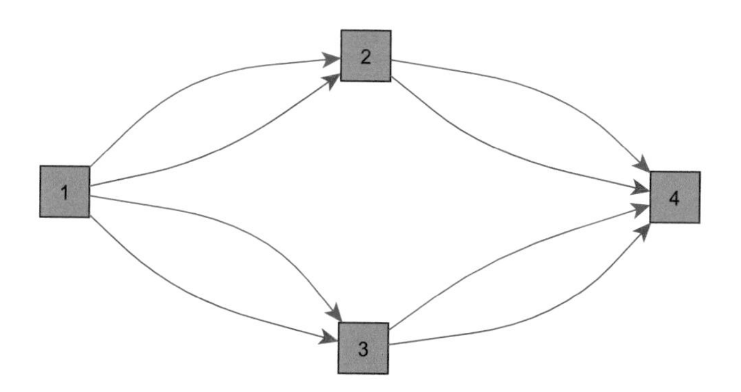

Fig. 3.1 Illustrative network

Table 3.1 Data from the network of Fig. 3.1

Edge	Origin→ Destination	Cost	Mode
1	1 → 2	4	1
2	1 → 2	5	2
3	1 → 3	3	1
4	1 → 3	5	2
5	2 → 4	3	1
6	2 → 4	2	2
7	3 → 4	4	1
8	3 → 4	4	2

The following are tests for other cost of mode change values.[1]

- Considering the cost of mode change equal to 0, Algorithm 1 found the shortest path $1 \longrightarrow^1 2 \longrightarrow^2 4$ with cost 6.
- Considering the cost of mode change equal to 0.5, the shortest path found is $1 \longrightarrow^1 2 \longrightarrow^2 4$ with a cost of 6.5.
- Considering the cost of mode change equal to 1, the shortest path found is $1 \longrightarrow^1 2 \longrightarrow^1 4$ with a cost of 7. In this case, there are other paths with the same cost: $1 \longrightarrow^1 3 \longrightarrow^1 4$ and $1 \longrightarrow^2 2 \longrightarrow^2 4$. The choice of path is made through the order that they appear in the execution of the algorithm, noting that in this example, we consider two labels per node (number of modes considered in the network).

This illustrative example shows us that as we alter the cost of mode change value, the shortest paths can change, but from a given value to the cost of mode change, the shortest path found is always the same, in this case, for values greater than or equal to 1.

Note also that when examining the node 4, we have eight paths arriving at such a node, which makes it necessary to limit the number of labels per node, as previously mentioned.

3.3.2.2 Instance 2

In this example, two transport modes are considered between each pair of nodes in the graph. This is a small network, containing four nodes and ten edges, Fig. 3.2.

The edges, the costs and the transport modes are defined in Table 3.2. The blue edges refer to mode 1 and the red edges refer to mode 2.

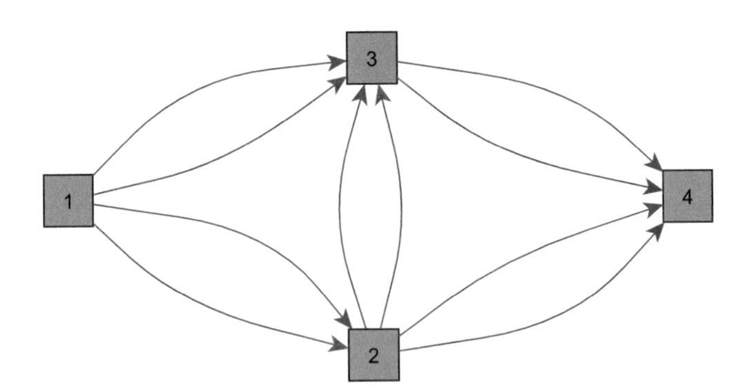

Fig. 3.2 Network with four nodes and two modes of transport

[1]The number above the arrow at each edge indicates the mode of transport used to traverse it.

Table 3.2 Data from the network of Fig. 3.2

Edge	Origin→ Destination	Cost	Mode
1	1 → 2	6	1
2	1 → 2	6	2
3	1 → 3	5	1
4	1 → 3	4	2
5	2 → 3	2	1
6	2 → 3	1	2
7	2 → 4	4	1
8	2 → 4	2	2
9	3 → 4	5	1
10	3 → 4	4	2

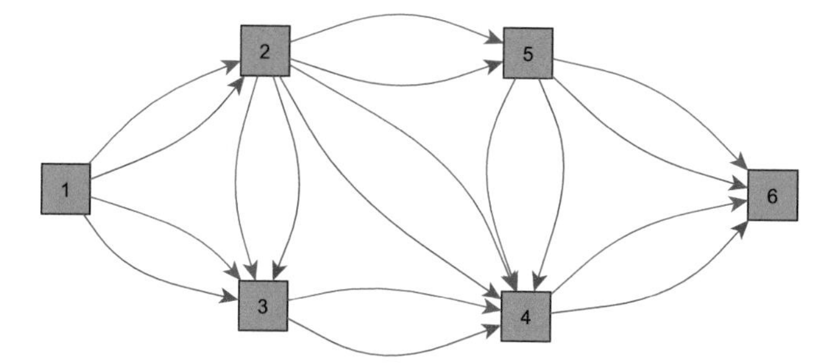

Fig. 3.3 Network with six nodes and two modes of transport

In this example, we also consider different values for the cost of mode change, but in all cases, Algorithm 1 found the shortest path: $1 \longrightarrow^2 2 \longrightarrow^2 4$ with cost 8.

3.3.2.3 Instance 3

In this example, two transport modes are considered between each pair of nodes in the graph. The network contains six nodes and 18 edges, Fig. 3.3.

The edges, the costs, and the transport modes are defined in Table 3.3. The blue edges refer to mode 1 and the red edges refer to mode 2.

Table 3.4 shows the results obtained through Algorithm 1 for different values of the cost of mode change.

When the cost of mode change is zero, the path $1 \longrightarrow^1 3 \longrightarrow^1 4 \longrightarrow^2 6$ also has a cost of 9. This happens because the edge $4 \longrightarrow 6$ has the same cost for the two transport modes considered. For change cost values greater than zero, the shortest path found is always the same, in this case $1 \longrightarrow^1 3 \longrightarrow^1 4 \longrightarrow^1 6$.

Note that it would not be necessary to test non-zero cost of mode change values because, as there was no change in mode at this cost, increasing would also have no shortest path with mode change.

Table 3.3 Data from the network of Fig. 3.3

Edge	Origin→ Destination	Cost	Mode
1	$1 \rightarrow 2$	4	1
2	$1 \rightarrow 2$	4	2
3	$1 \rightarrow 3$	2	1
4	$1 \rightarrow 3$	3	2
5	$2 \rightarrow 3$	3	1
6	$2 \rightarrow 3$	3	2
7	$2 \rightarrow 4$	6	1
8	$2 \rightarrow 4$	5	2
9	$2 \rightarrow 5$	5	1
10	$2 \rightarrow 5$	4	2
11	$3 \rightarrow 4$	5	1
12	$3 \rightarrow 4$	6	2
13	$4 \rightarrow 6$	2	1
14	$4 \rightarrow 6$	2	2
15	$5 \rightarrow 4$	5	1
16	$5 \rightarrow 4$	2	2
17	$5 \rightarrow 6$	1	1
18	$5 \rightarrow 6$	2	2

Table 3.4 Results obtained through Algorithm 1

Cost of mode change	Shortest path	Path cost
0	$1 \longrightarrow^1 3 \longrightarrow^1 4 \longrightarrow^1 6$	9
0.5	$1 \longrightarrow^1 3 \longrightarrow^1 4 \longrightarrow^1 6$	9
2	$1 \longrightarrow^1 3 \longrightarrow^1 4 \longrightarrow^1 6$	9

3.3.2.4 Instance 4

In this example, the multimodal network consists of 17 nodes and 52 edges, containing three modes of transport. The nodes 1 and 6 are origin nodes and the node 17 is the destination. The green edges refer to mode 1, the blue edges refer to mode 2, and the red edges refer to mode 3, Fig. 3.4.

The edges, costs, and modes of transport are defined in Table 3.5.

Tables 3.6 and 3.7 present the results obtained through Algorithm 1 for different values of the cost of mode change, considering nodes 1 and 6 as origin nodes.

Tests were carried out for other cost of mode change values and it was observed that for values greater than or equal to 8, the shortest path found is always the same.

Analyzing the results obtained in each instance, we conclude that the cost of mode change value interferes with the shortest path found, which is consistent with what happens in real problems, because users are generally unwilling to change mode, and often there is no other option.

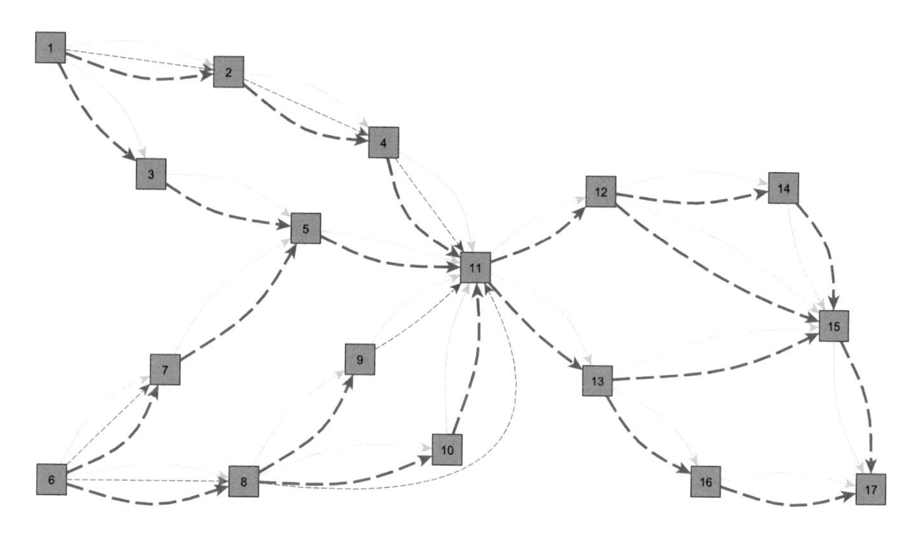

Fig. 3.4 Network with three modes of transport between some nodes

Table 3.5 Data of the network in Fig. 3.4

Edge	Origin→Destination	Cost	Modes
1	1 → 2	4	1
2	1 → 2	7	2
3	1 → 2	3	3
4	1 → 3	2	1
5	1 → 3	7	2
6	2 → 4	3	1
7	2 → 4	4	2
8	2 → 4	7	3
9	3 → 5	5	1
10	3 → 5	3	2
11	4 → 11	7	1
12	4 → 11	6	2
13	4 → 11	5	3
14	5 → 11	6	1
15	5 → 11	5	2
16	6 → 7	2	1
17	6 → 7	7	2
18	6 → 7	2	3
19	6 → 8	4	1
20	6 → 8	5	2
21	6 → 8	2	3
22	7 → 5	8	1
23	7 → 5	3	2

(continued)

Table 3.5 (continued)

Edge	Origin→Destination	Cost	Modes
24	8 → 9	3	1
25	8 → 9	4	2
26	8 → 10	3	1
27	8 → 10	5	2
28	8 → 11	13	3
29	9 → 5	4	1
30	9 → 5	1	2
31	9 → 11	5	1
32	9 → 11	2	3
33	10 → 11	3	1
34	10 → 11	6	2
35	11 → 12	12	1
36	11 → 12	2	2
37	11 → 13	11	1
38	11 → 13	4	2
39	12 → 14	4	1
40	12 → 14	5	2
41	12 → 15	10	1
42	12 → 15	12	2
43	13 → 15	5	1
44	13 → 15	7	2
45	13 → 16	8	1
46	13 → 16	9	2
47	14 → 15	4	1
48	14 → 15	4	2
49	15 → 17	5	1
50	15 → 17	3	2
51	16 → 17	3	1
52	16 → 17	2	2

Table 3.6 Results obtained through Algorithm 1 for the origin node 1

Cost of mode change	Shortest path	Path cost
0	$1 \longrightarrow^1 3 \longrightarrow^2 5 \longrightarrow^2 11 \longrightarrow^2 13 \longrightarrow^1 15 \longrightarrow^2 17$	22
0.5	$1 \longrightarrow^1 3 \longrightarrow^2 5 \longrightarrow^2 11 \longrightarrow^2 13 \longrightarrow^1 15 \longrightarrow^2 17$	23.5
2	$1 \longrightarrow^1 3 \longrightarrow^2 5 \longrightarrow^2 11 \longrightarrow^2 13 \longrightarrow^2 15 \longrightarrow^2 17$	26
4	$1 \longrightarrow^1 3 \longrightarrow^1 5 \longrightarrow^1 11 \longrightarrow^2 13 \longrightarrow^2 15 \longrightarrow^2 17$	31
8	$1 \longrightarrow^1 3 \longrightarrow^1 5 \longrightarrow^1 11 \longrightarrow^1 13 \longrightarrow^1 15 \longrightarrow^1 17$	34

3.4 Minimum Cost Flow Problem in Colored Graphs

This section introduces the multimodal transport network problem considering costs and capacities *crisp*. As in Sect. 3.3, we use the colored graphs to model the problem, where each transport mode considered is represented by a color. The costs

Table 3.7 Results obtained through Algorithm 1 for the origin node 6

Cost of mode change	Shortest path	Path cost
0	$6 \longrightarrow^{3} 8 \longrightarrow^{1} 9 \longrightarrow^{3} 11 \longrightarrow^{2} 13 \longrightarrow^{1} 15 \longrightarrow^{2} 17$	19
0.5	$6 \longrightarrow^{3} 8 \longrightarrow^{1} 9 \longrightarrow^{3} 11 \longrightarrow^{2} 13 \longrightarrow^{1} 15 \longrightarrow^{2} 17$	21.5
2	$6 \longrightarrow^{3} 8 \longrightarrow^{1} 10 \longrightarrow^{1} 11 \longrightarrow^{2} 13 \longrightarrow^{2} 15 \longrightarrow^{2} 17$	26
4	$6 \longrightarrow^{3} 8 \longrightarrow^{1} 10 \longrightarrow^{1} 11 \longrightarrow^{2} 13 \longrightarrow^{2} 15 \longrightarrow^{2} 17$	30
8	$6 \longrightarrow^{3} 8 \longrightarrow^{3} 11 \longrightarrow^{2} 13 \longrightarrow^{2} 15 \longrightarrow^{2} 17$	37

at the edges depend on the flow in them; thus, the formulation of this type of problem has a nonlinear objective function, as described in (3.2).

$$\min \ z = \sum_{(i,j)\in A} \sum_{l\in L} t_{ij}(x_{ij}) x_{ijl}$$

$$\text{s.t} \begin{cases} \sum_{j:(i,j)\in A} x_{ijl} - \sum_{j:(j,i)\in A} x_{jil} = b_{il}, \forall i \in N, \quad \forall l \in L \\ x_{ijl} \le u_{ijl}, \ \forall (i,j) \in A, \ \forall l \in L \\ x_{ij} = \sum_{l\in L} x_{ijl} \\ x_{ijl} \ge 0, \ \forall (i,j) \in A, \ l \in L \end{cases}, \tag{3.2}$$

where

- L is the set of all considered modes of transport.
- l is a mode of transport.
- x_{ijl} is the flow at the edge (i, j) using the mode l.
- x_{ij} is the total flow at the edge (I, j).
- b_{il} is the supply or demand from the node i into the mode l.
- $t_{ij}(x_{ij})$ is the travel time required to go through the edge (i, j) dependent on the flow at the edge (x_{ij}).
- u_{ijl} is the capacity of edge (i, j) using the mode l.

The travel time function is modeled as a cost function proposed by the BPR in 1964 [6]:

$$t_{ij}(x_{ij}) = t_0 \left[1 + \rho \left(\frac{x_{ij}}{u_{ij}} \right)^{\lambda} \right], \tag{3.3}$$

where

- $u_{ij} = \sum_{l\in L} u_{ijl}$.
- t_{ij} is travel time at the edge (i, j).
- t_0 is a free flow travel time.
- x_{ij} is the flow at the edge (i, j).
- ρ and λ are parameters of the model: $\rho = 0.15$ and $\lambda = 4$.

3.4.1 *Algorithm 2*

The second proposed algorithm is a flow incremental loading algorithm. Such an algorithm was chosen for its simplicity and efficiency and is a proposed solution for the multimodal transport network problem in which the shortest path is found through the algorithm described in Sect. 3.3.1. The following are the steps of the algorithm.

- Step 0 (Initialization): Set I (the number of increments). Let n be the number of iterations.
- Step 1: Find the colored shortest path using Algorithm 1.
- Step 2: Send flow incrementally:
  1. Let b the flow through the network. Send flow for the shortest path found in Step 1 incrementally $\left(\frac{b}{I}\right)$, respecting the capacity of each edge path.
  2. Update the costs at the edges through the function given by (3.3).
- Step 3: Stop criterion:
  1. If $n \leq I$ or there is a flow to transit, go to Step 1, do $n = n + 1$.
  2. Otherwise: the end.

3.4.2 *Examples*

In this section, the results obtained through Algorithm 2 presented, as described in Sect. 3.4.1. Three instances were used, Instances 1, 3, and 4 of Sect. 3.3.2, where tests were performed for different values in the number of increments.

The analysis of the results was made based on the flow sent by each path, the updated cost of each path, the total final cost, and the number of increments. The updated cost is calculated after the last portion of the flow has been sent.

In all examples presented, tests were performed with different cost of mode change values, but the results presented were obtained by considering the cost of mode change equal to 0.5. As this cost influences the shortest paths found in each iteration, and in this algorithm, the main objective is the incremental load of the flow, we chose to use only one cost of mode change value in the results described as follows.

3.4.2.1 Instance 1

The network considered is the same as in Fig. 3.1, where two modes of transport between each pair of nodes of the graph are considered. The demand from origin 1 to destination 4 is 20.

Table 3.8 Data of the network in Fig. 3.1

Edge	Origin→ Destination	Cost	Mode
1	$1 \to 2$	4	1
2	$1 \to 2$	5	2
3	$1 \to 3$	3	1
4	$1 \to 3$	5	2
5	$2 \to 4$	3	1
6	$2 \to 4$	2	2
7	$3 \to 4$	4	1
8	$3 \to 4$	4	2

Table 3.9 Flow sent for $I = 1$

Path	Flow sent	Initial cost	Updated cost
$1 \to^1 2 \to^2 4$	10	6.5	7.4
$1 \to^1 3 \to^1 4$	10	7	8.05

Table 3.10 Flow sent for $I = 2$

Path	Flow sent	Initial cost	Updated cost
$1 \to^1 2 \to^2 4$	10	6.5	7.4
$1 \to^1 3 \to^1 4$	10	7	8.05

Table 3.11 Flow sent for $I = 4$

Path	Flow sent	Initial cost	Updated cost
$1 \to^1 2 \to^2 4$	10	6.5	7.4
$1 \to^1 3 \to^1 4$	10	7	8.05

The edges, costs, and transport modes are defined in Table 3.8. The total capacity (maximum amount of flow to be carried to each edge independent of the mode of transport) of each edge is 10, that is, capacity is not defined by mode but by edge.

Algorithm 2 meets the network demand according to the number of increments. The tests were performed for different values of I (number of increments).

The flow sent to $I = 1$ occurred according to Table 3.9. The processing time was 0.1 s. The final total cost was 154.5.

The flow sent to $I = 2$ occurred according to Table 3.10. The processing time was 0.2 s. The final total cost was 154.5.

The flow sent to $I = 4$ occurred according to Table 3.11. The processing time was 0.4 s. The final total cost was 154.5.

In this example, the results obtained for the three values of I is the same, although it is not the ideal result, because, in flow incremental loading algorithms, the ideal is that as the I value increases, the solution found changes and the final total cost decreases. The equality of the results obtained in this example is justified by the fact that the path $1 \to^1 2 \to^2 4$ remains the shortest path after sending several parts of flow (for $I = 4$); thus, such a path is used until the flow reaches the capacity of the edges of the problem. When this path is full, the shortest path is $1 \to^1 3 \to^1 4$, and the remaining flow passes by this path. It is noteworthy that this fact occurred only in this example.

Table 3.12 Data of the network in Fig. 3.3

Edge	Origin → Destination	Cost	Mode
1	$1 \rightarrow 2$	4	1
2	$1 \rightarrow 2$	4	2
3	$1 \rightarrow 3$	2	1
4	$1 \rightarrow 3$	3	2
5	$2 \rightarrow 3$	3	1
6	$2 \rightarrow 3$	3	2
7	$2 \rightarrow 4$	6	1
8	$2 \rightarrow 4$	5	2
9	$2 \rightarrow 5$	5	1
10	$2 \rightarrow 5$	4	2
11	$3 \rightarrow 4$	5	1
12	$3 \rightarrow 4$	6	2
13	$4 \rightarrow 6$	2	1
14	$4 \rightarrow 6$	2	2
15	$5 \rightarrow 4$	5	1
16	$5 \rightarrow 4$	2	2
17	$5 \rightarrow 6$	1	1
18	$5 \rightarrow 6$	2	2

Table 3.13 Flow sent for $I = 1$

Path	Flow sent	Initial cost	Updated cost
$1 \rightarrow^1 3 \rightarrow^1 4 \rightarrow^1 6$	40	9	10.35
$1 \rightarrow^2 2 \rightarrow^2 5 \rightarrow^2 6$	20	10	10.0937

Table 3.14 Flow sent for $I = 3$

Path	Flow sent	Initial cost	Updated cost
$1 \rightarrow^1 3 \rightarrow^1 4 \rightarrow^1 6$	40	9	10.35
$1 \rightarrow^2 2 \rightarrow^2 5 \rightarrow^2 6$	20	10	10.0937

3.4.2.2 Instance 2

In this example, we use the network of Fig. 3.3. The edges, costs, and the transport modes are defined in Table 3.12. The total capacity at each edge is 40. The black edges refer to mode 1 and the blue edges refer to mode 2. The demand from origin 1 to destination 6 is 60. The cost of change is 0.5.

All demands were met according to the number of increments. Tests were performed for different values of I.

The flow sent to $I = 1$ occurred according to Table 3.13. The processing time was 0.1 s and the final total cost was 615.8740.

The flow sent to $I = 3$ occurred according to Table 3.14. The processing time was 0.2 s and the final total cost was 615.8740.

The flow sent to $I = 6$ occurred according to Table 3.15. The processing time was 0.4 s and the final total cost was 614.9960.

Table 3.15 Flow sent for $I = 6$

Path	Flow sent	Initial cost	Updated cost
$1 \to^1 3 \to^1 4 \to^1 6$	40	9	10.35
$1 \to^2 2 \to^2 5 \to^2 6$	10	10	10.0058
$1 \to^1 2 \to^1 5 \to^1 6$	10	10	10.0938

In this case, the results obtained results for $I = 1$ and $I = 3$ were the same. For $I = 6$ the results were different and as expected, they were better.

3.4.2.3 Instance 3

In this example, we use the network of Fig. 3.4, with 17 nodes, 52 edges, and three transport modes. The node 1 is the origin node and the destination is the node 17 with a demand of 60. The edges, the costs, and the transport modes are defined in Table 3.16. The total capacity at each edge is 50, with the exception of edge $15 \to 17$, which is 80.

All demands were met according to the number of increments. Tests were performed for different values of I.

The flow sent to $I = 1$ occurred according to Table 3.17. The processing time was 2 s and the final total cost was 1596.

The flow sent to $I = 2$ occurred according to Table 3.18. The processing time was 3 s and the final total cost was 1596.

The flow sent to $I = 5$ occurred according to Table 3.19. The processing time was 2 s and the final total cost was 1585.

In this case, the results obtained for $I = 1$ and $I = 3$ were the same. For $I = 5$, the results were different and as expected, the final total cost was lower.

3.5 Final Considerations

A smart transport system should be able to provide users with access to the various sectors of a city through different transport modes that connect them quickly, safely, and efficiently. To do this, it is necessary to adapt the capabilities of transport modes to passengers, and to develop an advanced traffic control system that is able to verify the integration and efficiency of the transport system as a whole.

To propose transformations and innovations for the transport system, several changes and investments are necessary, given the precariousness and slowness, mainly in large centers.

Smart cities depend heavily on the available transport systems, which is why they have been extensively studied and applied to solutions to real problems. As the complexity of such problems consumes a large amount of computational resources

Table 3.16 Data of the network in Fig. 3.4

Edge	Origin→ Destination	Cost	Mode
1	$1 \to 2$	4	1
2	$1 \to 2$	7	2
3	$1 \to 2$	3	3
4	$1 \to 3$	2	1
5	$1 \to 3$	7	2
6	$2 \to 4$	3	1
7	$2 \to 4$	4	2
8	$2 \to 4$	7	3
9	$3 \to 5$	5	1
10	$3 \to 5$	3	2
11	$4 \to 11$	7	1
12	$4 \to 11$	6	2
13	$4 \to 11$	5	3
14	$5 \to 11$	6	1
15	$5 \to 11$	5	2
16	$6 \to 7$	2	1
17	$6 \to 7$	7	2
18	$6 \to 7$	2	3
19	$6 \to 8$	4	1
20	$6 \to 8$	5	2
21	$6 \to 8$	2	3
22	$7 \to 5$	8	1
23	$7 \to 5$	3	2
24	$8 \to 9$	3	1
25	$8 \to 9$	4	2
26	$8 \to 10$	3	1
27	$8 \to 10$	5	2
28	$8 \to 11$	13	3
29	$9 \to 5$	4	1
30	$9 \to 5$	1	2
31	$9 \to 11$	5	1
32	$9 \to 11$	2	3
33	$10 \to 11$	3	1
34	$10 \to 11$	6	2
35	$11 \to 12$	12	1
36	$11 \to 12$	2	2
37	$11 \to 13$	11	1
38	$11 \to 13$	4	2
39	$12 \to 14$	4	1
40	$12 \to 14$	5	2

(continued)

Table 3.16 (continued)

Edge	Origin→ Destination	Cost	Mode
41	12 → 15	10	1
42	12 → 15	12	2
43	13 → 15	5	1
44	13 → 15	7	2
45	13 → 16	8	1
46	13 → 16	9	2
47	14 → 15	4	1
48	14 → 15	4	2
49	15 → 17	5	1
50	15 → 17	3	2
51	16 → 17	3	1
52	16 → 17	2	2

Table 3.17 Flow sent for $I = 1$

Path	Flow sent	Initial cost	Updated cost
$1 \rightarrow^1 3 \rightarrow^2 5 \rightarrow^2 11 \rightarrow^2 13 \rightarrow^1 15 \rightarrow^2 17$	50	23.5	26.4924
$1 \rightarrow^1 2 \rightarrow^2 4 \rightarrow^2 11 \rightarrow^2 12 \rightarrow^1 14 \rightarrow^2 15 \rightarrow^2 17$	10	26.0687	27.1477

Table 3.18 Flow sent for $I = 2$

Path	Flow sent	Initial cost	Updated cost
$1 \rightarrow^1 3 \rightarrow^2 5 \rightarrow^2 11 \rightarrow^2 13 \rightarrow^1 15 \rightarrow^2 17$	50	23.5	26.4924
$1 \rightarrow^1 2 \rightarrow^2 4 \rightarrow^2 11 \rightarrow^2 12 \rightarrow^1 14 \rightarrow^2 15 \rightarrow^2 17$	10	26.0687	27.1477

Table 3.19 Flow sent for $I = 5$

Path	Flow sent	Initial cost	Updated cost
$1 \rightarrow^1 3 \rightarrow^2 5 \rightarrow^2 11 \rightarrow^2 13 \rightarrow^1 15 \rightarrow^2 17$	48	23.5	26.2890
$1 \rightarrow^1 3 \rightarrow^2 5 \rightarrow^2 11 \rightarrow^2 12 \rightarrow^2 14 \rightarrow^2 15 \rightarrow^2 17$	2	25.8323	26.1474
$1 \rightarrow^3 2 \rightarrow^1 4 \rightarrow^2 11 \rightarrow^2 12 \rightarrow^1 14 \rightarrow^1 15 \rightarrow^2 17$	10	26.0687	27.1502

and, in most cases, these problems cannot be solved using deterministic tools that provide exact results, the search for other forms of resolution is of great importance.

In this work, two new approaches to the multimodal transport network problem are presented.

In the first, we deal with the shortest path problem in multimodal networks using colored graphs. When considering different values for the cost of mode change, different solutions have been found, a fact that occurs in real problems, because users are not usually willing to change their mode of transport, but make this change only when there is no other choice. The algorithm detects the mode changes through the coloring of the edges and a cost related to this change is added to the cost of the path. The use of coloring in graphs, in this case, makes the algorithm more intelligent and closer to reality, as different modes of transport, such as private vehicles, buses, and vans, share the same network and the coloration allows differentiation among them. It is worth remembering that this method is innovative, because in the literature we find only one work that proposes a similar algorithm, based on Dijkstra's classic algorithm.

In the second approach, we study the flow problem in multimodal transport networks, where we first find the shortest paths using the algorithm developed in the initial approach and then distribute the flow through the network incrementally, via the paths found. Through the incremental load of flow, it is possible to distribute the flow through the network in a balanced way, without congesting the shortest paths used. Another important aspect of this algorithm is the fact that the cost at the edges depends on the flow that passes through them, which makes the algorithm more adherent to reality. Finally, it should be remembered that this incremental flow loading algorithm is not restricted to multimodal transport network problems, and can be applied to other types of problems that can be modeled in the network form.

References

1. Abbaspour, R.A., Samadzadegan, F.: An evolutionary solution for multimodal shortest path problem in metropolises. Comput. Sci. Inf. Syst. **7**(4), 1–24 (2010)
2. Ahuja, T.L., Magnanti, R.K.: Network Flows. Prentice Hall, Philadelphia (1993)
3. Bellman, R.E.: On a routing problem. Q. Appl. Math. **16**, 87–90 (1958)
4. Bieli, M., Boumakoul, A., Mouncif, H.: Object modeling and path computation for multimodal travel systems. Eur. J. Oper. Res. **175**, 1705–1730 (2006)
5. Brito, J., Martínez, F.J., Moreno, J.A., Verdegay, J.L.: Fuzzy approach for vehicle routing problems with fuzzy travel time. In: International Conference on Fuzzy Systems, Barcelona (2010)
6. Bureau of Public Roads, Traffic assignment manual. Tech. Report, U.S. Department of Commerce, 1964
7. Dijkstra, E.W.: A note on two problems in connexion with graphs. Numer. Math. **1**, 269–271 (1959)
8. Dubois, D., Prade, H.: Fuzzy Sets and Systems: Theory and Applications. Academic, New York (1980)
9. Golnarkar, A., Alesheikh, A.A., Malek, M.R.: Solving best path on multimodal transportation networks with fuzzy costs. Iran. J. Fuzzy Syst. **7**(3), 1–13 (2010)
10. Heid, A., Galvez-Fernandez, C., Habbas, Z., Khadraoui, D.: Solving time-dependent multimodal transport problems using a transfer graph model. Comput. Ind. Eng. **61**(2), 391–401 (2010)
11. Lam, S.K., Srikanthan, T.: Accelerating the K-shortest paths computation in multimodal transportation networks. In: 5th International Conference on Intelligent Transportation Systems, Singapore (2002)
12. Lillo, F., Schmidt, F.: Optimal paths in real multimodal transportation networks: an appraisal using GIS data from New Zealand and Europe. In: Proceedings of the 45th Annual Conference of the ORSNZ (2010)
13. Loureiro, C.F.G.: Geração de colunas na solução de problemas de desenhos de redes multimodais de transportes. In: XVII Enc. Nac. de Eng. de Prod., Porto Alegre-RS (1997)
14. Lozano, A., Storchi, G.: Shortest viable path algorithm in multimodal networks. Transp. Res. **35**, 225–241 (2001)
15. Lozano, A., Storchi, G.: Shortest viable hyperpath in multimodal networks. Transp. Res. Part B **36**, 853–874 (2002)

16. Miller, H.J., Storm, J.D., Bowen, M.: GIS design for multimodal networks analysis. In: GIS/LIS '95 Annual Conference and Exposition Proceedings of GIS/LIS, Nashville, USA, pp. 750–759 (1995)
17. Modesti, P., Sciomachen, A.: A utility measure for finding multiobjective shortest paths in urban multimodal transportations networks. Eur. J. Oper. Res. **111**, 495–508 (1998)
18. Mouncif, H., Boulmakoul, A., Chala, M.: Integrating GIS-technology for modelling origin-destination trip in multimodal transportation networks. Int. Arab J. Inf. Tech. **3**, 256–263 (2006)
19. Mouncif, H., Rida, M., Boulmakoul, A.: An efficient multimodal path computation integrated within location based service for transportation networks system (multimodal path computation within LBS). J. Appl. Sci. **11**(1), 1–15 (2011)
20. Pandian, P., Natarajan, G.: A new method for finding an optimal solution of fully interval integer transportation problems. Appl. Math. Sci. **4**(37), 1819–1830 (2010)
21. Qu, L., Chen, Y.: A hybrid MCDM method for route selection of multimodal transportation network. Lect. Notes Comput. Sci. **5263**, 374–383 (2008)
22. Ramazani, H., Shafahi, Y., Seyedabrishami, S.E.: A shortest path problem in an urban transportation network based on driver perceived travel time. Sci. Iranica A **17**(4), 285–296 (2010)
23. Ramazani, H., Shafahi, Y., Seyedabrishami, S.E.: A fuzzy traffic assignment algorithm based on driver perceived travel time of network links. Sci. Iranica A **18**(2), 190–197 (2011)
24. Verga, J., Yamakami, A., Silva, R.C.: Multimodal transport network problem: classical and innovative approaches. In: Corona, C.C. (ed.) Soft Computing for Sustainability Science, pp. 299–332, 1st edn. Springer, New York (2018)
25. Viedma, F.E.L.: Coloured-edge graph approach for the modelling of multimodal networks. PhD. Thesis, Auckland University of Technology (2011)
26. Xin-Bo, W., Gui-Jun, Z., Zhen, H., Hai-Feng, G., Li, Y.: Modeling and implementing research of multimodal transportation network. In: The 1st International Conference on Information Science and Engineering, Nanjing, pp. 2100–2103 (2009)
27. Yu, H., Lu, F.: A multimodal route planning approach with an improved genetic algorithm. Int. Arch. Photogramm. Remote. Sens. Spat. Inf. Sci. **38**(2), 343–348 (2011)
28. Ziliaskopoulos, A., Wardell, W.: An intermodal optimum path algorithm for multimodal networks with dynamic arc travel times and switching delays. Eur. J. Oper. Res. **125**, 486–502 (2000)

Chapter 4
An Adaptive Large Neighborhood Search Heuristic to Solve the Crew Scheduling Problem

Leandro do Carmo Martins and Gustavo Peixoto Silva

Abstract This paper presents an adaptation of the adaptive large neighborhood search (ALNS) heuristic in order to solve the crew scheduling problem (CSP) of urban buses. The CSP consists in minimizing the total of crews that will drive a fleet in daily operation as well as the total overtime. The solution for the CSP is a set of duties performed by the crews throughout the day, and those duties must comply with labor laws, labor union agreements, and the company's operational rules. The CSP is a NP-hard problem and it is usually solved by metaheuristics. Therefore, an ALNS-like heuristic was developed to solve the CSP. Its implementation was tested with real data from a bus company which operates in Belo Horizonte, MG-Brazil, and it provided solutions quite superior to those both adopted by the company and the ones generated by other methods in the literature.

4.1 Introduction

The planning of the urban public transportation system is a complex and step-by-step process involving both the urban transport companies and the public authorities. This planning is made up of five phases: (1) definition of the buses routes, (2) definition of the trip timetable for the whole week, (3) scheduling of the vehicles for the trips, (4) daily schedule of the crews who will drive the vehicles, and (5) the crew roster of the crews over a giving planning period [2].

Among the planning phases, the companies must schedule the vehicles and the crews, and also their roster. Thus, the public transport companies must define the size of their fleets, the number of crews (one driver and one collector per bus), and their working and leisure times, so that all the trips may follow the rules with the

L. do Carmo Martins
IN3 Department of Computer Science, Open University of Catalonia, Barcelona, Spain

G. P. Silva (✉)
Department of Computer Science, Federal University of Ouro Preto,
Campus Universitário Morro do Cruzeiro, Ouro Preto, Brazil
e-mail: gustavo@iceb.ufop.br

V. N. Coelho et al. (eds.), *Smart and Digital Cities*, Urban Computing,
https://doi.org/10.1007/978-3-030-12255-3_4

lowest cost possible. This is a very complex task for which optimization methods can be used in their resolution [21].

The realization of the planned trips leads to the definition of a daily working schedule for drivers and collectors, called crew scheduling problem (CSP), which consists in determining the minimum number of crews and allocating to them the trips that must be performed on a particular work day, so that each trip is attributed to the duty of one crew, and that there is no overlap among the trips performed by the crews. The goal is to minimize their amount and the total overtime contained in the daily scale [2].

As input data, CSP has the vehicles scheduling previously defined for a particular kind of day (working day, Saturday, and Sunday/holiday). Vehicles scheduling defines the *vehicle blocks*, it means, the set of trips to be performed by each vehicle of the fleet. The blocks are split in *tasks*, which are a set of consecutive trips that must be performed by the same crew, once there are no enough conditions among those trips in order to change the crews. The necessary conditions to change two crews define a *relief opportunity*, which consists of a minimum time interval between two trips (*relief time*), and this interval time must occur in an appropriate location for that (*relief point*) [14, 17]. Such tasks are characterized by an initial and a final location, as well as a start and an ending time, which represents where and when the task must start and finish. Besides, the tasks are performed by an operating vehicle, previously defined, and its duration is due to the difference between its ending and starting times. A set of tasks performed by a crew is called *duty*. In this way, one crew is associated only to one duty and vice versa, and then they are considered synonymous from now on in this paper.

The duties can be classified as *straight duty* or *split duty*. In this last one, the work is divided into two pieces of work with an interval higher than 2 h between the first and second work piece. Split duty happens due to the peaks in trips demand, so common during the business days. This interval higher than 2 h is not considered as working time, so it is not paid to the crews. A straight duty does not present an interruption of this magnitude, and its duration is given by the difference between its finishing and starting time. On the other hand, the duration of a split duty is also given by the same difference, yet subtracting the interval higher than 2 h characterizes the job rules. Regardless of the kind of duty, the overtime is the amount of hours worked beyond the ordinary duty duration.

The duty construction must comply with a set of legal and operational restrictions, which makes the CSP hard to be solved. Fischetti et al. [5] show that CSP is a NP-hard problem. In the CSP resolution for the studied case, the following restrictions were taken into consideration: (1) the duties have a fixed remuneration for 6 h and 40 min of work, which defines the normal work duration; (2) the duties cannot exceed two overtime hours per day; (3) the interval between the end of one duty and its beginning in the next day must be 11 h or more; (4) the vehicles change, performed by a crew in operation, may only happen in predetermined places; (5) the quantity of vehicles change performed in one duty is limited; (6) the bus station change is allowed only between two tasks with an interval higher than 2 h (split duty).

This paper will present the CSP resolution by the adaptive large neighborhood search (ALNS) heuristic. ALNS is characterized by the *destruction* and *repair* of the current solution throughout several heuristics of *removal* and *insertion* used in the search. The heuristics used in each interaction are chosen by a dynamically adjusted weight based on the efficiency of each heuristic throughout the search. The developed implementation was tested with real data from a medium-sized company in Belo Horizonte, MG-Brazil, and the reached solutions outperformed both the solutions adopted by the company and the ones generated by other methods presented in the literature.

This paper is divided as follows: Sect. 4.2 presents a brief bibliographic review about the CSP. Section 4.3 describes how the ALNS was implemented in order to solve the CSP. Section 4.4 presents the results obtained from the computational experiments and the discussion of such results. Finally, Sect. 4.5 presents the conclusion of this paper.

4.2 Bibliographic Review

Usually, the CSP is formulated as a set covering or set partitioning problem and it used the technique of columns generation to solve it [14]. Lourenço et al. [7] formulate the CSP as a set covering problem, taking into account several objective-functions. The authors present multi-objectives metaheuristics to solve the CSP since, in practice, the problem presents different conflicting objectives. The metaheuristics base themselves on Tabu search and genetic algorithms (GA), and they have the mono-objective GRASP as internal procedure. For all considered problems, the multi-objective Tabu search and the multi-objective GA presented good results in a reasonable period of time.

Silva and Cunha [15] presented a new approach for the CSP based on the GRASP metaheuristic, using as local search the very large-scale neighborhood search (VLNS) [1]. In this approach, a neighbor solution is obtained by the reallocation and/or change of tasks between two or more duties, may even involving all the duties on the change. To search the neighborhood, an *improvement graph* is constructed. From this graph, there is a search for a *valid cycle*: a negative cycle that leads to the solution cost reduction. When a valid cycle is found, the changes in the cycle are performed and the graph is updated. The search is done when no negative cycle is found. The computational tests showed that the approach creates significant improvement comparing to the solution adopted by the company.

Reis and Silva [10] explore different methods of local search associated to the variable neighborhood search (VNS) metaheuristic in order to solve the CSP. The neighbor solutions of the current solution are obtained from the reallocation and/or tasks changes between two crews. The adopted methods of local search were the variable neighborhood descent (VND) and the VLNS [15]. It has been considered the different operational scenarios that allow one or two vehicles change per duty, and yet the scenario where the split duties are more expensive, in order to control its occurrence. Both techniques outperformed the solution of the company and the VNS-VLNS technique outperformed the classic VNS.

Chen and Shen [3] present a new strategy of columns generation to deal with the CSP. Traditionally, the columns to be added to the main problem are determined by the resolution of subproblems of computational high-cost. Therefore, the authors suggest that a set of duties reasonably large be considered. Such duties called shift-pool are precompiled with desired characteristics. During the process of columns generation, the subproblem is solved until there are no more columns with cost reduction on the shift-pool. This idea allows reducing the processing time to solve the CSP. Instances up to 701 trips, 55 vehicle blocks, and two (2) garages are considered in the experiments. The results show that the proposed algorithm outperforms the generation of traditional columns.

Souza [19] presents a hybrid model of integer linear programing along with the late acceptance hill climbing (LAHC) metaheuristic, in order to solve the CSP. It is presented a compact model composed of restrictions that allows solving the CSP directly from the tasks to be performed, unlike the classic models of set partitioning or set covering, where the duties must be generated and informed a priori. The LAHC aims to solve bigger problems. Therefore, the components of the solution are partitioned into smaller sets, which are submitted to the exact method. The model was tested with real problems and proved efficient in terms of solution generation, but in a high processing time.

Silva and Silva [16] apply the guided local search (GLS) metaheuristic to solve the CSP. GLS performs a local search in the current solution and it is different from the other metaheuristics by penalizing the indesirable characteristics of the current solution as a way to escape from local optimum. Thus, when the GLS finds a local optimum, it selects some attributes of the solution and penalizes them, making the solution no longer a local optimum and then moving the search to another local optimum. The process repeats until some stop criteria complies. The GLS was combined with the local search method variable neighborhood descent (VND), producing results similar to the ones in the literature.

Song et al. [18] approach the CSP with an improved GA with genes recombination, which is responsible for the duties reconstruction and capable of reducing the number of duties in the solution. Tabu and non-Tabu lists are used for the reconstruction of the initial solutions. The roulette method was adopted to select the individuals for the crossover characterized by two cut-off points. Mutation consists in the reconstruction of n duties randomly chosen. They are divided into segments that will be reconstructed and replaced. This method was capable of reducing over 30 duties when there are a large number of tasks, once it is a very fast algorithm.

The ALNS heuristic was proposed by Ropke and Pisinger [11] in order to solve the pickup and delivery problem with Time Windows. It starts from a feasible solution and applies destroy–reconstruct mechanisms to find better solutions. The removal and insertion heuristics are chosen from statistics gathered during the process. Three removal heuristics are used for that: Shaw, random, and worst; and two insertion ones: greedy and regret. Every iteration, one removal and one insertion heuristics are chosen independently, and then the pair is applied. All heuristics are used in the search, and the pair selection is made taking into account the individual weight of each heuristic. Those weights are dynamically adjusted due to the heuristics success. The ALNS was able to improve the best solutions known in the literature for over 50% of the problems, indicating the advantages of using concurrently several removal and insertion heuristics, instead of using only one pair of heuristics.

Later, the same authors integrated new removal heuristics to the ALNS. The three new removal heuristics: cluster, neighbor graph, and request graph were proposed and the problem started covering a class of vehicles routing problems (VRP) [12]. Lately, by covering distinct variants of the VRP, the authors proposed three new removal heuristics: time-oriented, node-pair removal, and historical request-pair [8].

Since the pickup and delivery problem with Time Windows is similar to the CSP, there is a hypothesis that ALNS can be efficient in its resolution. Thus, this paper presents a version of the ALNS heuristic focused on the CSP resolution, as well as the obtained results.

4.3 Adaptive Large Neighborhood Search Applied to the CSP

ALNS starts its search from an initial feasible solution and it works with different methods that destroy and repair the current solution. While a pair of those removal and insertion heuristics is adopted during the whole search in the large neighborhood search (LNS), several removal and insertion heuristics are used at the same search in the ALNS. A weight is initially attributed to each heuristic in order to find the most efficient pair for the problem. The weights are dynamically adjusted based on the efficiency of each heuristic throughout the search.

4.3.1 Solution Representation

For the CSP a solution is represented by a set of duties, and each duty has a sequence of tasks to be performed by the crew responsible for that. Such representation allows applying the removal and reinsertion mechanisms inherent to the ALNS.

4.3.2 Initial Solution

ALNS starts its search from an initial feasible solution. For CSP, it was generated in a greedy way. Due to a set of tasks to be performed in a particular day, the goal is to group them, respecting all the restrictions of the problem. A valid sequencing of tasks constitutes a duty and each task must be allocated to a unique duty.

The solution construction starts with an ordering of tasks by the increasing starting time. From the first task t_i in the ordered set of tasks, the tasks allocation begins. The task is included in the first duty. The next task t_j to be included is the one of least increase in the objective function, and it respects the following restrictions: (1) it belongs to the same vehicle of operation t_i; (2) the starting time of t_j is higher or equal to the finish time of t_i; and (3) the starting point of t_j is equal to the finishing point of t_i. The last task allocated becomes the new task t_i and the process repeats itself until the maximum time allowed for one duty gets reached. If the maximum time allowed for one duty exceeds the normal time, this allocation results in overtime for the duty. And if the maximum time allowed for one duty gets reached, a new duty begins and the process repeats itself until all tasks have been allocated.

4.3.3 Objective Function (OF)

Consider $d_i, i = 1, \ldots, n$, the duties of a solution s to be performed during a particular day. So, the daily cost C of the duties performance is calculated by

$$C(s) = \sum_{i=1}^{n} FixedCost_i + w_{ot} \times overtime_i + w_{sd} \times split\_duty_i. \tag{4.1}$$

The *FixedCost* is associated to the maintenance of a crew in the company's staff and is the same for all crews, $overtime_i$ is the amount of overtime, expressed in minutes, into duty d_i, and $split\_duty_i = 1$ if d_i is a split duty, otherwise $split\_duty_i = 0$. The constants w_{ot} and w_{sd} are adjusted to suit the solutions obtained according to the company's interests.

The cost of a CSP solution given by Eq. (4.1) was evaluated considering the following values for the parameters: $FixedCost = 10{,}000$, $w_{ot} = 4$, and $w_{sd} = 600$ and $w_{sd} = 5000$, so that the developed method can be compared to other authors who used the same database and costs in the objective function (OF). Thus, the crew cost has a high value so that the crews number reduction is the main optimization criteria, followed by the split duties reduction, and finally, the overtimes. The OF value for the solution adopted by the company was calculated by applying the same coefficients defined for the model.

4.3.4 Destructive Heuristics

Four destructive heuristics were implemented and three of them were adapted from those proposed by Ropke and Pisinger [11] and Shaw [13], and one of them was proposed in this paper. The tasks are the elements to be removed and reinserted in the CSP solution. Therefore, a destructive heuristic must remove a predefined quantity of q tasks of a given solution.

4.3.4.1 Random Removal

The random removal is characterized by the removal of randomly chosen q tasks. Initially a duty is drawn and a task, also randomly chosen, is removed from it. The procedure is over when q tasks have been removed from the solution.

4.3.4.2 Removal of Worst Position

In this kind of removal, the task that produces the highest difference between the OF value with the task and the one without it in the solution is removed, at every iteration. The cost of a task i in a solution s is given by $worst\_cost(i, s) = f(s) - f_{-i}(s)$, where $f_{-i}(s)$ is the cost of solution s without task i. This removal aims to allocate tasks that cause great impact in the OF. In other words, those tasks maximize the function $worst\_cost(i, s)$.

4.3.4.3 Shaw Removal

The idea of the Shaw Removal is to remove similar tasks, which allow them to be easily reinserted, generating then better solutions. The similarity of two tasks i and j is computed through a function of similarity $R(i, j)$. For CSP, two tasks are similar if they start and finish at the same place because, as they fit by the initial and final points of operation, it is impossible to perform the change of tasks which have different start and finish places. The similarity between tasks i and j is given by

$$R(i, j) = \alpha(|st_i - st_j| + |et_i - et_j|) + \beta(|(et_i - st_i) - (et_j - st_j)|). \quad (4.2)$$

In (4.2), st_i corresponds to the initial time and et_i to the final time of task i. The first portion of (4.2) considers the proximity of the time of beginning and end of the tasks, while the second portion considers the proximity of the length of the two tasks. Therefore, as lower the value of (4.2), higher is the similarity between the pair of tasks i and j. The weights α and β control the proportion of the tasks characteristics on the measurement.

The procedure begins randomly choosing a task to be removed. The next $q - 1$ tasks are chosen according to the similarity related to the task previously removed. Some tasks may not reach $q - 1$ similar tasks when there will be no $q - 1$ tasks that start and finish at the same place as the considered task. In this case, the quantity q of removed tasks becomes the possible quantity of removal.

4.3.4.4 Average Removal

Once one of the goals in solving the CSP is to minimize the amount of overtime, it is interesting that the duties have their duration approximately equal. Thus, in this paper is proposed the average removal method. In this kind of removal, tasks are removed from the duties in which their lengths exceed the average length. Removals from the first or last duty task are allowed and the choice is random at every removal. It is not interesting to remove a task contained in the duty interior, once it does not decrease the duty length and it can increase its internal idleness.

4.3.5 *Constructive Heuristics*

In order to repair the solutions partially destroyed, two constructive heuristics have been employed in the CSP. They were adapted from those proposed by Ropke and Pisinger [11] and Potvin and Rousseau [9]. In the CSP context, inserting a task in a position is the same as inserting a task in a duty, once it is not possible to allocate it in more than one position in the duty due to the temporal restriction which places the tasks in chronological order. When the insertion of new tasks in existing duties is not possible, a new duty is created from the task to be inserted.

During the duties reconstruction, the vehicles changes are allowed and this helps widening the searching. A maximum amount of vehicles changes in one duty is defined a priori. For the case studied, no more than two changes of vehicles per duty are allowed.

4.3.5.1 Greedy Insertion

The method of greedy insertion inserts a task in its cheaper position in the solution. Among the different possibilities of insertion, for each task i is chosen that one which results on the lowest cost. This position is called minimum cost position c_i. At the end, the task i is chosen among the ones not allocated and with the lowest value of c_i. The insertion in its minimum cost position is performed. The process continues until all tasks have been inserted.

4.3.5.2 Regret Insertion

The regret heuristics are based on the greedy heuristic and integrate a kind of evaluation like "look straight ahead" when selecting a task for insertion. Consider $f_\Delta(i, k)$ as the OF value change caused by the task i insertion on its kth valid cheaper position, so the value of insertion 2-regret is given by $c_i^* = f_\Delta(i, 2) - f_\Delta(i, 1)$. On each iteration, the task i that maximizes c_i^* is inserted. Therefore, is chosen for insertion the task that produces the highest impact in the OF if it is not allocated on its best position and has to go to its second best position.

4.3.6 Choosing the Heuristics of Removal and Insertion

After defining the heuristics that destroy and reconstruct a solution, the next step consists of choosing a pair of heuristics in order to modify the current solution. A unique pair of heuristics could be used during the whole search. However, it is interesting that all heuristics are used, once each one of them could be more suitable in a given moment of the search. There is a belief that this alternation results in a robust method [11].

The choice of the heuristics that form the pair is independent, and the selection of each one, during the search, is done by the principle of the roulette selection. At every heuristic, a weight is attributed and the sortition happens according to its choice probability. Considering a set of k heuristics with weights $w_i, i \in \{1, \ldots, k\}$, a heuristic j is selected with probability:

$$\frac{w_j}{\sum_{i=1}^{k} w_i}. \tag{4.3}$$

Suppose a pie chart where a space proportional to its weight is attributed to each heuristic. An outside roulette is placed over the graph and a heuristic is selected. As higher the heuristic weight, the higher its space in the graph, and then it is higher the chance to be chosen. The weight is straightly related to the heuristic efficiency in order to improve the solution.

4.3.7 Weights Adaptive Adjustment

The heuristics weights represent the efficiency of each one during the searching process provided by the ALNS heuristic. The weights are automatically adjusted by using statistics of the previous iterations. A score is kept for each heuristic that measures its performance in a particular number of iterations. As higher the heuristic score, the higher its efficiency.

The search is split into segments, which are split into iterations or processing time, then the criteria for algorithm stopping are defined. In the beginning of each segment, the score of each heuristic is defined as zero and it is increased during the search in three different situations. In the first case, σ_1 is increased in the score when the heuristic is able to find a new global solution s^*. σ_2 is increased in the score when the heuristic is able to find a solution s' not previously found and better than the current solution s. In the last case, σ_3 is increased in the score when the heuristic is able to find a solution s' not previously found and worse than the current solution s, but still accepted by the acceptance criteria though, to be described next.

Once the situation characterized by the σ_1 increase is more attractive than that one characterized by the σ_2 increase, which is more attractive than that one characterized by the σ_3 increase, it is comprehensible that $\sigma_1 \geq \sigma_2 \geq \sigma_3$. Even though the heuristics that improve the solution are preferable, it is also interesting to use heuristics which can diversify the search, like in the last case, which encourages heuristics to explore new regions of the search space.

One removal and one insertion heuristics are applied in each iteration of the ALNS. As both are responsible for the generation of a new solution, their scores are updated in the same amount because it is not possible to guarantee which one was the operation success factor. At the end of each segment, the heuristics weights are updated taking into account the score obtained in the current segment. Thus, as the scores start with the value zero for each heuristic, the respective weights start with value one. At the end of each segment, new weights are calculated using the saved scores. These scores are updated at the end of each iteration, and reset starting a new segment. Let $w_{i,j}$ be the weight of the heuristic i used in the segment j, the weight of the heuristic i to be used in the segment $j + 1$ is calculated as

$$w_{i,j+1} = w_{i,j}(1 - r) + r\frac{\pi_i}{\theta_i}. \tag{4.4}$$

In (4.4), π_i is the score obtained by the heuristic i during its last segment, and θ_i is the number of time that the heuristic i was used in the last segment. The reaction factor $r \in [0, 1]$ controls how much the next weights will be influenced by the weights of the previous segment. If $r = 0$, the weights of the previous segment are repeated next segment. If $r = 1$, the weights of the previous segment are totally discarded. A value between 0 and 1 for the reaction factor results in new weights that consider the heuristics performance in the previous and in the current segment.

4.3.8 Acceptance Criteria

Like in Ropke and Pisinger [11], it was used here the acceptance criteria of the simulated annealing metaheuristic, so that a solution s' will be accepted if it is better than s. If s' is worse than s, s' will be accepted with the probability $e^{-[f(s')-f(s)]/T}$, where $T > 0$ represents the current temperature. The temperature T starts in T_0 and decreases at every iteration by the expression $T = T \times c$, where $0 < c < 1$ is the *cooling rate*. A method to obtain the initial temperature that takes into account the characteristics of each instance, described by Talbi [20], was used as a good choice of T_0 , once it must be related to the problem. In this paper the initial temperature starts with a low value and increases at a rate of 10% until 70%, at least, of the neighbor solutions are accepted.

4.3.9 *Method to Minimize the Quantity of Duties*

The method used to minimize the duties amount in the CSP starts from a feasible initial solution s_i, with a particular quantity of duties. For each duty all of their tasks are removed from the solution and the greedy insertion heuristic tries to reinsert them on the other existing duties, creating then a new solution s'. If, at the end of the reinsertion, the quantity of duties of s' is lower than the one in s_i, s' is accepted as s_i, and the search restarts from the first duty of the new solution. Otherwise, it moves to the next duty. The procedure is finished when all duties have been considered in the process of destruction and reconstruction of the solution and the amount of duties has not been reduced.

The method of duties minimization is applied on the initial solution, before performing the ALNS search. Therefore, ALNS is expected to be able to reduce the variable costs of the solution, since the quantity of duties has already been minimized.

4.4 Computational Experiments

In order to test the ALNS, real data from a public transport company, which operates in Belo Horizonte city, were used and the experiments were performed in a computer with the following features: Intel Core i7-4770 processor, 8 GB RAM, Ubuntu 16.04, and the algorithm was developed in C++ programming language.

The data provided by the company are: the amount of tasks to be performed each working day, Saturday, and Sunday of the studied week, with their respective starting and ending time, initial and final points, and vehicle used in the operation. Besides vehicle scheduling, the company also provided their crew scheduling in operation. Thus, it was possible to calculate the length, the overtime quantity, and the characterization as straight and split duty. Table 4.1 contains, for each kind of day, the quantity of tasks to be performed, and input data that give the dimension of each problem.

Table 4.1 Quantity of tasks to be performed on working days, Saturdays, and Sundays

Days of week	Monday	Tuesday	Wednesday	Thursday	Friday	Saturday	Sunday
Number of tasks	705	674	814	872	787	644	345

4.4.1 Parameters Calibration

ALNS heuristic has a series of parameters to be defined. Firstly, it must be defined the quantity q of tasks to be removed and reinserted in the solution, which is controlled by the parameter ξ. In each iteration, a random number q that satisfies $0.008m \leq q \leq \xi m$ is chosen, where m represents the quantity of tasks of the problem. The acceptance criteria of a solution is controlled by an initial temperature T_0, cooled at a rate c. The Shaw removal heuristic is controlled by the parameters α, β, and p_{shaw}, while the removal of worst position heuristic is controlled only by the parameter p_{worst}. The parameters σ_1, σ_2, σ_3, and r, used in the weights adaptive adjustment, must also be defined. The ALNS stop criteria is defined by the quantity of segments, each one constituted by 100 iterations.

Among the considered parameters, T_0, ξ, and r were calibrated, while the rest of the cited parameters were defined according to Ropke and Pisinger [11]. After several computational tests and the statistical test of Kruskal–Wallis [6] and Dunn [4] have been performed, the final parameters setup was reached, represented by the vector $(\%T_0, \xi, c, \alpha, \beta, p_{shaw}, p_{worst}, \sigma_1, \sigma_2, \sigma_3, r) = (0.781, 0.025, 0.99975, 1, 1, 6, 6, 20, 10, 5, 0.8)$ for the business days and by $(\%T_0, \xi, c, \alpha, \beta, p_{shaw}, p_{worst}, \sigma_1, \sigma_2, \sigma_3, r) = (1.563, 0.05, 0.99975, 1, 1, 6, 6, 20, 10, 5, 0.8)$ for the weekend. These were the setups used to test the implemented ALNS.

4.4.2 Results

New tests were performed in order to measure the ALNS efficiency. Four distinct scenarios of tests were analyzed: the scenarios $1TV_{600}$ and $1TV_{5000}$ which allow no more than one vehicle change per duty when the split duties weigh are 600 and 5000, respectively; and also the scenarios $2TV_{600}$ and $2TV_{5000}$ which allow up to two vehicles change per duty in the two weightings of the split duties. For each problem and scenario, 10 computational tests were performed, each one limited to 1 h length. A solution of less OF value was selected for comparing different resolution approaches. The results generated by ALNS are compared to the ones obtained by the methods GRASP [15], VNS-VLNS [10], and GLS [16]. The different resolution approaches considered in each scenario of tests are justified by the scenario setups adopted by the authors of the mentioned methods.

Tables 4.2 and 4.3 present the results obtained by ALNS and other resolution approaches on the scenarios of weights 600 and 5000, respectively, for the split duties. The average OF value, the percentage of standard deviation (SDev%), the OF value of the best solution (Best OF) obtained, and its detailing in terms of duties quantity (#duties), split duties (#SD), and overtimes (OT) are all presented for ALNS. For the rest of the approaches, only the best solutions and their details are presented. The gap is presented in the last line: GAP = $(S_{alns} - S_{best})/S_{best}$, where S_{alns} is the OF value of the best solution obtained by ALNS and S_{best} is

Table 4.2 Results obtained by ALNS, GRASP, VNS-VLNS, and GLS on the scenario $1T\,V_{600}$

Method	Item	Monday	Tuesday	Wednesday	Thursday	Friday	Saturday	Sunday
ALNS	Average OF	1,220,118	1,172,848	1,400,409	1,523,416	1,442,174	1,063,718	572,884
	SDesv%	0.943	0.535	0.747	0.645	0.447	0.919	0.451
	Best OF	**1,204,500**	**1,162,832**	1,385,804	**1,505,896**	**1,433,608**	**1,050,576**	565,812
	#Crews	117	113	135	147	141	101	55
	#SD	21	16	18	16	14	28	10
	OT	91:15	96:48	104:11	109:34	105:02	99:04	40:53
GRASP	Best OF	1,264,836	1,265,880	1,462,384	1,564,420	1,510,520	1,131,272	590,716
	#Crews	124	125	144	154	149	111	58
	#SD	14	7	11	12	10	15	7
	OT	68:29	48:40	65:46	71:45	60:30	51:08	27:09
VNS-VLNS	Best OF	1,223,556	1,191,952	**1,380,000**	1,517,168	1,435,332	1,089,764	**562,444**
	#Crews	120	117	135	149	141	107	55
	#SD	19	14	19	17	14	10	7
	OT	50:39	56:28	77:30	70:42	70:33	57:21	34:21
GLS	Best OF	1,253,532	1,222,920	1,417,292	1,557,100	1,474,588	1,104,096	581,468
	#Crews	123	120	139	153	145	108	57
	#SD	23	23	14	22	19	17	9
	OT	40:33	38:00	78:43	57:55	54:57	57:54	74:14
Company	Best OF	1,364,404	1,322,420	1,518,460	1,651,256	1,576,564	1,253,100	686,468
	#Crews	134	130	149	162	155	124	68
	#SD	6	3	5	4	1	0	0
	OT	86:41	85:55	106:05	120:14	108:11	54:35	26:57
GAP	Best OF	−1.582%	−2.504%	+0.421%	−0.749%	−0.120%	−3.730%	+0.599%

Table 4.3 Results obtained by ALNS, VNS-VLNS, and GLS on the scenario $1TV_{5000}$

Method	Item	Monday	Tuesday	Wednesday	Thursday	Friday	Saturday	Sunday
ALNS	Average OF	1,280,396	1,217,594	1,447,686	1,584,688	1,484,026	1,140,896	603,150
	SDesv%	0.713	0.601	0.357	0.783	0.402	0.820	0.053
	Best OF	**1,256,628**	**1,202,692**	**1,440,876**	**1,565,016**	1,471,920	**1,127,876**	602,836
	#Crews	120	115	138	151	141	105	57
	#SD	7	6	7	6	7	11	5
	OT	90:07	94:33	107:49	104:14	112:10	95:19	32:39
VNS-VLNS	Best OF	1,270,628	1,213,176	1,471,176	1,583,644	**1,471,100**	1,152,552	**601,296**
	#Crews	120	114	140	148	139	109	57
	#SD	11	11	11	17	12	10	5
	OT	65:07	75:44	67:24	77:41	87:55	52:18	27:35
GLS	Best OF	1,276,280	1,216,340	1,455,988	1,581,920	1,478,832	1,146,440	601,560
	#Crews	120	116	139	150	142	110	57
	#SD	12	8	10	13	8	7	5
	OT	67:50	68:05	66:37	70:30	78:28	47:40	27:20
Company	OF value	1,390,804	1,335,620	1,540,460	1,668,856	1,580,964	1,253,100	686,468
	#Crews	134	130	149	162	155	124	68
	#SD	6	3	5	4	1	0	0
	OT	86:41	85:55	106:05	120:14	108:11	54:35	26:57
GAP	Best OF	−1.102%	−0.864%	−1.038%	−1.069%	+0.056%	−1.619%	+0.256%

the OF value of the best solution known for a particular problem and scenario of tests. A negative GAP (green) means that ALNS outperformed the best solution in a particular percentage. A positive GAP (red) means that ALNS generated a worst solution in a particular percentage. The values in bold represent the best solutions found per problem.

By analyzing the first scenario of tests, the ALNS heuristic generated the best solutions for the problems of Monday, Tuesday, Thursday, Friday, and Saturday. For Wednesday and Sunday, the solutions obtained by ALNS were at most 0.6% worse than the best solutions generated by Reis and Silva [10].

Also, most of the solutions generated by ALNS are characterized for presenting an inferior number of duties related to the solutions generated by the other methods. This reduction is explained by the overtime increase in the solution, which represents the problem trade-off: by decreasing the amount of duties, the overtimes naturally increase, once there will be fewer crews performing the same tasks.

Considering the second tests scenario, presented in Table 4.3, there is a split duty reduction in the generated solutions when they are compared to the ones generated in the previous scenario. That was the strategy adopted so that the heuristics could generate solutions containing the quantity of split duties next to the company solution. Analyzing the presented solutions, ALNS generated the best results for the problems of Monday, Tuesday, Wednesday, Thursday, and Saturday. For Friday and Sunday the solutions obtained by ALNS were at most 0.256% worse than the best solutions also generated by Reis and Silva [10].

In practice, it is acceptable that crews perform, at most, up to two vehicles change per duty. This change process is undesirable once it is hard to control the damages in the vehicles, caused by the crews use. Due to this factor, the companies usually do not consider more than one vehicles change per duty. However, this scenario was studied in order to verify if this way of flexibility can bring some economy to the companies. The results are presented on Tables 4.4 and 4.5.

By analyzing Table 4.4, ALNS generated the best solutions for all business days and Saturdays in this scenario. About Sunday, the solution generated by ALNS was just 0.563% worse than the best solution generated by Reis and Silva [10]. Similarly to the $1TV_{600}$ scenario, the reduction in duties generated by ALNS, related to the other methods, is justified by the overtime increase in each solution.

Table 4.5 shows that ALNS generated the best results for the problems of Monday, Tuesday, Wednesday, Thursday, and Saturday. For Friday and Sunday, the solutions obtained by ALNS were at most 0.736% worse than the best solutions generated by Reis and Silva [10].

By analyzing the quality of the solutions obtained by ALNS on the four scenarios, only Saturdays' problem, on the scenarios $1TV_{600}$, presented an amount of split duties superior to 20%, which is undesirable by the public transport companies. The rest of the solutions are composed by a quantity of split duties inferior to 20% of the total duties, highlighting the applicability of the solutions generated in practical and real environments. Another factor to be noticed is the small percentage of the standard deviation in all solutions, showing the heuristics robustness. Besides, on

Table 4.4 Results obtained by ALNS, GRASP, and VNS-VLNS in the scenario $2TV_{600}$

Method	Item	Monday	Tuesday	Wednesday	Thursday	Friday	Saturday	Sunday
ALNS	Average OF	1,224,303	1,161,067	1,395,434	1,520,991	1,436,904	1,065,081	5,69,930
	SDesv%	0.678	0.336	0.409	0.485	0.484	0.602	0.739
	Best OF	**1,210,492**	**1,152,376**	**1,389,968**	**1,510,252**	**1,425,480**	**1,055,260**	565,608
	#Crews	118	112	136	148	139	102	55
	#SD	17	17	13	13	15	20	10
	OT	84:33	92:24	92:22	93:33	110:20	96:55	40:02
GRASP	Best OF	1,264,386	1,265,916	1,457,172	1,563,532	1,508,884	1,131,828	591,060
	#Crews	124	125	143	154	149	111	58
	#SD	14	7	12	11	9	14	8
	OT	66:32	48:49	83:13	70:33	56:11	55:57	26:05
VNS-VLNS	Best OF	1,223,556	1,189,408	1,393,920	1,515,196	1,437,060	1,089,764	**562,444**
	#Crews	120	117	137	149	141	107	55
	#SD	19	10	15	17	17	10	7
	OT	50:39	55:52	62:10	53:36	56:18	43:12	34:21
Company	OF value	1,364,404	1,322,420	1,518,460	1,651,256	1,576,564	1,253,100	686,468
	#Crews	134	130	149	162	155	124	68
	#SD	6 3	5	4	1	0	0	
	OT	86:41	85:55	106:05	120:14	108:11	54:35	26:57
GAP	Best OF	−1.079%	−3.214%	−0.284%	−0.327%	−0.812%	−3.270%	+0.563%

Table 4.5 Results obtained by ALNS and VNS-VLNS in the scenario $2TV_{5000}$

Method	Item	Monday	Tuesday	Wednesday	Thursday	Friday	Saturday	Sunday
ALNS	Average OF	1,270,998	1,208,713	1,446,612	1,587,650	1,486,715	1,125,628	603,987
	SDesv%	0.907	0.500	0.524	0.662	0.624	0.911	0.450
	Best OF	**1,252,288**	**1,200,932**	**1,435,604**	**1,573,848**	1,471,984	**1,108,708**	5,99,496
	#Crews	121	116	138	152	141	104	56
	#SD	4	4	6	6	7	9	6
	OT	92:52	87:13	106:41	99:22	112:26	98:47	39:34
VNS-VLNS	Best OF	1,264,724	1,220,112	1,449,952	1,580,160	**1,461,288**	1,141,916	**597,496**
	#Crews	121	118	139	150	139	110	58
	#SD	12	8	11	14	10	11	6
	OT	51:34	58:25	58:19	58:33	66:05	49:39	27:12
Company	OF value	1,390,804	1,335,620	1,540,460	1,668,856	1,580,964	1,253,100	686,468
	#Crews	134	130	149	162	155	124	68
	#SD	6 3	5	4	1	0	0	
	OT	86:41	85:55	106:05	120:14	108:11	54:35	26:57
GAP	Best OF	−0.993%	−1.597%	−0.999%	−0.401%	+0.736%	−2.995%	+0.335%

Table 4.6 Better solutions comparison of different scenarios

	$1TV_{600}$	$2TV_{600}$	$1TV_{5000}$	$2TV_{5000}$
Monday	1,211,836	**1,203,568**	1,261,280	**1,257,688**
Tuesday	1,169,944	**1,153,616**	1,201,956	**1,197,704**
Wednesday	**1,384,688**	1,388,460	**1,431,336**	1,439,944
Thursday	1,514,668	**1,513,188**	**1,564,068**	1,575,324
Friday	**1,431,040**	1,431,724	1,475,376	**1,469,816**
Saturday	**1,053,000**	1,055,320	1,126,256	**1,117,452**
Sunday	565,224	**564,696**	599,212	**594,592**

the four studied scenarios, the proposed ALNS outperformed the solution adopted by the company at every single analyzed day.

Another characteristic analyzed in the obtained results was the flexibility of allowing, at most, just one or two vehicles change per duty. Table 4.6 presents the costs (OF value) of the best solutions obtained when solving the problems, considering the weights 600 and 5000 for the split duties on the scenarios that allow, at most, one and two vehicles change per duty, respectively. The values in bold indicate which one of the scenarios resulted in the solution with the lowest OF value for each day.

Considering the first block of Table 4.6 ($w_{sd} = 600$), it is observed that the flexibility caused by the vehicles change did not lead to a significant impact on the cost of the solutions. For three out of seven problems, cheaper solutions were obtained by permitting up to one vehicle change per duty, while the solutions that allow performing up to two changes stood out for the remaining problems. This fact can be explained by the flexibility that the smallest weight for the split duties provides, once its number tends to increase, considering the low value of this kind of duty. On the other hand, by analyzing the second block ($w_{sd} = 5000$), the influence that the possibility of performing more vehicles changes during a duty may influence on the OF value is notable. In this case only two out of the seven problems presented better solutions, allowing up to one vehicle change per duty, while the remaining five problems had better solutions when considering up to two changes. As the value of the split duties is higher in this case, the search tends to explore the most economic situations like those that incorporate more vehicles changes in the duties.

4.5 Conclusions

This paper presents an implementation of the adaptive large neighborhood search heuristic in order to solve the crew scheduling problem of the public transportation system. Some existing removal and insertion heuristics were adapted to the CSP context, and a new heuristic of task removal was proposed: the average removal.

A set of parameters used by ALNS was calibrated, and the best setup was selected from statistical tests for the resolution of problems belonging to the CSP for further performances. In the tests instances, the ALNS was able to generate good results and it outperformed the most competitive approaches in most of the problems, proposed by Reis and Silva [10] and Silva and Silva [16]. At least 71% and, at most, 86% of the best solutions were generated by considering the set of the seven solved problems.

Concluding, the ALNS is an efficient resolution mechanism of a class of problems belonging to the CSP. Its superiority by solving these problems can be explained by its adaptive characteristic, which makes it robuster and adaptable to the instances of distinct characteristics.

Acknowledgements The authors thank and appreciate CAPES (Coordination for the Improvement of Higher Level Personnel), CNPq (National Council for Scientific and Technological Development), FAPEMIG (Foundation for Research Support of Minas Gerais State), and UFOP (Federal University of Ouro Preto) for all the support received during this paper development.

References

1. Ahuja, R.K., Orlin, J.B., Sharma, D.: Very large-scale neighborhood search. Int. Trans. Oper. Res. **7**, 301–317 (2000)
2. Bodin, L., Golden, B., Assad, A., Ball, M.: Routing and scheduling of vehicles and crews: the state of the art. Comput. Oper. Res. **10**, 63–211 (1983)
3. Chen, S., Shen, Y.: An improved column generation algorithm for crew scheduling problems. Int. J. Inf. Comput. Sci. **10**, 175–183 (2013).
4. Dunn, O.J.: Multiple comparisons using rank sums. Technometrics **6**, 241–252 (1964)
5. Fischetti, M., Martello, S., Toth, P.: The fixed job schedule problem with spread-time constraints. Oper. Res. **35**, 849–858 (1987)
6. Kruskal, W.H., Wallis, W.A.: Use of ranks in one-criterion variance analysis. J. Am. Stat. Assoc. **47**, 583–621 (1952)
7. Lourenço, H.R., Paixão, J.P., Portugal, R.: Multiobjective metaheuristics for the bus driver scheduling problem. Transp. Sci. **35**, 331–343 (2001)
8. Pisinger, D., Ropke, S.: A general heuristic for vehicle routing problems. Comput. Oper. Res. **34**, 2403–2435 (2007)
9. Potvin, J.-Y., Rousseau, J.-M.: A parallel route building algorithm for the vehicle routing and scheduling problem with time windows. Eur. J. Oper. Res. **66**, 331–340 (1993)
10. Reis, A.S., Silva, G.: Um estudo de diferentes métodos de busca e a metaheurística VNS para otimizar a escala de motoristas de ônibus urbano. Transporte em Transformação XVI—Trabalhos Vencedores do Prêmio CNT de Produção Acadêmica **1**, 45–64 (2012)
11. Ropke, S., Pisinger, D.: An adaptive large neighborhood search heuristic for the pickup and delivery problem with time windows. Transp. Sci. **40**, 455–472 (2006)
12. Ropke, S., Pisinger, D.: An unified heuristic for a large class of vehicle routing problems with backhauls. Eur. J. Oper. Res. **171**, 750–775 (2006)
13. Shaw, P.: A new local search algorithm providing high quality solutions to vehicle routing problems. APES Group, Department of Computer Science, University of Strathclyde, Glasgow, Scotland. Citeseer (1997)
14. Smith, B.M., Wren, A.: A bus crew scheduling system using a set covering formulation. Transp. Res. **22A**, 97–108 (1988)
15. Silva, G.P., Cunha, C.B.: Uso da técnica de busca em vizinhançe grande porte para a programação da escala de motoristas de ônibus urbano. Transportes **18**, 37–45 (2010)

16. Silva, T.A., Silva, G.P.: O uso da metaheurística guided local search para resolver o problema de escala de motoristas de ônibus urbano. Transportes **23**, 1–12 (2015)
17. Silva, G.P., Souza, M.J.F., von Atzingen, J.: Métodos exatos para resolver o problema de programação da tripulação. Transportes **14**, 25–32 (2006)
18. Song, C., Guan, W., Ma, J., Liu, T.: Improved genetic algorithm with gene recombination for bus crew scheduling problem. In: Mathematical Problems in Engineering. Hindawi Publishing Corporation, Cairo (2015)
19. Souza, D.S.: Uma abordagem híbrida para resolver o problema da escala de motoristas de ônibus urbano. Master thesis, Federal University of Ouro Preto, Ouro Preto (2014)
20. Talbi, E.G.: Metaheuristics: From Design to Implementation. Wiley, Hoboken (2009)
21. Wren, A.: Scheduling vehicles and their drivers—forty years experience. Technical Report, School of Computing Studies, Leeds University, Leeds (2004)

Chapter 5
Proposal for Analysis of Location of Popular Residential Using the p-Median

Bruno F. de Azevedo and Nelio D. Pizzolato

Abstract The location of projects in the My House My Life Program (PMCMV) is of great importance because it is not only a business relationship between buyers and developers, but also a contribution to the social transformation. The use of modern tools that help the launching plan of Real Estate companies, financiers, and buyers is a constant challenge to the sector, which requires accurate decisions. This article proposes a new tool for the planning of PMCMV residential real estate projects, using Operations Research techniques such as the p-median to locate candidate areas to receive ventures as close as possible to the existing potential demands. In order to illustrate the proposed method, a detailed case study was carried out in the Municipality of Rio de Janeiro. The results obtained in the case study demonstrate, in some cases, large variations between the positions of higher demand and the launches of developments in the PMCMV format promoted by the market.

5.1 Introduction

In recent years, real estate market has faced a troubled scenario in sales of residential housing units. Without an international scenario, the so-called crisis of subprimes originated without financing of property the person without income. In the national scenario, for a variety of reasons, often the international financial crisis, there was a reduction in demand, leading to a serious internal crisis of unemployment and low demand, especially in the area of construction. Of course, as the business-building cycle is rather long, it cannot be better.

Since 2009, with the launch of the My Home My Life (PMCMV) Program, a number of market launches have been spread throughout the country to meet

B. F. de Azevedo (✉)
UNESA Productivity Research Program Scholarship, Universidade Estácio de Sá-UNESA, Petrópolis, RJ, Brazil

N. D. Pizzolato
Universidade Católica de Petrópolis, Petrópolis, RJ, Brazil
e-mail: nelio.pizzolato@ucp.br

V. N. Coelho et al. (eds.), *Smart and Digital Cities*, Urban Computing,
https://doi.org/10.1007/978-3-030-12255-3_5

this very large real estate demand, and keep it going. The location of real estate developments is, in fact, a fundamental factor in the succession of popular enterprises, which carry in themselves also an attempt to correct a historical social problem, especially in urban areas.

This research presents a proposal with the purpose of assisting entrepreneurs in the decision-making for the development of PMCMV residential real estate projects. The proposal consists of applying Operational Research tools to locate candidate points for the implementation of residential real estate projects for sale, according to the criteria of the program, thus promoting a reduction in the use of intuition in the purchase or definition of land, adding new techniques to many aspects of the analysis of viability of the business, to provide a more secure vision of its investments. In this work, a case study was developed in the city of Rio de Janeiro, which has particular characteristics in relation to population distribution. This municipality presents a diversity of demand in a large part of its area, with real estate regions marketed to buyers with high purchasing power, close to regions occupied by low-income families. This particularity of Rio de Janeiro becomes an element considered no case study, which can contribute to and influence the urban planning of the city.

In addition to this Sect. 5.1, this work is composed as follows: Sect. 5.2 presents the My Home My Life Program, describing its most recent evolution. Section 5.3 presents a Methodology, in addition to presenting a brief review of the literature. Sect. 5.4 describes a practical application to the municipality of Rio de Janeiro, while Sect. 5.5 evaluates the results and in Sect. 5.6 are made as conclusions of the study.

5.2 My Home My Life Program

Launched in 2009 by the Federal Government, the My Home My Life Program (PMCMV) aims to allow access to the home for low- and middle-income families. In partnership with States, Municipalities, companies, and nonprofit entities, the program is directly linked to the National Housing Secretariat of the Ministry of Cities. The program also has the assistance of the banks Caixa Econômica Federal and Banco do Brasil S. A. in the granting of credit for real estate financing.

In the first phase of the program, year 2009, families were divided into three income brackets for access to program benefits, as described below:

- Range 1—Families with gross income up to R$ 1600.00;
- Range 2—Families with gross income between R$ 1600.01 and R$ 3275.00;
- Range 3—Families with gross income between R$ 3275.01 and R$ 5000.00.

In the second phase of the program, year 2011, the three bands had their values changed, as described below, in order to adjust prices to the market:

- Range 1—Families with gross income up to R$ 1600.00;
- Range 2—Families with gross income between R$ 1600.01 and R$ 3600.00;

- Range 3—Families with gross income between R$ 3600.01 and R$ 5000.00.

In 2016, continuing the program, the third phase of the program began, with price readjustment and the creation of a new intermediate band, as described below:

- Range 1—Families with gross income up to R$ 1800.00;
- Range 1.5—Families with gross income between R$ 1800.01 and R$ 2350.00;
- Range 2—Families with gross income between R$ 2350.01 and R$ 3600.00;
- Range 3—Families with gross income between R$ 3600.01 and R$ 6500.00.

For ranges 1 and 1.5, the Program grants subsidy of up to 90% of the value of the property and, as the bands increase, the subsidy and interest rates undergo changes, however, always below traditional real estate financing modalities. In these first two lanes, credit is facilitated and income verification and risk analysis requirements are waived, with families having to register in City Halls or organizers to benefit from the program by lot, which does not occur in other bands, where the buyer can choose the property that wishes to acquire.

In order to benefit from the program, families must:

- Own family income compatible with the Program, as explained above;
- The provision of the financing cannot exceed 30% of the family income;
- The property must be used as housing of the owner of the financing;
- The owner of the financing cannot have other real estate financing or real estate property taken out on his behalf;
- The financing holder may not have used the FGTS for real estate financing in the last 5 years;
- In lanes 2 and 3, the holder cannot have credit restriction on his behalf;
- The property cannot be sold before the end of the financing;
- The age of the older tenderer along with the term of the financing cannot exceed 80 years, 5 months, and 29 days;
- The property must be situated in the same place as the current residence or work of the owner, or where one intends to live or work.

Here, we chose to describe only the benefits and rules of the Program in the view of the buyer, as there are other parties involved in the program that have specific rules, such as builders/developers. The purpose of this description is to understand where the public with the capacity to acquire these properties and where these ventures have been launched.

5.3 Methodology and Literature Review

The use of models to locate facilities, supported by geographical information systems (GIS), has become a powerful tool for decision-making processes. The graphical interface brought by the GIS, associated with the georeferenced database, allows the public power to prepare location plans on a certain area, taking into account the configuration of the road network, the geographical and topological barriers, as well as the distribution of the population by age, family income, family size, socioeconomic aspects, and possible restrictions.

The first urban studies involved the location of public schools, and a celebrated and pioneering work was published by Tewari and Jena [12]. Using the meager resources of the time, the authors studied the problem of the location of secondary schools over a vast geographic district in India, containing two cities, hundreds of clusters, over a million and a half inhabitants, and succeeded in developing a proposal taking into account issues and peculiarities such as the language spoken at school and the distance of 8 km from each student to reach the school. This study stimulated the appearance of many others, such as Molinero [5], Beguin et al. [3], Pizzolato and Silva [6], Barcelos et al. [2], and Teixeira and Antunes [11], among others.

In a more recent study, White et al. [13] examine the use of optimization techniques to locate public facilities in developing countries, highlighting studies related to health and education. According to this study, Rahman and Smith [9] study the use of locational analysis of health units in developing countries; Galvo et al. [4] verified the location of maternity hospitals in Rio de Janeiro; Pizzolato et al. [7] consider the location of public schools in urban areas; Yasenovsky and Hogdon [14] study the location of health units in rural Ghana; and even more recent studies can be cited. Thus, Pizzolato et al. [8] review studies based on the p-median and its extensions, while Smith et al. [10] apply hierarchical location studies directed to health posts; and analysis of location of real estate developments of Azevedo [1].

In addition to studies on the location of schools and health facilities, as discussed above, there are a number of similar applications, such as the location of warehouses, shopping malls, hospitals, libraries, police stations, and rescue teams, among others.

Regarding the location of real estate projects, this work proposes the following methodology, following the five steps listed below and discussed below:

- Step 1: Data Collection—Consists of the step of obtaining data that will aid in the foundation of the demand to be studied;
- Step 2: Network Vertex Marking—This stage deals with the development of vertex of the network to be studied;
- Step 3: Vertex Demand Calculation—The demand for the vertex is then calculated according to the data obtained in Step 1;
- Step 4: Application of the p-Median Model—This step consists of the direct application of the p-median to the studied problem;
- Step 5: Comparing Results with Market Launches—With the results obtained in Step 4 and the real estate market data, this step has the purpose of comparing the computational results with the behavior of market launches.

5.3.1 Step 1: Data Collection

In this study, we chose to concentrate the study in the city of Rio de Janeiro, which is composed of a large variety of real estate projects launched in recent years and has market information available for comparison of results.

In order to analyze the existing demand in the city, the information of the 2010 Census was used, carried out by the Brazilian Institute of Geography and Statistics (IBGE). In order to carry out the Census, IBGE develops a division of all the cities of the national territory into census tracts, each composed in urban areas for approximately 300 (three hundred) dwellings, in order to favor the later work of the enumerator. As this study was intended to study the entire city of Rio de Janeiro, the space division was used in Areas of Weighting.

These areas are composed of regions with clusters of census tracts so that inferences can be made about the region. The IBGE provides microdata of the research that contain answers of a determined sample of the population, according to the Area of Weighting. The city of Rio de Janeiro has 200 (two hundred) Areas of Weighting.

5.3.2 Step 2: Network Vertex Marking

The geometric centers of the Weighting Areas were adopted as vertex of the network, a division made by IBGE, as noted above, for the dissemination of Census data. Through the IBGE website, maps were acquired in shapefile files, which can be opened in GIS.

The use of the geometric centers of the Weighting Areas is an approximation made in this work, considering that all the demand of the region is concentrated in this point.

To use the files made available by IBGE, the SIG QGIS 2.8.2 was adopted, which is a free program, available for download at http://www.qgisbrasil.org. The maps of the territorial division of the Municipality of Rio de Janeiro, in Areas of Weighting, were imported into the QGIS 2.8.2. And, later marked the geometric centers of each Weighting Area, according to Fig. 5.1.

5.3.3 Step 3: Vertex Demand Calculation

To apply the p-median model, it was necessary to find the demand in each vertex determined previously. The IBGE provides the microdata in a text file, where each line represents a search performed, with the responses described in codes. This text file can be imported into Microsoft Excel, a software that allows better use of the data.

Fig. 5.1 Marking network vertex

In this paper, it was decided to determine as demand the combination of information below, taken from the Census:

- Weighting Area Code—Variable code: V0011;
- Persons in charge of the private household who have informed that they do not reside in their own properties—Variable code: V0201;
- Monthly household income—Variable code: V6529;
- Age of the head of the household, being considered in the demand only those responsible between 18 and 50 years—Variable codes: V6036 and V0502;

In spite of the PMCMV Differentiated Ranges, in this work we opted to analyze the first Band, since it represents the band with greater housing demand and with worse housing conditions. Since the PMCMV was launched in 2009 and was revised in 2011, it was decided to adopt the parameters of 2011, in the division of the income brackets, that is, households with a gross income of R$ 1600.00.

In the Census survey, the sample data presented fractions, corresponding to the universe of the Census survey (the fractions are also provided by the IBGE), so we chose to divide the values found in the sample by the fraction corresponding to the universe. If the fraction represents part of the universe, when dividing the sample by the fraction, one can obtain the universe of the area. In this way, it was possible to find a demand closer to the real in the Areas of Weighting.

5.3.4 Step 4: Application of the p-Median Model

The p-median model consists of grouping N vertex into p sets, where an enterprise should exist. During this process microregions are produced, called

$C1, C2, \ldots, Cp$, whose medians are the locations and each of the other vertex belongs to one of the regions whose median is the nearest.

In the model proposed in this paper, for locating candidate sites to receive residential projects, there is the basic premise that people wish to reside near their current locations. This premise was fundamental in the application of this model.

The p-median model corresponds to the following binary linear programming model:

$$\sum_{i=1}^{n}\sum_{j=1}^{n} q_i d_{ij} x_{ij} \tag{5.1}$$

Subject to

$$\sum_{i=1}^{n} x_{ij} = 1; \ J \in N \tag{5.2}$$

$$\sum_{j=1}^{n} x_{ij} = p \tag{5.3}$$

$$x_{ij} \leq x_{jj}; \ i, j \in N \tag{5.4}$$

$$x_{ij} \in \{0, 1\}; \ i, j \in N \tag{5.5}$$

At where,

N is the set of vertex of the network, $N = \{1, \ldots, n\}$;

d_{ij} is a symmetric matrix of distances between vertex with $d_{ii} = 0, i \in N$;

p is the number of medians to be located;

q_i represents the weight (demand) of vertex i;

x_{ij} are the decision variables, with $x_{ij} = 1$ if the vertex i is allocated to the vertex j, and $x_{ij} = 0$ otherwise, and $x_{jj} = 1$ if vertex j is a median and $x_{jj} = 0$ otherwise, to $i, j \in N$.

The objective function (5.1) is to minimize the weighted distances of each vertex to the nearest median; the constraints (5.2) and (5.4) require that each vertex i be allocated to a single vertex j, which must be a median. The restriction (5.3) determines the exact number p of medians to be located and (5.5) indicates the conditions of integrality. This model will be applied to the proposed problem.

It was possible to apply the p-median model using the Geographic Information System (GIS), SPRING 5.3, developed by the Image Processing Division (DPI), after marking the vertex of the network and surveying the demands at each vertex of the National Institute of Space Research (INPE). We chose this specific GIS, since it already has integrated the system of location of medians (based on linear programming discussed above).

With the vector files (shapefile) generated in QGIS 2.8.2, this material was imported into SPRING 5.3. With the two layers imported to SPRING 5.3, one with the Weighting Areas and another with the vertex of the network, it was possible to apply the median localization function, available in the software.

In order to carry out a more detailed study, it was decided to assign the demands to the problem. In this way, it was possible to analyze the problem for Range 1 of PMCMV.

The vertex demands were inserted in SPRING 5.3 as attributes (data tables), corresponding to the code of the Weighting Area. These data are embedded in the vertex layer of the network, that is, the demand data is part of the vertex and can be used in several analyses.

In SPRING 5.3, the "Median Location" function was used. This function applies the p-median to a particular layer, according to the desired criteria. In this window was selected the desired object in the analysis, in this case it was the vector file that represents the vertex of the network. Within the "Location of Medians" function, it is also possible to determine an attribute as demand at each point.

In the analysis done in this work, we decided to determine a number of medians for the demand, according to the number of neighborhoods where the My Home My Life (PMCMV) projects were launched in the Municipality. It is believed that this way it would be possible to analyze and compare the actual market launches with the results of the study.

There were 11 (eleven) medians, the same number of market launches for Range 1 of PMCMV.

5.3.5 *Step 5: Comparing Results with Market Launches*

The data referring to the launch of PMCMV real estate projects were taken from a spreadsheet provided by the City Hall of Rio de Janeiro, which includes the ventures of the program that were obtained until 2014. It is intended here to use this information as compared to the results of the p-median for each program range.

5.4 Results

The results obtained in the steps described above will be presented below for Range 1 of the program. These results will be demonstrated in a figure with the arcs linking the demand areas to the proposed sites for implementation of the projects, a table containing the results of each median and a comparative table between the market launches and the result of the p-median in descending order.

The tables containing the results of the medians are composed of the weighting area codes and their respective denominations (reported by the IBGE), number of

Fig. 5.2 Result medians location of the municipality of Rio de Janeiro in SPRING 5.3 for range 1

Table 5.1 Median results for range 1

Pondaration area code	Local	Vertex	Demand
3304557005008	Cosmos 2	25	17,153
3304557005031	Mar 1	18	17,240
3304557005049	Leblon 2	16	17,197
3304557005084	Inhama	22	17,229
3304557005096	Encantado, Abolio e gua Santa	1	689
3304557005106	Colgio	25	16,845
3304557005108	Cascadura	1	953
3304557005144	Santa Teresa	31	17,176
3304557005150	Taquara 1	24	15,966
3304557005169	Jardim Sulacap, Vila Militar, and Deodoro e Campo dos Afonsos	16	17,277
3304557005187	Senador Camar 1	21	15,965

vertex served by each median, and the number of residential units of demand served by the medians.

In SPRING 5.3, the presented results comprise households with income of up to R$ 1600.00, who reside in nonowned properties, and with the age of the person in charge of the household up to 50 years are presented in Fig. 5.2, Tables 5.1, and 5.2.

5.5 Discussion of Results

Analyzing the data obtained through the market launch information, as presented in Table 5.2, the region of the West Zone of the Municipality of Rio de Janeiro concentrates most of the launches.

Table 5.2 Comparison between market launches and simulation results

Market launches region	Quant	Simulation results region	Quant
Santa Cruz	5586	Jardim Sulacap, Vila Militar, and Deodoro e Campo dos Afonsos	17,277
Rocha	2240	Mar 1	17,240
Senador Camar	2133	Inhama	17,229
Cosmos	1945	Leblon 2	17,197
Barros Filho	1260	Santa Teresa	17,176
Estcio	998	Cosmos 2	17,153
Cidade de Deus	996	Colgio	16,845
Paciłncia	902	Taquara 1	15,966
Campo Grande	735	Senador Camar 1	15,965
Jacarepagu	460	Cascadura	953
Santssimo	423	Encantado and Abolio e gua Santa	689

Comparing these launches with the information obtained through the application of the methods presented in the location proposal described in Sect. 5.3, it is possible to make a detailed analysis of the supply and demand of real estate in the city.

In the results found for Range 1, with a household income of up to R$ 1600.00, it is noticed that one of the medians determined in the computational simulation is located in the south zone in one of the most valued areas of the city (Leblon 2). This area still has one of the biggest demands on total distribution. This result demonstrates the particularity of Rio de Janeiro, which has a large number of communities with lower-income populations around the southern and central regions. Although the existence of this demand is clear to the residents of the city, a plan of action to offer an alternative solution to housing without adequate conditions of use has not yet been effective enough to meet these demands.

Figure 5.3 shows the differences between the market launches and the simulation result. The red stars represent the neighborhoods where launches occurred, with the quantities of units launched. The blue circles represent the medians found in the simulation results. There is a clear preference for market launches in the western region as mentioned above. While launches near the south and central zones have fewer units, even if they are in high demand.

Another interesting aspect is also the uniformity between the number of demands met in each median, around 15,000–17,000, which may represent a good capacity to absorb this demand in all the regions studied. Obviously, this demand for the p-median problem is not really "Housing Demand," because the latter is linked to home formation, since the demand studied also contains families that do not have the interest or ability to purchase of a property. However, the problems of housing for families in this income range need to be treated with greater care, since many households in this range are unable to provide security to the resident families.

Also, noteworthy in this study is the demand found in neighborhoods in the center and north of Rio de Janeiro. Some of these neighborhoods have had launches in

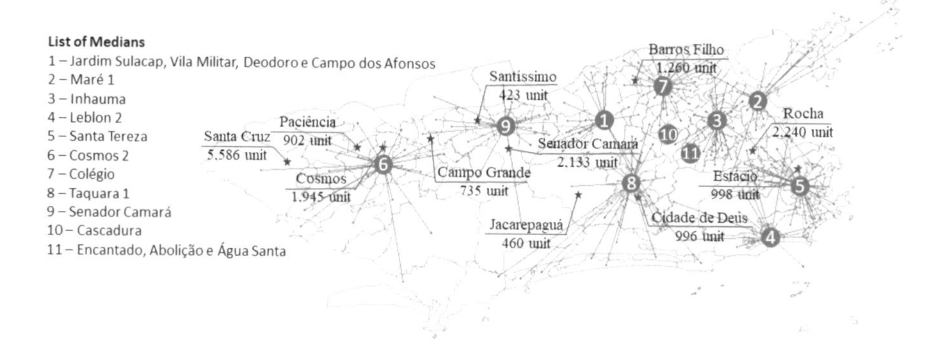

Fig. 5.3 Comparative of market launches with medians

recent years, such as Estácio, which appears in sixth place in the ranking of releases, but still the number of launches is much lower than the data obtained from the IBGE.

This concentration of launches in a certain region, and lack of launches in other regions with great demand, is linked to the urban planning of the city. This distribution of housing also represents a social challenge, since these launches make availability in regions where migration will not occur, mainly from the south and central zones. This discussion takes into account the serious problem of socio-spatial segregation that exists in Brazil, a topic that needs to be seriously debated to be understood with the search for adequate solutions.

5.6 Conclusions

The results presented in this article demonstrate the possibilities in the use of Operational Research tools to assist in the decision-making process of real estate entrepreneurs and public managers in the area of urban planning. The determination of the location of projects and the definition of an adequate launch plan are constant challenges for the success of companies in the segment and an adequate distribution of housing by the public managers.

The proposal presented is not a substitute for entrepreneurs' experience, but it demonstrates a powerful tool to improve the quality of the strategic decisions of the companies in the segment. Small- and medium-sized companies, which do not have the resources to conduct consistent surveys, may use the IBGE data and the proposals presented herein to improve their performance in the market.

In relation to the results found, in comparison with the market launches, there are large differences between demand and units launched in Range 1, showing an opportunity for investment in this modality, which still finds some restriction on the part of the developers.

Another fact that calls attention is the distribution of demand in every municipality, very different from the market launches, concentrated in certain regions. Note the need to explore neighborhoods near the center and south and center of the Municipality, which often does not occur due to differences in land costs between these regions.

In addition to the applications presented, this proposal represents great possibilities, using Census data with the possibility of parameterizing the research according to the need and strategic planning of the entrepreneur to build a method of its own and to recognize the capacity of a given business in a less intuitive way.

References

1. Azevedo, B.F.: Análise de localização de empreendimentos imobiliários: Um estudo de caso do Município do Rio de Janeiro. Petrópolis (2016)
2. Barcelos, F.B., Pizzolato, N.D., Lorena, L.N.: Localização de escolas de ensino fundamental com modelos capacitado e não-capacitado: Caso de Vitória/ES. Pesqui. Oper. **24**(1), 133–149 (2004)
3. Beguin, H., Deconnink, J., Peeters, D.: Optimizer la localization des ecoles primaires: Le cas de Mouscron, Belgique. Rev. Econ. Reg. Urbaine **5**, 795–806 (1989)
4. Galvão, R.D., Espejo, L.G.A., Boffey, B.: A hierarchical model for the location of perinatal facilities in the Municipality of Rio de Janeiro. Eur. J. Oper. Res. **138**, 495–517 (2002)
5. Molinero, C.M.: Schools in Southhampton: a quantitative approach to school location, closure and staffing. J. Oper. Res. Soc. **39**(4), 339–350 (1988)
6. Pizzolato, N.D., Silva, H.B.F.: The location of public schools: evaluation of practical experiences. Int. Trans. Oper. Res. **4**(1), 13–22 (1997)
7. Pizzolato, N.D., Barros, A.G., Barcelos, F.B., Canen, A.G.: Localização de Escolas Públicas: Síntese de Algumas Linhas de Experiłncias no Brasil. Pesqui. Oper. **24**(1), 111–131 (2004)
8. Pizzolato, N.D., Raupp, F.M.P., Alzamora, G.S.: Revisão de Desafios Aplicados em Localização com Base em Modelos da p-mediana e suas Variantes. Rev. Pesqui. Oper. Desenvolv. (PODES) **4**(1), 13–42 (2012)
9. Rahman, S., Smith, D.K.: Use of location-allocation models in health service development planning in developing nations. Eur. J. Oper. Res. **123**, 437–452 (2000)
10. Smith, D.K., Harper, P.R., Potts, C.N.: Bicriteria efficiency/equity hierarchical location models for public service application. J. Oper. Res. Soc. **64**, 500–512 (2013)
11. Teixeira, J.C., Antunes, A.P.: A hierarchical location model for public facility planning. Eur. J. Oper. Res. **15**(1), 92–104 (2008)
12. Tewari, V.K., Jena S.: High school location decision making in rural India and location-allocation models. In: Ghoshand, A., Rushton, G. (eds.) Spatial Analysis and Location-Allocation Models. Van Nostrand Rheinhold, New York (1987)
13. White, L., Smith, H., Curie, C.: OR in developing countries: a review. Eur. J. Oper. Res. **208**, 1–11 (2011)
14. Yasenovsky, V.S., Hodgson, M.J.: Hierarchical location-allocation with spatial choice interaction modeling. Ann. Assoc. Am. Geogr. **97**, 496–511 (2007)

Further Reading

15. Azevedo, B.F.: Método de localização de áreas candidatas a receberem empreendimentos residenciais com utilização da p-mediana: Um estudo do Município do Rio de Janeiro. 16ł Conferência Internacional da LARES (2016)
16. Azevedo, B.F.: Proposta de análise de demanda residencial para terrenos com base nos dados do IBGE Um estudo de caso no Município do Rio de Janeiro. 16ł Conferência Internacional da LARES (2016)
17. Lorena, L.A.N., Senne, E.L.F., Paiva, J.A.C., Pereira, M.A.: Integração de Modelos de Localização a Sistemas de Informações Geogŕficas. Gest. Prod. **8**(2), 180–195 (2001)

Chapter 6
An Ant Colony System Metaheuristic Applied to a Cooperative of Recyclable Materials of Sorocaba: A Case Study

Gregory Tonin Santos, Luiza Amalia Pinto Cantão, and Renato Fernandes Cantão

Abstract In the last decade, selective waste collection and the posterior reinsertion of the recycled materials into the productive chain have become not only an important economic activity but also essential for the reduction of the environmental impacts associated with landfills. The collection step of the recycling process is, in essence, a transportation operation, with costs directly coupled with those of fuel. Thus, fuel consumption reduction can increase profit margins with a pleasant side effect of reducing greenhouse gas emissions. In this scenario, Green Vehicle Routing Problems (GVRP) are instrumental as enablers of more economic routes in the sense of fuel consumption. Therefore, in this work we propose the application of a GVRP-based model with the objective of fuel consumption reduction and its application with data from CORESO, a recycling cooperative from the city of Sorocaba. Solutions for the model were obtained through the application of a tuned Ant Colony System metaheuristic associated with Tabu Search. We show that these solutions are on par with the exact ones and are calculated in a tiny fraction of the time spent by CPLEX. The results of this work were partially integrated in the cooperative routine for a period of tests.

6.1 Introduction

The populational growth experienced in Brazil during the last two decades [21, 22] understandably resulted in an increased production of solid waste. Along with the move towards a more industrialized economy, the waste composition is no longer predominantly organic, but synthetic, with a slower degradation rate in the

G. T. Santos · L. A. P. Cantão (✉)
São Paulo State University – UNESP, São Paulo, Brazil
e-mail: luiza.amalia@unesp.br

R. F. Cantão
Federal University of São Carlos – UFSCar, São Carlos, Brazil
e-mail: rfcantao@ufscar.br

V. N. Coelho et al. (eds.), *Smart and Digital Cities*, Urban Computing,
https://doi.org/10.1007/978-3-030-12255-3_6

environment. In that scenario, recycling becomes an important tool to mitigate the negative impacts related to this increase in waste production [40].

Recycling is an umbrella term encompassing all processes enabling the reuse of discarded materials, reinserting that in the production cycle. The benefits come both from environmental and social points of view: natural resources are spared and jobs are created.

The whole process begins by the separation of waste in its organic and recyclable parts, in residences and industries alike. Later, the recyclable part is collected by cooperatives, garbage pickers or recycling companies for suitable treatment and preparation for reuse [33]. The separation and collection phases are referred together as "selective waste collection".

In Brazil, Law 12.305/2010 [4], Solid Waste National Policy[1] (SWNP) establishes a hierarchy for the management of solid waste, stating that landfills are the last resort for its destination. That way, selective waste collection became a fundamental municipal policy because in general landfills are managed by counties. Furthermore, an extra incentive is given for cities that implement the selective waste collection through recycling cooperatives or other forms of garbage pickers associations: these cities have priority in Federal funding.

In Brazil, the basic working units of the recycling cycle are scrappers, industries and garbage pickers. Scrappers are usually favoured over garbage pickers because they often have the infrastructure needed to collect and sell more and better recycled materials. On the other hand, garbage pickers usually work alone in unsanitary conditions, sweeping streets or landfills, resulting in smaller amounts of material to be sold [40]. It should be mentioned that *recyclable and reusable waste collector* is officially recognized as an occupation by the Ministry of Labor and Employment [31].

Following [40], several studies validate the role of selective waste collection cooperatives as a decisive factor in reducing the environmental impact related to solid waste. Nevertheless, the same author points the obstacles garbage pickers face when trying to organize themselves in cooperatives, even with private and public sectors help. The "Cooperativa de Reciclagem de Sorocaba—SP (CORESO)" (Fig. 6.1) is one of these cooperatives and information about it will be instrumental in this work.

In this sense, costs associated to transportation are pivotal in the economic planning. According to [43], for an actual shipping company from Shangai, China, fuel costs sum up to 67.5% of all transportation costs, making fuel consumption an ideal candidate for cost reduction.

Directly proportional to fuel consumption (and cost), there is the emission of polluting gases, including greenhouse effect ones. Thus, fuel consumption reduction not only brings economical benefits for the cooperative but also environmental and public health at large [13].

[1]From the original in Portuguese, "Política Nacional de Resíduos Sólidos".

Fig. 6.1 CORESO cooperative. (**a**) Mechanical treadmill transporting material for initial triage. (**b**) Initial triage, sorting out, for instance, metal from plastic. (**c**) Sorting out different types of plastic. (**d**) Compaction, reducing transportation requirements

Many authors [41] consider only carbon dioxide (CO_2) given its preponderance among the greenhouse gases, ease of detection and significative proportion when compared to other gases from fossil fuel combustion. For instance, in 2014 only, the transportation industry was responsible for 43.5% of all CO_2 emissions in Brazil [34].

One possible approach to reduce fuel consumption is the creation of optimal transportation routes. Within this proposal, we have the *Vehicle Routing Problem (VRP)*, an NP-hard [15] integer programming problem modelling a fleet of vehicles that must satisfy predetermined demands of a set of clients, respecting scenario-specific constraints [13].

VRP was first developed by Dantzig and Ramser, in 1959 [42], with the intention of improving the logistics of fuel distribution for gas stations. This seminal model spawned a whole class of ever more sophisticated models and solution methods, as the heuristic developed by Clark and Wright, in 1964, with a substitution algorithm replacing expensive arcs for cheaper ones, thus effectively creating better routes [30]. The work of [39] presents a VRP with time windows, while [26] introduces the concept of capacitated VRP, in which the total transported weight cannot surpass the gross load capacity of the vehicle.

A natural development of the VRP was its application to logistic problems with environmental characteristics, giving rise to a new class of problems and algorithms grouped under the name *Green Vehicle Routing Problem (GVRP)*. GVRP is an extension of the standard VRP, with objectives and constraints gearer towards the minimization of environmental impacts. The work of [15], for instance, minimizes the traveled distance for alternative fuel vehicles, routing them through convenient gas stations so they never run out of fuel.

The model proposed by Xiao et al. [43] aims to reduce fuel cost by reducing its consumption. Consumption is, in this case, considered as a function of traveled distance, load weight, fuel consumption rate and fuel price, as well as some other indirect factors like cruise speed, taxes, vehicle depreciation and so on.

The solution of NP-hard problems like the VRP presents yet another challenge, demanding approximate solution methods (exact ones can be obtained only for small problems) [26]. Thus, research has been focused in metaheuristics [8], a class of methods that tries to obtain approximate solutions near the optimum for complex combinatorial problems in reasonable computational time [9].

The metaheuristic implemented for the VRP proposed in this work is the *Ant Colony System (ACS)*, introduced by Colorni et al. [8] simply as Ant System—first applied to the Travelling Salesman Problem (TSP)—and further enhanced by [14], foundation for the algorithm used here. Following [28], ACS is based on real ants behaviour: despite their short-sightedness, they are able to choose the shortest path among several sources of food and their colony. Ants leave a chemical trail of pheromones that tends to evaporate in time. In ideal conditions, therefore, shortest paths will evaporate less, effectively marking them for the next ants.

That said, ACS as proposed by Dorigo and Gambardella [14] implements "artificial ants" emulating the behaviour of their biological counterparts, resulting in the same search effect observed in real world, but this time towards the problem solution. In [28], some differences between the artificial and real ants are highlighted, as the introduction of a visibility feature inversely proportional to distance, a form of *heuristic information*.

Mazzeo and Loiseau [29], Bullnheimer et al. [5], Gambardella et al. [17], Bell and McMullen [3] and Gambardella et al. [16] are pioneers in the application of ACS to VRP. Mazzeo and Loiseau [29], for instance, show that this metaheuristic can produce good results when applied to capacited VRP instances with more than 50 nodes.

In this sense, this brief literature review demonstrates that GVRP is an active research field, both in its application in different scenarios and in the development of new algorithms and heuristics.

Thus, the present work aims to present ways to make selective waste collection more efficient from the point of view of the cooperatives. This will be accomplished minimizing the consumption of fuel, given the central position of the transportation activities to selective waste collection.

The general objective is to create a convenient VRP formulation modelling the minimization of fuel consumption—and consequently the emission of greenhouse gases, observing the constraints related to truck capacity, travel time and availability of vehicles, resulting in more economic routes for the work week (5 days).

In Sect. 6.2, we introduce the mathematical model and a brief description of the used metaheuristic. Section 6.3 explains the methods used to adjust model parameters, while Sect. 6.4 will analyse the results. Section 6.5 weaves the final conclusions. The work ends with references.

6.2 Mathematical Formulation

6.2.1 *Vehicle Routing Problem with Fuel Consumption Minimization Objective*

In this work, we have chosen to use a mathematical model based on that of Xiao et al. [43] given the amount of documentation about it, innovative approach and wide applicability, as exposed by Lin et al. [27]. As described in Sect. 6.1, this model considers transportation speed as a constant. The main factor related to fuel consumption is the total transported weight, as proposed by Zhang et al. [44] and Tiwari and Chang [41].

The model was adapted for the case of CORESO, a recycling cooperative located at the city of Sorocaba-SP, using the ACS heuristic and crafting a test scenario based in [38].

Arenales et al. [2] establish that the *classic Vehicle Routing Problem* can be seen as an oriented complete graph $G = (N, E)$, where $N = C \cup \{0, n+1\}$, $C = \{1, \ldots, n\}$ is a set of nodes or vertices (each one is a client), and 0 and $n+1$ are nodes representing the warehouse. The set $E = \{(i, j) : i, j \in N, i \neq j, i \neq n+1, j \neq 0\}$ are edges connecting these nodes. No edge ends at node 0 and no edge starts from node $n+1$. Costs are associated to each edge $(i, j) \in E$ while to each client i there is a demand D_i.

The first step in order to understand the model is to describe collection points (visited streets), location of these points (geographical coordinates), respective demands and distance among them. The next step is the development of an FCR (Fuel Consumption Rate) function, relating the basal vehicle autonomy (with no load) and load weight. Following [43], we have used a linear function in the load weight:

$$\rho_{ij} = \rho\left(Q_{ij}\right) = \rho_0 + \alpha Q_{ij}, \qquad \alpha = \frac{\rho^\star - \rho_0}{Q} \tag{6.1}$$

where α is the slope of the linear FCR function, and ρ_0 and $\rho^\star$ are parameters of the empty and fully loaded truck, respectively. Variable Q_{ij} is the weight transported from node i to node j. FCR calculation can be made more precise through statistical inference, information directly from truck manufacturers [43] or from fiscalization agencies such as INMETRO [23] (Instituto Nacional de Metrologia, Qualidade e Tecnologia) in Brazil.

In order to fully describe the model, we set

- c_0: fuel price (Brazilian reals per liter);
- d_{ij}: distance between nodes i and j;
- D_i: demand for node i;
- F: fixed transportation cost;
- y_{ij}: weight transported from i to j;
- x_{ij}: binary decision variable. If $x_{ij} = 1$, the vehicle travels from i to j, otherwise $x_{ij} = 0$.

A few other constraints must be posed for the VRP:

- Total transported load cannot exceed the vehicle capacity;
- Each client can be serviced only once;
- Routes start and end at node 0 (warehouse).

Thus, a GVRP based on [43] and minimizing fuel consumption can be written as:

$$
\begin{aligned}
&\min \sum_{j=1}^{n} F x_{0j} + \sum_{i=0}^{n}\sum_{j=0}^{n} c_0 d_{ij}\left(\rho_{ij}x_{ij} + \alpha y_{ij}\right) && && \text{(a)}\\
&\text{Subject to } \sum_{i=0}^{n} x_{ij} = 1 && \forall\, j = 1, \ldots, n && \text{(b)}\\
&\sum_{j=0}^{n} x_{ij} - \sum_{j=0}^{n} x_{ji} = 0 && \forall\, i = 0, 1, \ldots, n && \text{(c)}\\
&\sum_{j=0,\, j\neq i}^{n} y_{ij} - \sum_{j=0,\, j\neq i} y_{ji} = D_i && \forall i = 1, \ldots, n && \text{(d)}\\
&y_{ij} \leq Q x_{ij} && \forall\, i, j = 0, 1, \ldots, n && \text{(e)}\\
&x_{ij} \in \{0, 1\} && \forall\, i, j = 0, 1, \ldots, n && \text{(f)}
\end{aligned}
\qquad (6.2)
$$

Equation (a) is the cost function that we seek to minimize. Constraint (b) imposes that a client can be visited only once and (c) that a vehicle visits a client and leaves thereafter. Constraint (d) models the increase in load as the demand of that node, after visitation. Constraint (e) limits the vehicle load, sending it back to the warehouse when full or close to it. Constraint (f) states the binary condition of variable x_{ij}.

When compared to [43], our model changes the signs of the summations in Eq. (d). The reference is based on water trucks starting fully loaded and returning empty to the warehouse, the exact opposite of our situation, hence the sign change. The objective function was also simplified, disregarding the fixed transportation cost F.

Equation (6.3) simplifies Eq. (a) from (6.2), with ρ_{ij} the FCR for each edge i, j with the slope α already taken in account:

$$\sum_{i=1}^{n}\sum_{j=1}^{n} c_0 \rho_{ij} d_{ij} x_{ij} \tag{6.3}$$

The basic difference between Equation (a) from (6.2) and (6.3) is that the former considers a homogeneous fleet, while the latter can be used for different vehicles. Several authors adopt homogeneous fleets, like [24, 25, 37] and [35].

6.2.2 Ant Colony System (ACS)

ACS is suitable for problems modelled through graphs, where one node (named warehouse) is the starting point of a closed circuit encompassing all other nodes, respecting some set of constraints. The implementation of the ACS heuristic follows the ones proposed by Dorigo and Gambardella [14] and Lopes et al. [28].

In order to avoid the ACS coming up with repeated solutions, a memory is attached to the original algorithm in the form of a Tabu Search (TS) [18]. De la Cruz et al. [11] shows that the combination of ACS and TS presents a very competitive option when compared to notable algorithms from the literature. Another interest point of view comes from Zhang et al. [44]: TS heuristic helps to control the random characteristic of ACS, leading to better solutions.

Figure 6.2 shows a flowchart resuming the main blocks of our implementation. The first block is named “GVRP”, responsible for problem initialization, ants and parameters setting. Tabu Search and evaluation of objective function are also implemented here. The next block is called “Probability”, where the probability of an ant following a given trail is obtained. The block known as “Best” updates the objective function value. Finally, we have “Pheromone”, responsible for the update of the pheromone concentration after each iteration. There are two extra blocks representing stopping criteria (maximum number of iterations) and solution presentation. Arrows represent the order of execution for the blocks.

6.3 Methodology

6.3.1 Materials

The main implementation of the Ant Colony System with Tabu Search, as well as all instances tested, was made in Matlab version R2010a. Microsoft Excel® Spreadsheet version 1610 helped in the treatment of raw data coming from the cooperative (collection stations geographical coordinates and the number of houses

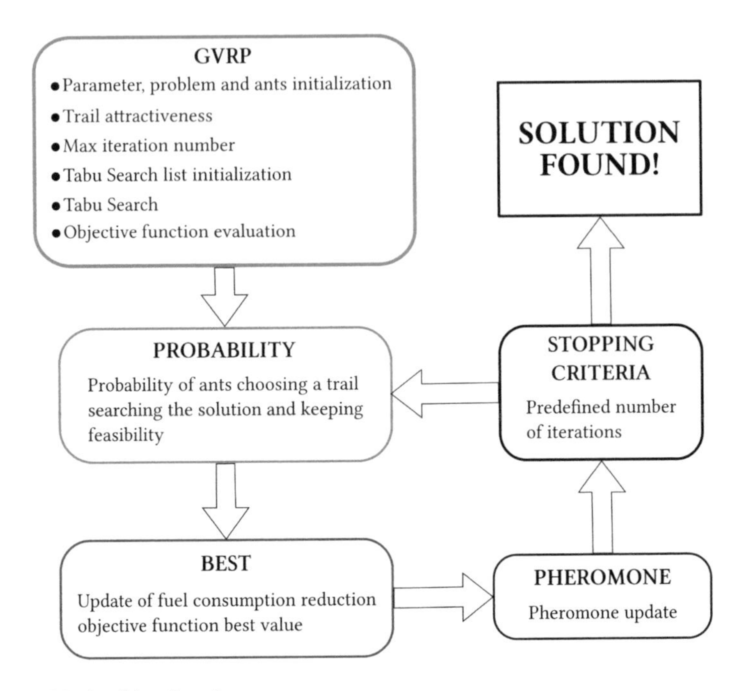

Fig. 6.2 ACS algorithm flowchart

per street). ArcGis 10.3 was used to tabulate georeferenced data for further treatment on a spreadsheet, while Google Earth Pro provided coordinates for each street, as well as the number of houses on each street (used to assess the demand).

Geographical information about the limits of the city of Sorocaba were downloaded for free from the collection of the Environmental spatial data infrastructure of São Paulo State (DATAGEO) [10].

Everything was implemented and tested on a desktop computer with an Intel i7, second generation, 3.4 GHz, with 8 GB of RAM memory running Microsoft Windows 7®.

6.3.2 Methods

A problem from [38] and three instances from [6]—M-n101-k10, M-n121-k7 and M-n151-k12 (n is the number of nodes and k the number of ants)—downloaded from the website of Pontifícia Universidade Católica do Rio de Janeiro (PUC-Rio) [36]—were chosen to help benchmarking the metaheuristics and adjust some parameters from the ACS. Reference [38] also brings exact solutions obtained with GAMS 24.3 and CPLEX.

The ACS parameters are: the importance of the trail α, importance of viability β, trail quality q_0 and pheromone rate ρ. Three sets of parameters, A, B and C were adjusted, being $A(\alpha = 0.5; \beta = 6; \rho = 0.3; q_0 = 40)$, $B(\alpha = 0.3; \beta = 6; \rho = 0.3; q_0 = 60)$ and $C(\alpha = 0.3; \beta = 6; \rho = 0.28; q_0 = 60)$. The total number of iterations for the ACS was set to 3000.

Distances between pairs of collection points (including the warehouse) were calculated with an Earth curvature-corrected Euclidean distance [20], as seen in Eq. (6.4). Taking $L_k \frac{90-Lat_k}{180}\pi$ and $M = \frac{Lon_i - Lon_j}{180}\pi$, we have

$$d_{ij} = 6371 \arccos\left(\cos L_j \cos L_i + \sin L_j \sin L_i \cos M\right), \tag{6.4}$$

where Lat and Lon are latitude and longitude, respectively. Works by Demir et al. [12] and Ćirović [7] make use of the same formulate for the distance.

6.3.2.1 Case Study

Operational data from the cooperative CORESO were compiled and updated in bimonthly meetings, pinpointing the visited streets, daily schedule and monthly collected weight, among others.

Operational data for CORESO cooperative was discussed and compiled along the project duration in bimonthly meetings. This data comprised the streets targeted for waste collection (turned into nodes on the graph), daily schedule and total collected weight per month. The meeting were also a place to discuss the application of our results in the daily cooperative activities in the form of new routes and schedule.

Compiled data has shown that the CORESO concentrates its efforts in the East zone of Sorocaba during workdays, totaling 230 streets/nodes with about 9000 collection points, when considered both residences and collective generators as industries or apartment complexes. Geographical coordinates for every street were obtained with Google Earth Pro and through Eq. (6.4) distances could then be calculated. Equation (6.4) proved to be in good agreement with the "Measure" tool from Google Earth Pro.

The demand was averaged for each street as the maximum of the monthly collected weight divided by the total number of residences visited by CORESO multiplied by the number of residences on that particular street. Great generators as condominiums or apartment buildings were inquired about the total weight of reciclables they generated. If the information was not available, the number of domiciles entered in the average as explained. Houses were counted with Google Earth Pro.

CORESO works with a different subset of all visited streets per workday, resulting in five different schedules. In order to effectively implement the optimized routes, each subset of streets was tested with the best ACS parameters (A and C).

CORESO owns two trucks with a capacity of 4 tons each (homogeneous fleet). Greenhouse gas emissions were estimated for CO_2 (carbon dioxide), CO (carbon monoxide), NO_x (nitrogen oxides) and NMHC (hydrocarbons) based on data available from INMETRO [23] using a similar[2] vehicle, the Citroën Jumper motor 2.3 16V.

According to CORESO, both trucks have an average autonomy of 7.5 km/L, leading us to adopt 9.0 km/L and 6.0 km/L for the empty and fully load vehicle, respectively. Using these values in the FCR, we obtain $\rho_0 = \frac{1}{9}$ L/km = 0.1111 L/km and $\rho^\star = \frac{1}{6}$ L/km = 0.1667 L/km. Equation (6.1) gives $\alpha \cong 1 \times 10^{-5}$, resulting in the FCR function $\rho = 1 \times 10^{-5} Q + 0.1111$.

The trucks run on diesel, with a cost of R$ 2.999 (Brazilian real) per liter, according to ANP (National Agency for Petroleum, Natural Gas and Fuel) on 03/01/2017 for Petrobrás S.A. distributor at Sorocaba [1].

Optimal routes were further represented spatially based on the georeferenced collection points, over the cartographic sheets for the city of Sorocaba (scale 1:10,000), for São Paulo state (scale 1:50,000) and for the country (scale 1:2,500,000) [10].

All results were presented to the cooperative management team and the truck drivers in the form of itineraries to be followed on each workday. Fuel reduction estimates were also presented as a motivation factor towards the adoption of the new scheduling.

6.4 Results and Discussion

Table 6.1 summarizes the solutions obtained with the given parameters for the test instances. Columns are, in order, ACS parameter set, instance name, solution from the ACS and percentual error in relation to the best solution for that instance. Instances M-n101-k10, M-n151-k12 M-n121-k7 have used 10, 12 and 7 ants, respectively, while the instance from [38], 8 ants.

It was not possible to identify a single best parameter set, given that *A* performed best with 100 nodes (M-n101-k10), *B* with 150 nodes (M-n151-k12) and *C* with both the GVRP [38] and the instance with 120 nodes (M-n121-k7).

Table 6.2 shows a comparison between the metaheuristic solution and the exact solution using GAMS 24.3 for the scenario in [38] for the three sets of parameters. The total cost for the exact solution is R$ 74.49 (Brazilian real) with a computational time of 7199.82 s.

From Table 6.2, it is possible to identify the computational efficiency of the metaheuristic when applied to GVRP. A solution differing by 1.38% from the exact one obtained within 0.32% of the time (parameter set *A*).

Parameter *C* resulted in the solution with the smallest deviation from the exact one, mere 0.55% within 0.33% of the computational time, showing the efficiency and applicability of ACS with Tabu Search, confirming the hypothesis that this metaheuristic can be a very competitive solution method, both in solution quality and computational time.

[2]No data was available on the actual trucks.

Table 6.1 Instance solution for each parameter set

Parameters	Instances	Results	Difference with best (%)
A	M-n101-k10	937.98	7.67
	M-n151-k12	1257.64	21.76
	M-n121-k7	1371.96	32.68
	[38]	75.52	1.38
B	M-n101-k10	1067.43	22.53
	M-n151-k12	1121.2	8.55
	M-n121-k7	1152.53	11.46
	[38]	76.65	2.90
C	M-n101-k10	985.82	13.16
	M-n151-k12	1164.65	12.75
	M-n121-k7	1062.06	2.71
	[38]	74.90	0.55

Table 6.2 Cost and computational time comparison for the exact GAMS and ACS metaheuristic solutions

	Route cost (R$)	Comp. time (s)	Dif. cost (%)	Perc. of time (%)
ACS parameter *A*	75.52	**22.83**	1.38	**0.32**
ACS parameter *B*	76.65	24.09	2.90	0.33
ACS parameter *C*	**74.90**	23.47	**0.55**	0.33

Bold face indicate best results

Table 6.3 Costs (R$) per workday for parameters *A*, *B* and *C*

	A	*B*	*C*
Monday	**7.46**	7.55	7.61
Tuesday	**4.27**	4.29	4.28
Wednesday	**4.11**	**4.11**	**4.11**
Thursday	4.96	4.73	**4.64**
Friday	**4.00**	4.09	4.09
Total	24.80	24.77	24.73

Table 6.3 presents the costs for each workday for the three sets of parameters, with the smallest ones featured in bold face.

Although the best solution for the problem in [38] is related to the parameter set *C*, parameter set *A* performed better when applied to the test instances from Table 6.1 and in four of the five workdays. This is expected given the *A* is tuned to graphs with less than 100 nodes as is the case (the largest number of nodes is Monday, with 60). This can be seen as a sensibility regarding the ACS parameters and reinforcing the need of more than one parameter set.

Table 6.4 illustrates the best routes and associated costs for Problem (6.2) using parameter set *C* on Thursday and *A* for the other workdays.

There are two routes per workday, one for each truck, totalizing R$ 24.50 per week. Considering a mean autonomy of 7.5 km/L at R$ 2999 per liter [1], the total fuel consumption is 8.17 L. Table 6.5 shows the emission of greenhouse gases per workday based on data from [23].

Table 6.4 Routes for (6.2)

Workday		Route	Load (kg)	Cost (R$)
Monday	1	0 - 15 - 16 - 17 - 14 - 20 - 19 - 18 - 6 - 5 - 4 - 3 - 7 - 26 - 25 - 24 - 13 - 22 - 12 - 11 - 10 - 23 - 9 - 21 - 8 - 33 - 34 - 1 - 27 - 39 - 28 - 38 - 35 - 32 - 29 - 30 - 37 - 36 - 2 - 31 - 54 - 58 - 57 - 56 - 0	3986.20	5.77
Monday	2	0 - 53 - 55 - 52 - 51 - 59 - 48 - 49 - 47 - 46 - 50 - 40 - 45 - 41 - 42 - 44 - 43 – 0	1326.00	1.69
Tuesday	3	0 - 29 - 30 - 31 - 41 - 28 - 25 - 26 - 24 - 27 - 23 - 22 - 21 - 7 - 48 - 44 - 46 - 43 - 5 - 2 - 3 - 4 - 6 - 9 - 13 - 14 - 10 - 11 - 15 - 12 - 16 - 42 - 45 - 47 - 0	3995.90	2.31
Tuesday	4	0 - 40 - 20 - 19 - 17 - 18 - 35 - 32 - 33 - 1 - 36 - 37 - 39 - 34 - 38 - 8 – 0	1316.30	1.97
Wednesday	5	0 - 43 - 44 - 45 - 37 - 36 - 35 - 34 - 40 - 39 - 28 - 41 - 42 - 29 - 27 - 30 - 26 - 24 - 25 - 23 - 1 - 4 - 22 - 5 - 6 - 8 - 7 - 21 - 20 - 9 - 10 - 11 - 16 - 15 - 14 - 17 – 0	3984.20	2.56
Wednesday	6	0 - 19 - 13 - 18 - 12 - 31 - 32 - 3 - 2 - 33 - 38 - 0	1328.10	1.56
Thursday	7	0 - 33 - 29 - 26 - 28 - 27 - 18 - 11 - 7 - 10 - 9 - 6 - 4 - 3 - 2 - 1 - 12 - 5 - 8 - 15 - 14 - 13 - 21 - 22 - 17 - 25 - 24 - 23 - 38 – 0	3992.10	2.65
Thursday	8	0 - 16 - 19 - 20 - 37 - 36 - 34 - 35 - 30 - 32 - 31 – 0	1320.10	1.97
Friday	9	0 - 35 - 25 - 26 - 28 - 30 - 31 - 14 - 33 - 34 - 29 - 40 - 39 - 37 - 36 - 13 - 10 - 4 - 12 - 11 - 1 - 9 -6 - 8 - 5 - 7 - 3 - 2 – 0	3994.10	2.33
Friday	10	0 - 15 - 22 - 19 - 20 - 17 - 21 - 18 - 38 - 23 - 16 - 27 - 24 – 0	1318.10	1.68

Parameter set *C* was used for Thursday and *A* for the other workdays

Table 6.5 Greenhouse gases emission (g) per workday

Workday	CO_2	NO_x	CO	NMHC
Monday	4910.487	5.396	0.878	0.373
Tuesday	2813.470	3.092	0.503	0.214
Wednesday	2706.708	2.974	0.484	0.206
Thursday	3051.932	3.354	0.545	0.232
Friday	16,116.636	17.710	2.880	1.226

Figures 6.3 and 6.4 represent graphically the routes. Dots are collection points, grouped by workday (230 nodes), as informed by CORESO. Figure 6.4 helps visualizing the pairs of routes per workday, while Figures 6.5, 6.6, 6.7, 6.8 and 6.9 show the routes separately for each workday.

After the tests conclusion, we scheduled a meeting on June 2017 with CORESO's managers and truck drivers to present the results and to propose their application on the cooperative routine. Drivers agreed to cooperate during 3 weeks to assess the

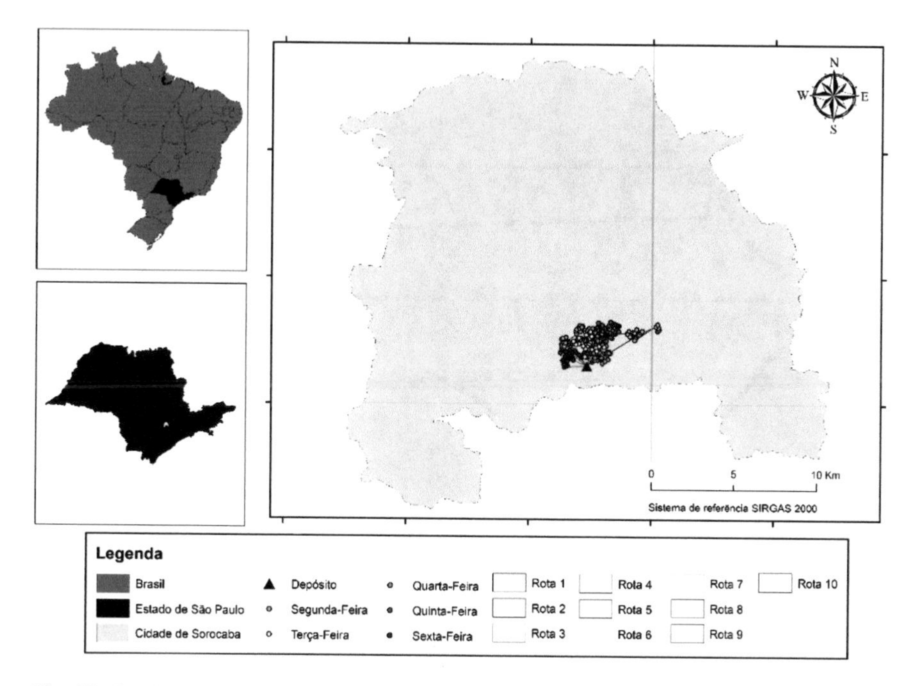

Fig. 6.3 Spatial representation of the routes from Table 6.4

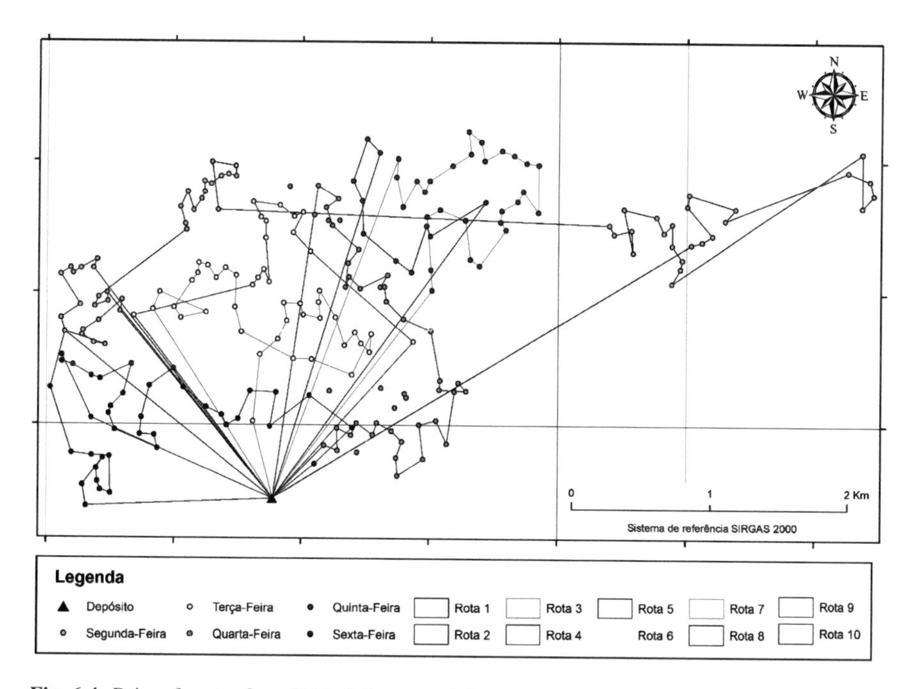

Fig. 6.4 Pairs of routes from Table 6.4, per workday

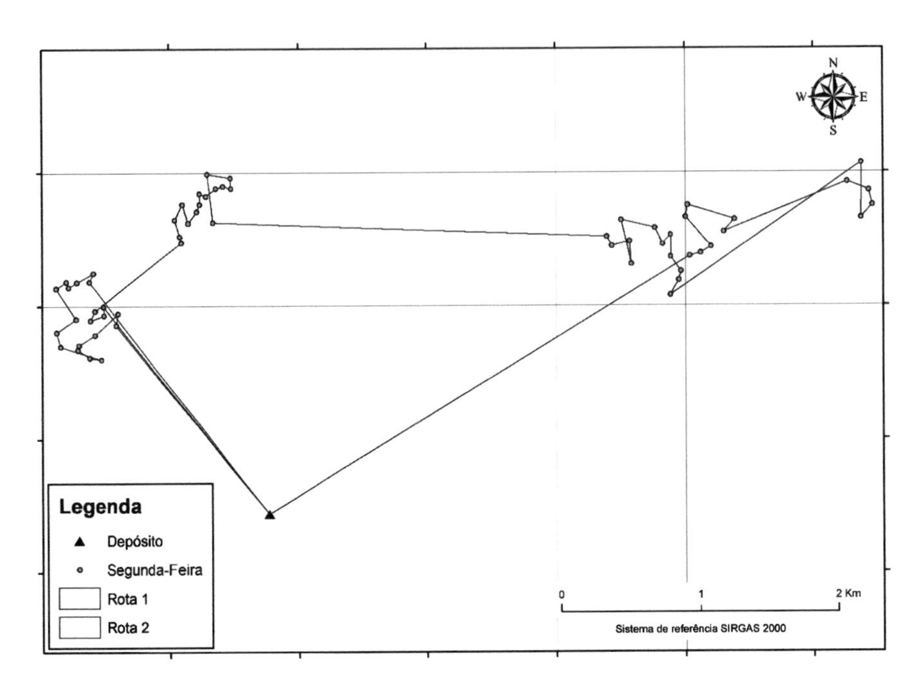

Fig. 6.5 Routes from Table 6.4, Monday

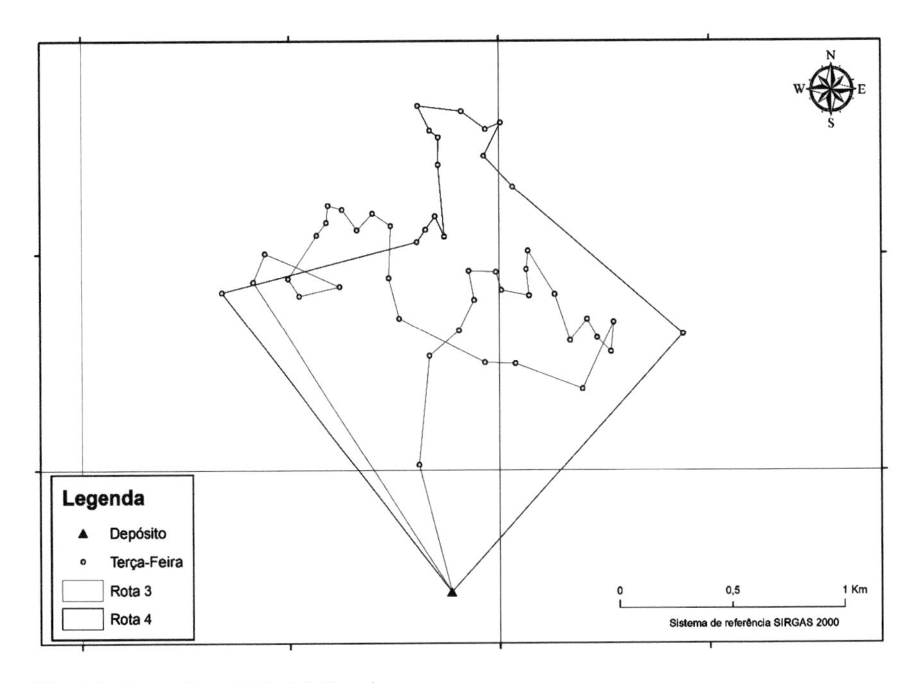

Fig. 6.6 Routes from Table 6.4, Tuesday

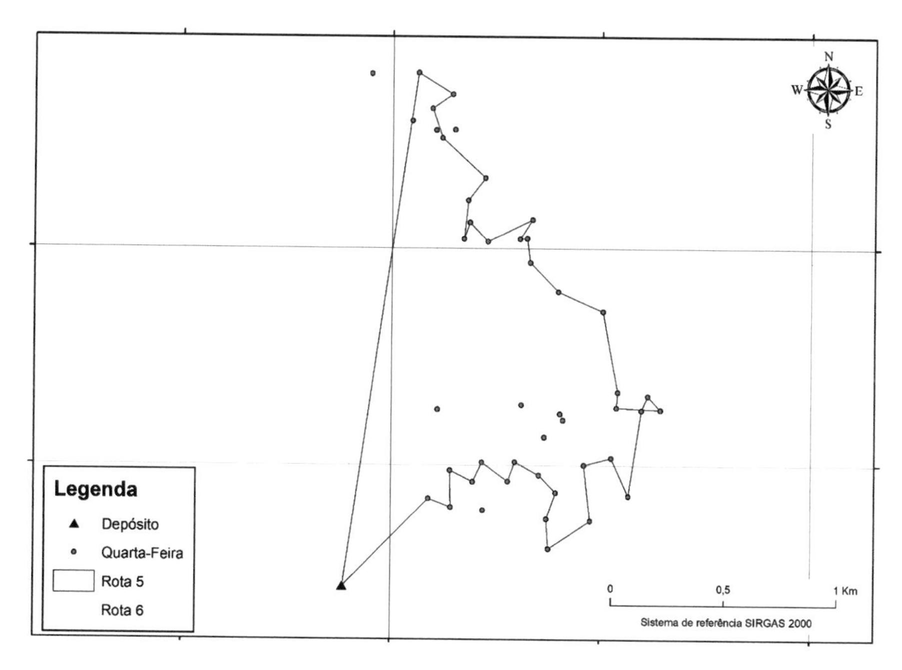

Fig. 6.7 Routes from Table 6.4, Wednesday

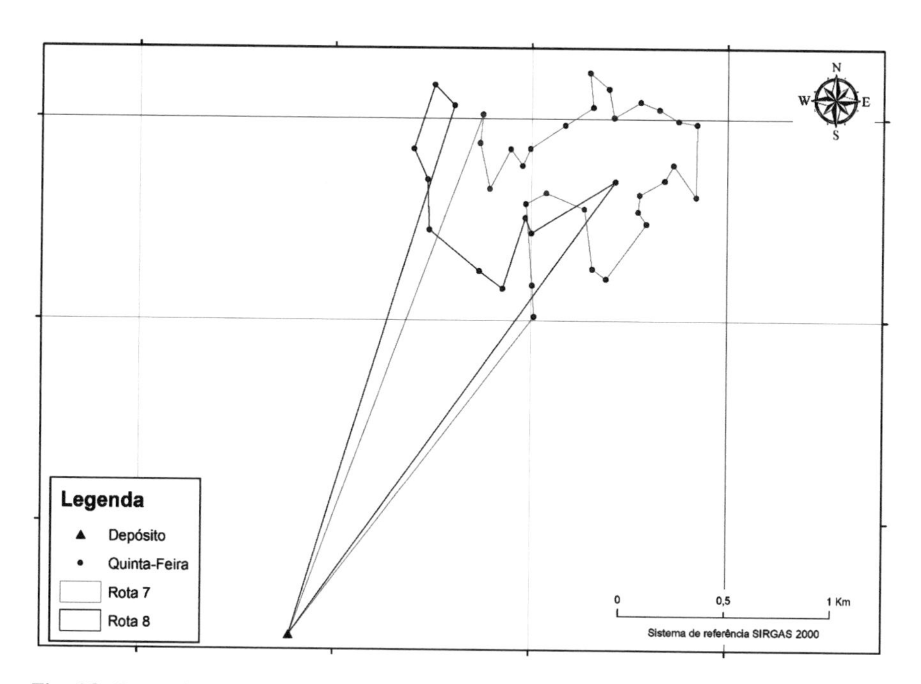

Fig. 6.8 Routes from Table 6.4, Thursday

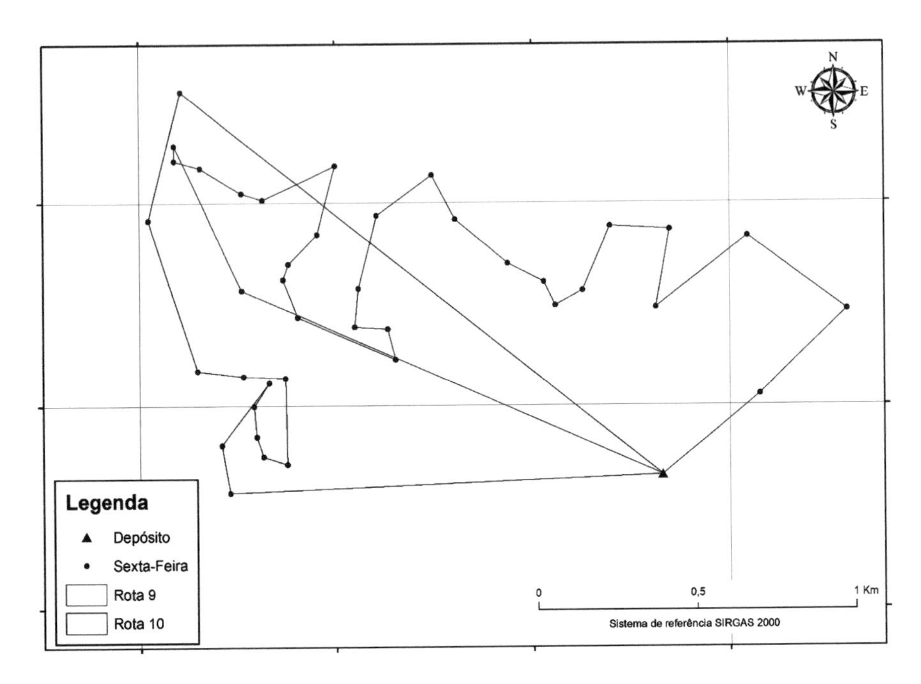

Fig. 6.9 Routes from Table 6.4, Friday

effectiveness of our routes. Unfortunately, some unforeseen events prevented the application during all days. Nevertheless, drivers reported an improvement related to traveled distance, although not mentioning a reduction on fuel consumption.

CORESO's trucks do not come equipped with a GPS making it difficult for the drivers to follow a route traced by streets. Indeed, the drivers are used to create their routes by neighbourhood using a notebook, thus diminishing the model benefits. They suggested the application of the model per neighbourhood, but we concluded that it would be a less economic option, given the wild variation in the number of streets. Some have as many as 5 streets while others, 28, making the solution dependent on the neighbourhood tour order.

A new field test would require the acquisition of two GPS with a few months of adaptation of use for the drivers until they become comfortable following a route designed by streets and not by neighbourhoods.

In the last round of conversations with the CORESO's managers, they reinforced their interest in applying the model, but progressively, giving the driver time to adapt.

6.5 Conclusion

In this work, we presented an application of the Green Vehicle Routing Problem (GVRP) to the case of a recycling cooperative—CORESO—situated in Sorocaba. We proposed an Ant Colony System (ACS) with Tabu Search (TS) approach to the solution of the GVRP.

The ACS metaheuristic has shown enormous potential applied to GVRP, obtaining solutions within an error margin of only 0.55% when compared to exact solutions obtained with an industry standard solver as GAMS. It should be also pointed that the solution process took a mere 0.33% of the computational time of GAMS. Parameter setting is a crucial step in this successful application of the method.

Some indirect benefits for the cooperative can emerge from the adoption of a GVRP model based scheduling system. Greenhouse gases emission assessment, especially carbon dioxide (CO_2), can help the cooperative enter the carbon credits market (Quioto Protocol (1997) [32]). Different scenarios can quickly be tested and the cooperative productivity can be increased while reducing furthermore the environmental impacts.

From a managerial point of view, the adoption of a methodology based on GVRP with metaheuristic solutions can bring agility into the decision process. Routes and scheduling can be readily altered based on changes on the setup: increase or decrease on the number of streets, fluctuations on fuel price, acquisition of more trucks and so on.

Despite a few unforeseen events, the optimal routes were partially implemented by the cooperative, resulting in a slight improvement related to traveled distances. That was a huge step forward, but clearly there is room for improvements towards the full application of the model.

From a wider perspective, Sorocaba is a participant of the "Programa Município VerdeAzul (PMVA)" a state government initiative intended to provide tools to quantify and support effective environmental management practices. A software tool that can be instrumental on reducing greenhouse gases emissions can earn Sorocaba a few positions up on the environmental state ranking [19], thus granting priority on funds from the "Fundo Estadual de Controle de Poluição (FECOP)" and more investments in environmental preservation.

Acknowledgements To Fundação de Amparo a Pesquisa do Estado de São Paulo (FAPESP), process number 2016/12488-6 for funding this research and CORESO recycling cooperative for their trust and support.

References

1. ANP – Agência Nacional do Petróleo, Gás Natural e Biocombustíveis: Sistema de levantamento de preços (2017). http://anp.gov.br/preco/prc/Resumo_Por_Municipio_Index.asp. Accessed 25 May 2017
2. Arenales, M.N., Armentano, V., Morabito, R., Yanasse, H.: Pesquisa Operacional. Elsevier, Rio de Janeiro (2007)

3. Bell, J.E., McMullen, P.R.: Ant colony optimization techniques for the vehicle routing problem. Adv. Eng. Inf. **18**(1), 41–48 (2004)
4. Brasil: Lei n. 12.305, dispõe sobre a política nacional de resíduos sólidos. Diário Oficial da República Federativa do Brasil (2010). http://planalto.gov.br/ccivil_03/_ato2007-2010/2010/lei/ll2305.htm. Accessed 05 Jul 2017
5. Bullnheimer, B., Hartl, R.F., Strauss, C.: Applying the ant system to the vehicle routing problem. In: de Recherche en Informatique et Automatique, I.N., 1525 LPCU (eds.) 2nd International Conference on Metaheuristics – MIC97 (1997)
6. Christofides, N., Mingozzi, A., Toth, P.: The vehicle routing problems. In: Christofides, N., Mingozzi, A., Toth, P., Sandi, C. (eds.) Combinatorial Optimizations. Wiley, New York (1979)
7. Ćirović, G., Pamučar, D., Božanić, D.: Green logistic vehicle routing problem: routing light delivery vehicles in urban areas using a neuro-fuzzy model. Expert Syst. Appl. **41**(9), 4245–4258 (2014)
8. Colorni, A., Dorigo, M., Maniezzo, V.: Distributed optimization by ant colonies. In: Varela, F.J., Bourgine, P. (eds.) Proceeding of the First European Conference on Artificial Life, pp. 134–142. Elsevier Publishing, Cambrige (1992)
9. Dalboni, F.L.: Algoritmos evolutivos eficientes para um problema de roteamento de veículos. PhD thesis, Federal University Fluminense (2003)
10. DATAGEO – Infraestrutura de dados espaciais ambientais do Estado de São Paulo: Base territorial ambiental unificada (2017). http://datageo.ambiente.sp.gov.br. Accessed 05 Jul 2017
11. De la Cruz, J.J., Paternina-Arboleda, C.D., Cantillo, V., Montoya-Torres, J.R.: A two-pheromone trail ant colony system—tabu search approach for the heterogeneous vehicle routing problem with time windows and multiple products. J. Heuristics **19**(2), 233–252 (2013)
12. Demir, E., Bektaş, T., Laporte, G.: The bi-objective pollution-routing problem. Eur. J. Oper. Res. **232**(3), 464–478 (2014)
13. Demir, E., Bektaş, T., Laporte, G.: A review of recent research on green road freight transportation. Eur. J. Oper. Res. **237**(3), 775–793 (2014)
14. Dorigo, M., Gambardella, L.M.: Ant colony system: a cooperative learning approach to the traveling salesman problem. IEEE Trans. Evol. Comput. **1**(1), 53–66 (1997)
15. Elhedhli, S., Merrick, R.: Green supply chain network design to reduce carbon emissions. Transp. Res. D Transp. Environ. **17**(5), 370–379 (2012)
16. Gambardella, L.M., Taillard, E., Agazzi, G.: MACS-VRPTW: a multiple ant colony system for vehicle routing problems with time windows. Technical report, Istituto Dalle Molle Di Studi Sull Intelligenza Artificiale (1999)
17. Gambardella, L.M., Rizzoli, A.E., Oliverio, F., Casagrande, N., Donati, A.V., Montemanni, R., Lucibello, E.: Ant colony optimization for vehicle routing in advanced logistics systems. In: Proceedings of the International Workshop on Modelling and Applied Simulation, pp. 3–9 (2003)
18. Gomes, A.: Uma introduça ao à busca tabu. Departamento de Ciência da Computação, Instituto de Matemática e Estatística, Universidade de São Paulo, São Paulo (2009). http://www.ime.usp.br/~gold/cursos/2009/mac5758/AndreBuscaTabu.pdf. Acessado em 20 Jun 2017
19. Governo do Estado de São Paulo: Município verdeazul (2017). http://www.ambiente.sp.gov.br/municipioverdeazul/o-projeto/. Accessed 05 Jul 2017
20. Hartung, G.E.: Distâncias entre dois pontos – UCA (2011). http://portaldoprofessor.mec.gov.br/fichaTecnicaAula.html?pagina=espaco%2Fvisualizar_aula&aula=30254&secao=request_locale=es. Acessado em 12 Aug 2015
21. IBGE – Instituto Brasileiro de Geografia e Estatística: Estimativas populacionais para os municípios brasileiros em 01.07.2011 (2017). http://downloads.ibge.gov.br/downloads_estatisticas.htm?caminho=/Estimativas_de_Populacao/Estimativas_2001/. Accessed 05 Jul 2017
22. IBGE – Instituto Brasileiro de Geografia e Estatística: Estimativas populacionais para os municípios e para as unidades da federação brasileiros em 01.07.2016 (2017). http://www.ibge.gov.br/home/estatistica/populacao/estimativa2016/estimativa_dou.shtm. Accessed 05 Jul 2017

23. INMETRO – Instituto Nacional de Metrologia Qualidade e Tecnologia: Programa brasileiro de etiquetagem - PBE (2016). http://www.inmetro.gov.br/consumidor/pbe/veiculos_leves_2016.pdf. Accessed 20 Jun 2017
24. Kim, B.I., Kim, S., Sahoo, S.: Waste collection vehicle routing problem with time windows. Comput. Oper. Res. **33**(12), 3624–3642 (2006). Part Special Issue: Recent Algorithmic Advances for Arc Routing Problems
25. Koç, Ç., Karaoglan, I.: The green vehicle routing problem: a heuristic based exact solution approach. Appl. Soft Comput. **39**, 154–164 (2016)
26. Laporte, G.: The vehicle routing problem: an overview of exact and approximate algorithms. Eur. J. Oper. Res. **59**(3), 345–358 (1992)
27. Lin, C., Choy, K., Ho, G., Chung, S., Lam, H.: Survey of green vehicle routing problem: past and future trends. Expert Syst. Appl. **41**(4, Part 1), 1118–1138 (2014)
28. Lopes, H.S., de Abreu Rodrigues, L.C., Steiner, M.T.A. (eds.): Meta-Heurísticas em Pesquisa Operacional, 1st edn. Omnipax, Curitiba, PR, chap Otimização por Colônia de Formigas (2013)
29. Mazzeo, S., Loiseau, I.: An ant colony algorithm for the capacitated vehicle routing. Electron. Notes Discrete Math. **18**, 181–186 (2004). Latin-American Conference on Combinatorics, Graphs and Applications
30. Miura, M.: Resolução de um problema de roteamento de veículos em uma empresa transportadora. Trabalho de Conclusão de Curso. Escola Politécnica da Universidade de São Paulo, São Paulo (2003)
31. MMA – Ministério do Meio ambiente: Catadores de materiais recicláveis (2017). http://www.mma.gov.br/cidades-sustentaveis/residuos-solidos/catadores-de-materiais-reciclaveis. Accessed 20 Jun 2017
32. MMA – Ministério do Meio ambiente: Mudança do clima (2017). http://www.mma.gov.br/clima/convencao-das-acoesunidas/protocolo-de-quiotos. Accessed 20 Jun 2017
33. MMA – Ministério do Meio Ambiente: Reciclagem (2017). http://www.mma.gov.br/informma/item/7656-reciclagem. Accessed 20 Jun 2017
34. MME – Ministério de Minas e Energia: Resenha energética brasileira do exercício 2014 (2014)
35. Montoya, A., Guéret, C., Mendoza, J.E., Villegas, J.G.: A multi-space sampling heuristic for the green vehicle routing problem. Transp. Res. C Emerg. Technol. **70**, 113–128 (2016)
36. PUC-Rio – Pontifícia Universidade Católica do Rio de Janeiro: Capacitated vehicle routing problem library—CVRPLIB (2017). http://vrp.atd-lab.inf.puc-rio.br/index.php/en/. Accessed 20 Jun 2017
37. Qian, J., Eglese, R.: Fuel emissions optimization in vehicle routing problems with time-varying speeds. Eur. J. Oper. Res. **248**(3), 840–848 (2016)
38. Santos, G.T., Cantão, L.A.P, Cantão, R.F.: Vehicle routing problem with fuel consuption minimization: a case study. In: Maturana, S. (ed.) Prodeedings of the XVIII Latin-Iberoamerican Conference on Operations Research, CLAIO 2016, pp. 802–809 (2016). ISBN: 978-956-9892-00-4
39. Solomon, M.M.: Algorithms for the vehicle routing and scheduling problems with time window constraints. Oper. Res. **35**(2), 254–265 (1987)
40. Souza, M.T.S.d., Paula, M.B.d., Souza-Pinto, H.d.: O papel das cooperativas de reciclagem nos canais reversos pós-consumo. Rev. Adm. Empresas **52**, 246–262 (2012)
41. Tiwari, A., Chang, P.C.: A block recombination approach to solve green vehicle routing problem. Int. J. Prod. Econ. **164**, 379–387 (2015)
42. Toth, P., Vigo, D.: The Vehicle Routing Problem. Society for Industrial and Applied Mathematics (2002). http://epubs.siam.org/doi/pdf/10.1137/1.9780898718515
43. Xiao, Y., Zhao, Q., Kaku, I., Xu, Y.: Development of a fuel consumption optimization model for the capacitated vehicle routing problem. Comput. Oper. Res. **39**(7), 1419–1431 (2012)
44. Zhang, S., Lee, C., Choy, K., Ho, W., Ip, W.: Design and development of a hybrid artificial bee colony algorithm for the environmental vehicle routing problem. Transp. Res. D Transp. Environ. **31**, 85–99 (2014)

Chapter 7
Multilevel Optimization Applied to Project of Access Networks for Implementation of Intelligent Cities

Márcio Joel Barth, Leandro Mengue, José Vicente Canto dos Santos, Juarez Machado da Silva, Marcelo Josué Telles, and Jorge Luis Victória Barbosa

Abstract Studies about network infrastructure have been realized and applied in a high variety of service-based industries, these studies are currently being used to design the network infrastructure in smart cities. However, planning network infrastructure in different levels is a big problem to be solved, because, generally, literature presents solutions where just one level is processed and the problems are solved individually. Planning the distribution and connection of equipment at various levels of a network infrastructure is an arduous task, it is necessary to evaluate the quantity and the best geographical distribution of equipment at each level of the network. This research presents a metaheuristic inspired by the concepts of the genetic algorithms. The proposed paper can search for solutions to plan the network infrastructure of multilevel capacitated networks, solving the network planning problem and obtaining results that are 20% better at cost when compared with other solutions.

7.1 Introduction

Globalization, urbanization, and industrialization have been recognized as three major factors driving human civilization in the twenty-first century. According to the Organization for Economic Co-operation and Development (OECD), approximately 70% of the world's population will soon live in urban areas. In this context, many cities have created initiatives to become more intelligent in functional terms, adopting Information and Communication Technology (ICT) as a way to revitalize economic opportunities and to strengthen their global competitiveness, consequently providing greater comfort to its habitants. The technologies aim to

M. J. Barth (✉) · L. Mengue · J. V. C. dos Santos · J. M. da Silva · M. J. Telles · J. L. V. Barbosa
University of Vale do Rio dos Sinos—UNISINOS, São Leopoldo, RS, Brazil
e-mail: marcio-barth@procergs.rs.gov.br; leandro@mengue.com.br; jvcanto@unisinos.br; jbarbosa@unisinos.br

V. N. Coelho et al. (eds.), *Smart and Digital Cities*, Urban Computing,
https://doi.org/10.1007/978-3-030-12255-3_7

facilitate the accomplishment of routine tasks, to offer new services, to automate actions, and to promote improvements in the daily life of the individuals and in the administration of the cities. The big challenge that arises is which infrastructure will support all the connectivity necessary to make the information available to the various services, since a great network of "things" must be formed and consequently will create a big "cloud of things" to store all the information generated. So, data communication networks have become as important as other services such as power and water supply to users, giving the feeling that this service should be available in the same way as the power supply to turn on a lamp. In this way, data communication networks need to have the same capillarity as electricity networks, for example, to be able to serve users in the same way. In addition, with the increasing demand for connection of several types of networked devices such as the implementation of *Smart Grid*, *IoT*, and *Smart City* [16, 17, 20], an accurated planning is required in order to provide access to communication networks in the best way and at an appropriate cost.

In the literature, there are already several works [2, 13, 21, 22] indicating the best application of methods for the elaboration of network projects in isolation form, but the purpose of this work is to design the system in an integrated way in order to provide access to data communication network to several habitants/consumers/services connecting *(Smart City)* connecting various types of devices such as sensors, *Smart Grid*, and end consumers, optimizing the best cost to network installation. The objective of this article is to introduce a computational method that can design access networks to solve the problem of network designs by exploring the technique known as the problem of the Multilevel Capacitated Minimum Spanning Tree (MCMST).

7.2 Basic Concepts

This chapter provides a brief introduction on the theories necessary for the understanding of the work.

7.2.1 Smart Cities

Cities are certainly the greatest social constructions of mankind, not only because of their complexity, but because of their ability to bring people together and facilitate common interests. The problems of cities are known, often multidisciplinary and complex solutions. Problems in meeting demands related to education, health, urban mobility, and sanitation, among others, can be listed as priorities for public managers. Those problems are also the main focus of companies involved in making solutions and managing problems in urban environments.

As suggested by Lee et al. [10], a model proposal for intelligent city analysis may be listed as: open government, service innovation, partnership formation, urban proactivity, intelligent city governance, and intelligent city infrastructure integration. This last topic meets the objective of this research, since it specifically addresses the ICT infrastructure to support the initiatives of intelligent cities, propitiating a high-speed communication network and providing the connectivity of several complementary devices.

7.2.1.1 Examples of Need for Connectivity in Smart Cities

The idea of an access network project for smart cities from the ICT point of view is to offer citizens applications and services that will improve their life quality. Examples of applications, like Smart Grid, Smart Parking, Traffic Video Monitoring, can also be seen in Fig. 7.1.

7.2.2 Access Network Topologies

Access network is a term that describes the communication network part that provides the connectivity between end users (terminal nodes) and central unit of the network. End users can also be signal relays for subnets. Next Generation Network (NGN) refers to the convergence trend of fixed and mobile communication networks, not leaving aside the currently used access technologies. The key factor to achieve this goal is the encapsulation of traffic in the *IP* protocol. On the other hand,

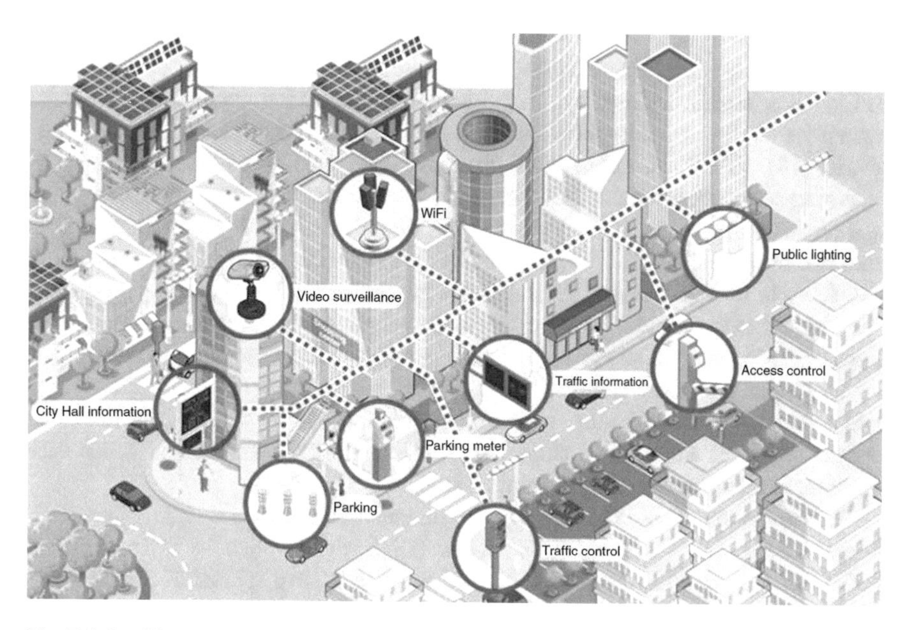

Fig. 7.1 Intelligent systems framework of a city

Internet of things for smart cities is a new concept and not deeply researched, that has been conducted so far, as necessitating a dedicated communication network to accommodate these applications. The combination of the aforementioned facts led to research for solutions that provide high rates of transmission and high availability for users.

7.2.3 *Fiber Optic Network Technologies*

Optical fiber is lightweight and flexible, made of glass (silica) or plastic, used in telecommunication networks to carry light signals with a greater bandwidth than any other wired technology. Due to low attenuation, the signals are transmitted over long distances [11]. The optical fiber can be single mode or multimode. Multimode fibers are characterized by a large core diameter (50–62 μm). Due to this fact, in multimode fibers there is a risk of greater dispersion which limits the distances with this type of fiber [1]. The single-mode fiber has a smaller diameter than the multimode (5–10 μm), which allows the use of larger distances. Advantages of using fiber optics: (a) high data transfer rates, on the order of terabits per second in laboratory conditions; (b) use of long distances due to low attenuation; and (c) no external electromagnetic field, providing network security.

7.2.3.1 Passive Optical Networks

With the growing need for data transfer by companies and home users using fiber-optic telecommunication networks, the architecture of Passive Optical Networks (PON) has emerged. This enabled telecom companies to invest less financial resources in the infrastructure network to serve customers in the access networks [18]. The PON architecture allows a single fiber to be divided into more fibers, serving several end users. This is possible with a passive component called *splitter*, which is a passive fiber optic coupler that divides the “light” of a single fiber into two or more fiber channels. This makes a connection in the telecommunication exchange to be able to serve several customers in the last mile, making it possible for companies to provide the use of fiber optic communication to a large part of the population (residential and/or commercial users) with less financial investments.

The signal of a fiber can be divided into 2^n fibers, currently n varies from 2 to 64 (it can also be said that the separation ratio is 2^n). Figure 7.2 illustrates how the splitter separates a fiber optic cable into a various fiber optic cables, in which case the fiber is capable of serving several customers.

The divisions of a PON are shown in the network of Fig. 7.3. Clients are connected to either an Optical Network Unit (ONU) or an Optical Network Terminal (ONT) of a fiber, which has been divided into the system splitter of the Network Optical Distribution (ODN) of the Optical Distribution Network, which is connected to the Optical Line Terminal (OLT). This shows that it is possible to connect multiple clients with only one fiber in the OLT.

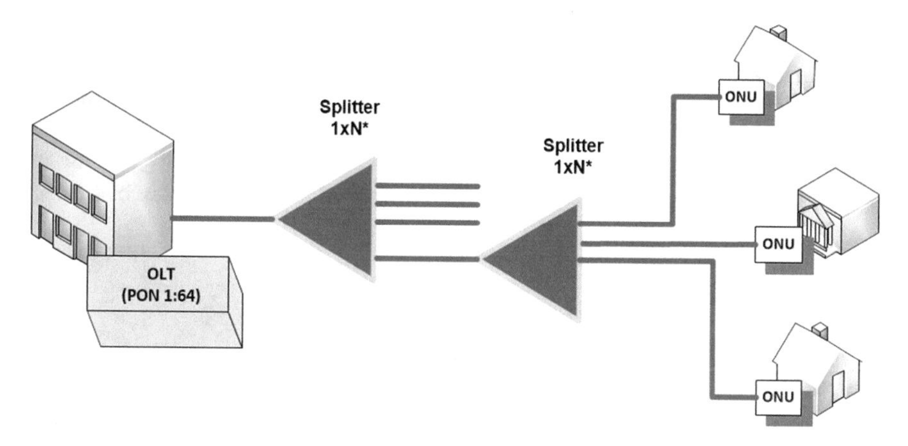

Fig. 7.2 Visual splitter scheme

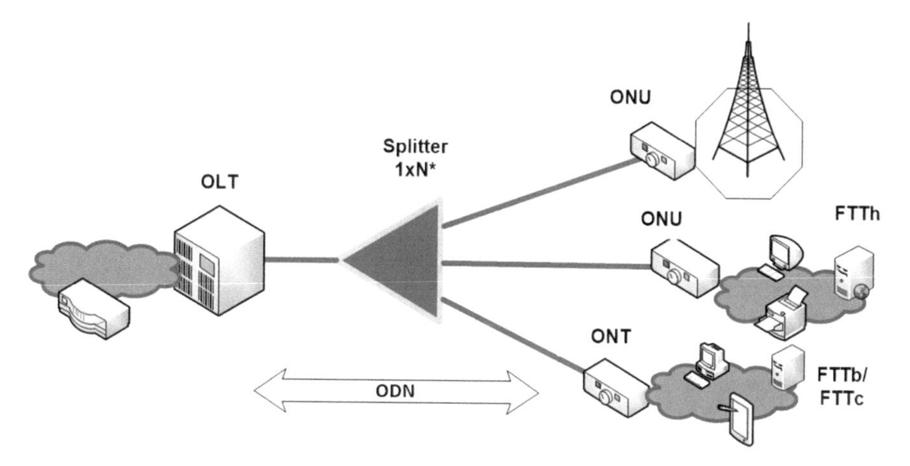

Fig. 7.3 Passive optical network example

The PON network is an evolution of the point-multipoint network architecture, whose main characteristic is the nonuse of electric energy through its physical environment [1]. Therefore, it takes the name of passive optical network, that is, its elements responsible for the distribution of the signal (splitters) are totally passive.

7.2.4 Graphical and Numerical Representation of Telecommunications Networks

The graphical representation of telecommunications networks has been used for the schematic representation of graphs, which facilitate the understanding of the explanations having the advantage of the easy apprehension by the human global perception with regard to some of its topological aspects [14]. On the other hand, the schematic representation is not suitable for providing a computer with data on a graph structure. The data related to a graph will always need an internal numerical representation, with which the computer can work. Next, the concept of graph theory and its form of numerical representation for application in this work will be treated.

7.2.4.1 Theory of Graphs

Graphs are important mathematical tools with applications in several areas of knowledge, successfully employed in solving computational problems [8]. A graph is mathematically represented as a tuple:

$$G = (V, A, \psi_G), \tag{7.1}$$

where V represents the vertices (or nodes), A represents the edges, also called links between vertices, and ψ_G the function that associates each edge of the graph G at a pair of vertexes. Figure 7.4 shows the example of a graph represented by the set $G = (V, A)$, where $V = \{1, 2, 3, 4, 5\}$ is the set that represents the nodes, whereas the edges are represented by the set $A = \{a, b, c, d, e, f\}$.

The mathematical representation of a graph, obtained from its geometric representation, can be idealized by at least three forms [5]:

- Adjacency Matrix,
- Incidence Matrix,
- Linked lists.

In this work, the mathematical representation of graphs is presented in the form of adjacency matrices, since this form expresses in a simple way how the vertices are related (adjacencies) in a graph and also a smaller computational effort to access the information [5]. According to [14], a telecommunications network can be represented using Graph Theory. In his work, he presented the nodes as a representation of the routers:

$$V = \{1, 2, \ldots, ||V||\} \tag{7.2}$$

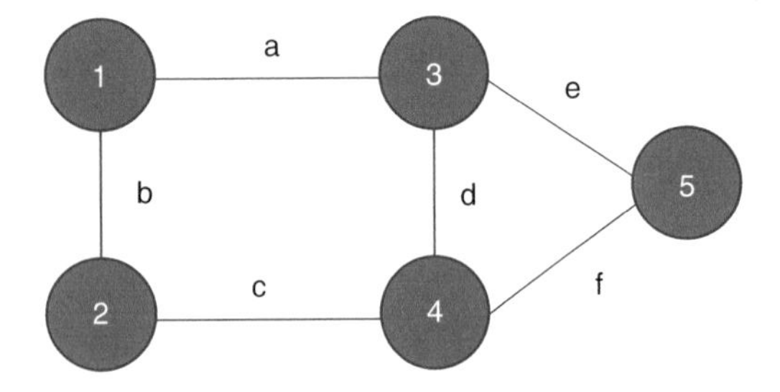

Fig. 7.4 Example of a graph

The representation of the links between the nodes is represented by the edges:

$$\mathbf{A} = a_{ij}, \tag{7.3}$$

where $\mathbf{A}$ is a binary matrix, in which i and j represent the indexes, rows, and columns, respectively, which contains the values 0, for absence of link, or 1, for the presence of connection between two vertices.

In the work developed by Martins [12], we also use graph theory to represent a communication network for energy meters.

For this work, simple and nonoriented graphs representing the data communication network will be considered, since all equipment involved in the infrastructure are transceivers.

7.2.4.2 Representation of a Graph by Adjacency Matrix

It is a simple representation where a graph is expressed by a matrix $\mathbf{A} = a_{ij}$ through the nodes and relations between the edges. The elements of these arrays are usually represented by 0-1 (Boolean) [12]. The binding function ψ_G, by definition, can be expressed by the equation:

$$\psi_G = \begin{cases} a_{ij} = 1, & \text{if there is a connection } (i, j), \\ a_{ij} = 0, & \text{if there is no connection } (i, j). \end{cases} \tag{7.4}$$

As an example, the communication network, represented in Fig. 7.5, is assumed as a network established between routers, which can be represented by a simple graph, as shown in Fig. 7.4.

From the graph formed by the observation of the network, it is quickly possible to identify the adjacency matrix formed by the bonding function of the nodes ψ_G,

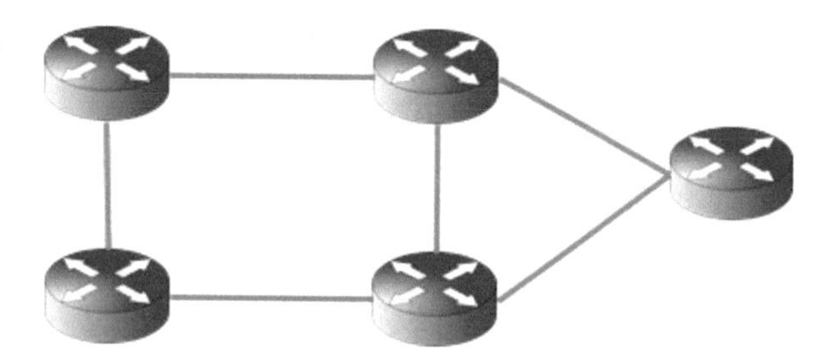

Fig. 7.5 A communications network and its links

exemplifying in Eq. (7.5):

$$\mathbf{A} = \begin{bmatrix} 0 & 1 & 1 & 0 & 0 \\ 1 & 0 & 0 & 1 & 0 \\ 1 & 0 & 0 & 1 & 1 \\ 0 & 1 & 1 & 0 & 1 \\ 0 & 0 & 1 & 1 & 0 \end{bmatrix} \tag{7.5}$$

7.2.5 The Minimum Spanning Tree Problem

The minimum spanning tree problem (MSTP) is to find, given a graph with weighted edges, a connection structure (tree) in which all nodes (spanning) connect (directly or indirectly) to each other. This structure must have the lowest possible weight, where the weight is given by the sum of the weights of the chosen edges (minimum).

Let $G = (V, A)$ be a nondirected graph and connected to a set of nodes $V = \{0, 1, \ldots, v\}$ and a set of edges A. Each node $i \in V$ has a weight $b_i \geq 0$ with $b_0 = 0$. The weight of the node can be interpreted as a required demand. Each edge $(i, j) \in A$ has a cost c_{ij} associated with its use in the spanning tree. The Capacitated Minimum Spanning Tree Problem (CMSTP) is to determine a minimal cost spanning tree on G, centered on node 0 (root), with the additional restriction that the sum of the weights of the nodes of any subtree connected to the root cannot be greater than a constant Q.

When all weights of nodes are equal, we have the homogeneous version of the problem that can be treated as a unit demand problem. For $2 < Q < n/2$, in the work of [15] it was shown that the problem is of the NP-difficult class.

7.2.5.1 Multilevel Capacitated Minimum Spanning Tree Problem

The multilevel capacitated minimum spanning tree problem (MCMSTP) is a generalization of the CMSTP, being the first research presented on the subject by Gamvros et al. [4]. The difference between the CMSTP and the MCMSTP is that in the second problem the installation of facilities with different capacities is allowed.

In the MCMSTP, a graph $G = (V, A)$ is given, with a set $N = \{0, 1, 2, \ldots, n\}$, where node 0 represents the central terminal from which the flow should exit and the others are the consumers, and a set of arcs A, b_i being the traffic demand (or weight) of the node i to be transported from node 0, features of type $1, 2, \ldots, L$ with capacities $Z_1 < Z_2 < \cdots < Z_L$ and a cost function c^l_{ij} denoting the cost of a facility of the type l installed between the nodes i and j. The goal then, like MCMSTP, is to find a minimum cost network to carry the required traffic, where the flow over each facility cannot be greater than its capacity.

In the flow-based models for the MCMSTP, the flow goes from the central node towards the terminal nodes. In the model presented by Gamvros et al. [4], the flow leaves the end nodes towards the central node. Each terminal node has a product that must be transported to the central node. The origin of the product k is the K node. The third variable, f_{ij}^{k}, indicates the flow of the product k through the arc (i,j).

7.3 Target Issue and Proposed Methodology

In this section, the focus problem about network infrastructure design that is used by telecommunication companies is detailed, a present challenge in different business areas and utility companies such as energy, gas, and water distribution. The problem is often the same: to optimize the installation of equipment in a geographic region in order to reduce costs. This problem has been modeled and studied in the literature as a Network Design Problem (NDP). Thus, for the network planning problem, the Multilevel Capacitated Minimum Spanning Tree (MCMST) model will be used; And, some points need to be observed: (a) number of clients to be served (demand nodes) of the network; (b) number of network levels; (c) quantity and location of equipment at each level (facilitators); and (d) the path between the nodes considering the geographical area.

7.3.1 Target Issue

With the growth in traffic demand driven by the universalization of services, the need for high-speed telecommunication access networks and capacity is increasingly needed, especially with regard to the implementation of smart city concepts, where sensors must be strategically distributed for the collection of information in order to supply with information the applications available to managers and citizens. In many cities, fiber optic networks already exist, but they still cannot have maximum coverage to provide the services of an intelligent city. In order to carry out the planning of communication network projects to meet the requirements of an intelligent city, the following section presents the proposed method, where the strategies that allow the elaboration of projects for access communication networks are applied through computational methods.

7.3.2 Proposed Methodology

This section has the purpose of presenting the mathematical modeling of the issue, the strategies adopted in the solution search, and the proposed architecture for the computational system.

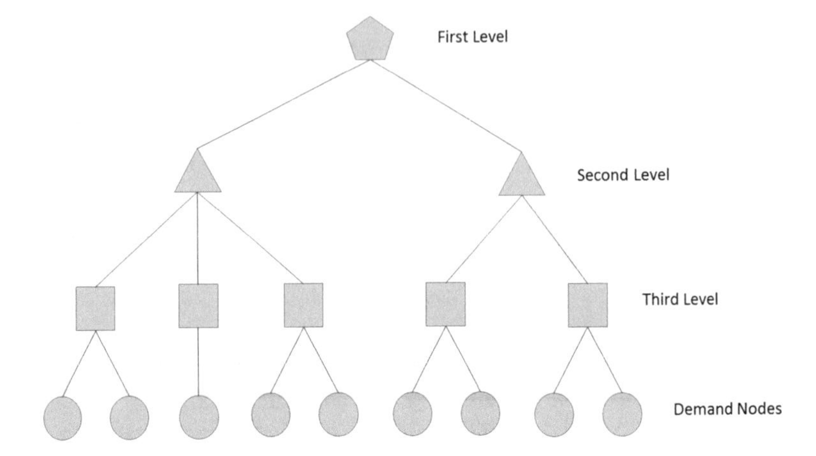

Fig. 7.6 Exemplification of the levels of a network with 3 levels

7.3.2.1 Mathematical Modeling of the MCMST Problem Applied in this Work

This section presents the mathematical formulation (part of the contribution of this paper) that represents the problem of multilevel network infrastructure. The representation makes it possible to identify the various levels of the network (MCMSTP) by maintaining a unique path between the central node (OLT—Optical Line Terminal) and the client nodes, as can be seen in Fig. 7.6. Taking into account some assumptions that impact the cost of the project.

The optimization model is given by:

Minimize:

$$\sum_{i=1}^{m}\sum_{j=1}^{n} X_{ij}\left(C_c D_{ij}\right) + \sum_{l=1}^{V-1}\sum_{i=1}^{n}\sum_{j=1}^{n} Z_{lij}\left(C_l D_{ij}\right) + \sum_{l=1}^{V}\sum_{i=1}^{n} W_{li} E_l \tag{7.6}$$

Restrictions:

$$\sum_{l=1}^{V-1}\sum_{j=1}^{n} Z_{lij} = 1, \forall i \tag{7.7}$$

$$\sum_{j=1}^{n} X_{ij} = 1, \forall i \tag{7.8}$$

$$\sum_{l=1}^{V} W_{li} \leq 1, \forall i \tag{7.9}$$

where:

- m: Number of customers;
- n: Number of nodes eligible to receive equipment;
- X_{ij}: Connection between the client i and the equipment j of level 1;
- V: Number of network levels;
- Z_{lij}: Connection between nodes i and j to level l;
- C_c: Cost of customer connection at the level 1;
- D_{ij}: Distance between nodes i and j to level l;
- C_l: Level l connection cost;
- W_{li}: Equipment installed at level l of node i;
- E_l: Cost of an equipment at the level l

The objective function represents the total cost of the network consisting of the installation cost (cables and installation services) plus the cost of the equipment used at the network nodes. The cost of fiber optic cables, other materials, and the service is obtained by a simple function using the length of the cables installed. The monetary value of the variables associated with costs is stored in auxiliary arrays and the cost of installing an equipment is considered ten times greater than the previous level (100, 1000, 10,000, ...). This relationship of values is used to maintain compatibility with the values applied in the implementation of different levels of infrastructure. The value ratio used for each level takes into account the cost differences between the equipment. Another cost considered in the function is the distance of the route between the equipment and between equipment and customers. Generally, the cost of this distance is the monetary cost of the cables used plus the installation cost. For the experiments, one (1) monetary unit is considered for the first level, two (2) monetary units for the second level, three (3) monetary units for the third level, and so on.

The constraint (7.7) defines the binding between nodes for each level, where this binding can only have a one-level destination. While the constraint (7.8) shows that each client can only be connected to a point on the first level. And, the constraint (7.9) identifies that each point can only have one type of equipment installed.

7.4 Methodology of the Experiments

The NDP solution consists of two steps, which together aim to optimize the design of the network by minimizing connection and installation costs. These steps are:

- Creation of an adjacency array from the problem database;
- Execution of the optimization routine using genetic algorithm.

The adjacency matrix contains the minimum distances between points in the network and is created using the Floyd–Warshall [3] algorithm, because it has excellent performance and has a very simple implementation. Thus, the array is created quickly. The purpose of assembling the adjacency matrix at the beginning

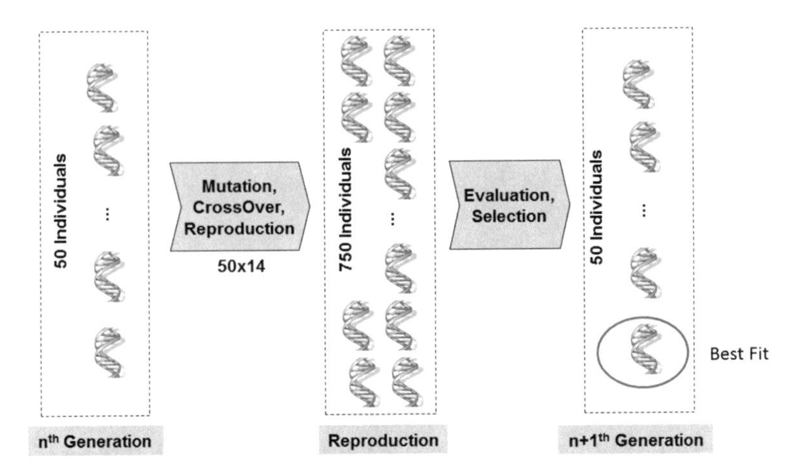

Fig. 7.7 Overview of the proposed genetic algorithm

of the solution is to avoid the need to perform minimum path routines in the genetic algorithm evaluation routine, as well as to store the minimum paths between the pairs of points.

The model used to obtain NDP solutions uses a genetic algorithm [6, 7], as it is a widely studied metaheuristic that yields good results for NP-Difficult problems such as NDP [9]. Figure 7.7 gives an overview of how the genetic algorithm works, having its main idea based on the evolution of species in nature, where the fittest individuals are more likely to survive and transmit their genetic characteristics to the new Generations.

The algorithm starts with a randomly generated population and through the process of mutation and crossing new individuals are generated. A selection process, where the fittest are more likely to survive, is responsible for reducing the population again to the beginning of a new generation and restarting the process from this new population. During the processing of the generations, individuals are evaluated and the one with better fitness is stored as the best solution.

Representation of Individuals The genetic algorithm uses a chromosome to constitute an individual, which represents a viable solution to the problem. The size of the chromosome is equal to the number of network points where it is possible to install an equipment. In this, sizes are not included in the demand points, because in our model there is no need to include them in the chromosome. In the chromosome, each part (gene) represents a point on the network where there is the possibility of installing an equipment. Each gene can assume a value of 0 at the level limit, where 0 means that the point will not be used, 1 means that it will be used by a level 1, 2 by a level 2 equipment, and so on up to the limit of working levels, which can be seen in Fig. 7.8.

Objective Function For each individual obtained, it is necessary to extract the index representing their quality or aptitude. This is done through the objective

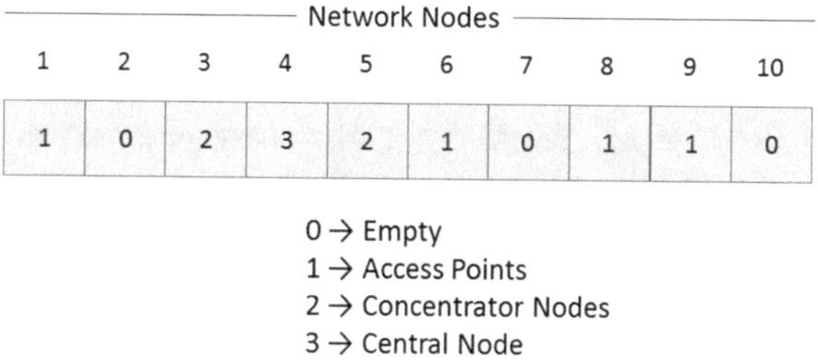

Fig. 7.8 Example of a chromosome with a level limit equal to 3

Network Nodes

1	2	3	4	5	6	7	8	9	10
1	0	2	3	2	1	0	1	1	0

0 → Empty
1 → Access Points
2 → Concentrator Nodes
3 → Central Node

function, which in this work is the sum of the costs of each connection and each installation. These values are obtained after the processing of the evaluation routine of the individuals.

Population and Initialization The model was implemented with a population of 50 individuals, which are randomly initialized. At initialization, the parameter Pv was implemented, which represents the probability that a chromosome gene will be started with zero value (empty network point). Thus, Pv = 1 indicates that the probability of having empty points is the same as the other types, Pv = 2 indicates double the probability, and so on. The motivation of using this parameter is very common in NDP problems, where an optimized network with more empty points is desired than with equipment. Thus, the genetic algorithm is already part of a better solution, besides the fact that the execution of the generations becomes faster, since the processing time of the evaluation function is proportional to the number of points with equipment. In the proposed experiments, the value used for Pv was 2.

Reproduction and Selection In each new generation, 700 new individuals are created by combining each individual with the 14 most apt of the population. Thus, out of a total of 750 individuals, only 50 are selected for the next generation. The creation of the 14 new individuals (for each one already existent) is made using the mapped partial cross-over method (PMX) and the mutation method through internal exchange in the chromosome (Swap operator). The PMX crossover method requires an adjustment where, after performing the crossover, a check is made to ensure that all levels are represented on the chromosome with at least one occurrence. The crossover percentage is 90% and the mutation rate is variable, being determined randomly between 0.1% and 10% every 10 generations. The selection of the 50 surviving individuals for the next generation is made by composing the best individual (elite selection) plus another 49 selected through the selection roulette method, where the probability of selection is given by the inverse of the index generated by the objective function.

Instances Used After the definition of the model and its coding, the program was tested with two distinct instances, one with 3 and other with 5 levels for each of the three databases in order to obtain the optimization results.

Implementation The implementation of the proposed model is basically divided into two distinct parts:

- Routine of control for evolutionary part: responsible for the processing of the generations, maintenance of the population, crosses between individuals, mutations of the chromosomes, application of selection roulette to select the new generation, and verification of the criterion of stop, which in our case will be the limit of 5000 generations.
- Routine of evaluation of the individuals: responsible for the transformation of the chromosome in map containing the connections of the demands and of the nodes of connection until the central node. This step is performed in three distinct parts:
  - Connecting the demands to the first level nodes: each demand is connected to the nearest first level node, among those available according to the indication on the chromosome. The choice of the nearest level 1 point occurs with the aid of the adjacency matrix generated at the beginning of the optimization process.
  - Connection of the intermediate nodes to the nodes of the next level: each node of the level n is connected to the node of the nearest level $n + 1$, among those available according to the indication of the chromosome. The choice of the closest level point $n + 1$ occurs with the aid of the adjacency matrix generated at the beginning of the optimization process.
  - Sum of the costs of each connection and each installation, thus generating the value of the objective function.

7.5 Tests and Results

The genetic algorithm able to work with the proposed model was coded in Pascal language and each test instance was executed 10 times. For the execution of the program, a computer with Windows 8.1 operating system, 2.4 GHz Intel Core i7-4500U processor, and 8 GB of memory was used, where only one physical processing core was used exclusively for execution (without parallel processing). Three databases were used according to the information contained in Table 7.1, whose results are compared with the work presented by Silva et al. [19] and shown below.

Table 7.1 Georeferenced databases

Data base	Demand nodes	Coordinates (local candidates for installation of facilities)	Area (km^2)
Base 1	50	390	2.1
Base 2	105	583	2.5
Base 3	405	1624	6.1

Instance 1 Results Table 7.2 presents the results obtained by the proposed genetic algorithm. In the table, the results are separated into rows for each level. Each column represents a database that accounts for: (a) the number of demands to be met; (b) the number of facilities found to meet the demands at each level; and (c) the cost of the solution in monetary units. In Fig. 7.9, the map with the result of base 1 with 3 levels is presented. The last line of the table brings the percentage improvement result in relation to the other algorithms.

Table 7.2 Instance 1 results

Algorithm	Data base	Base 1	Base 2	Base 3
	Demands	50	105	405
GA [19]	Level 1	28	46	132
	Level 2	6	10	81
	Level 3	1	1	75
	Execution time (s)	5589	17,385	213,287
	Total cost	52,250	73,526	963,382
GA [19]	Level 1	14	23	67
	Level 2	4	4	15
	Level 3	1	1	3
	Execution time (s)	5326	13,922	136,583
	Total cost	46,476	56,666	184,987
GA (proposed)	Level 1	7	12	87
	Level 2	2	2	9
	Level 3	1	1	1
	Execution time (s)	120	180	3600
	Total cost	35,615	47,608	149,274
	% improvement	**23.4%**	**16.0%**	**19.3%**

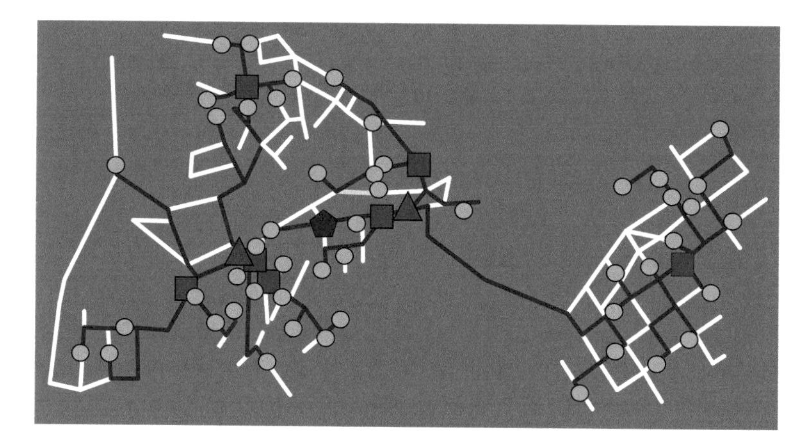

Fig. 7.9 Map with the result for base 1 of 3 levels

Table 7.3 Instance 2 results

Algorithm	Data base	Base 1	Base 2	Base 3
	Demands	50	105	405
GA [19]	Level 1	24	46	126
	Level 2	6	11	87
	Level 3	1	1	75
	Level 4	1	1	68
	Level 5	1	1	68
	Execution time (s)	7339	26,328	390,967
	Total cost	1,152,755	1,172,926	75,688,821
GA [19]	Level 1	20	23	65
	Level 2	6	6	16
	Level 3	1	1	3
	Level 4	1	1	2
	Level 3	1	1	2
	Execution time (s)	7339	19,232	220,884
	Total cost	1,150,260	1,158,810	2,394,322
GA (proposed)	Level 1	16	19	398
	Level 2	3	3	222
	Level 3	1	1	1
	Level 4	1	1	1
	Level 5	1	1	1
	Execution time (s)	442	1200	8000
	Total cost	1,138,735	1,150,620	1,915,742
	% improvement	**1.0%**	**0.7%**	**20.0%**

Instance 2 Results Although it is common for telecommunications networks to have three levels of equipment, in some situations, such as PON, it may be necessary to plan networks with more levels. Table 7.3 presents the results with 5 levels. In Fig. 7.10, the map with the result of base 1 with 5 levels is presented. The last line of the table brings the percentage improvement result in relation to the other algorithms.

7.6 Conclusion and Future Work

Telecommunications networks are becoming very important for society as a whole and especially for smart cities. The availability of means of communication brings with it the need for better network planning. In this paper, the NDP was modeled as an optimization problem. The expected result of this work was reached with the best network planning at a more affordable cost, taking into consideration the comparison with the work of [19] presenting better results due to the decision to elaborate the adjacency matrix containing the minimum distances between network

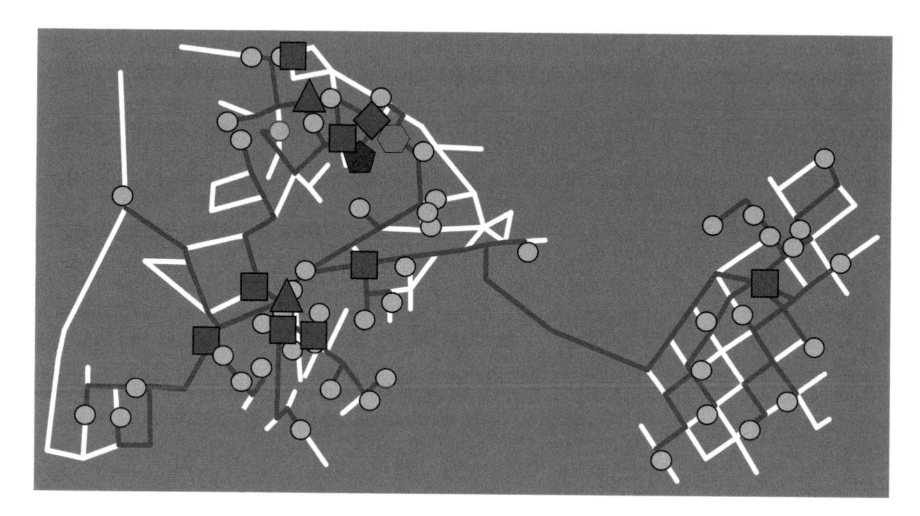

Fig. 7.10 Map with the result for base 1 of 5 levels

points before starting the optimization with the genetic algorithm, since the results obtained showed an improvement of approximately 20% for the 3-level and 20% instance for base 3 in the 5-level instance.

As future work, more restrictions will be added and analyzed beyond those presented in this paper, such as: (1) the maximum length of a cable is 2000 m; (2) the normal attenuation of an optical fiber is 0.25 dB/km, the attenuation for each fusion welding is 0.05 dB/splicing, and the safety attenuation is 5.2 dB; (3) the maximum tolerated attenuation value is 28 dB; and (4) the maximum value of splits is 32. These technical restrictions are of vital importance because if the level of attenuation and the maximum distance between the client nodes and the central point of the network is not respected it will not work in perfect conditions. And, also the development to effectively find the best configuration of the hybrid network (optical—wireless) of a smart city in a simplified way, in order to provide: (a) to meet all the data communication demands necessary to operationalize a Smart city; (b) the refinement and expansion of the hybrid network, regarding the issue of redundancy; and (c) the comparative analysis between CAPEX $\times$ OPEX aiming at the sustainability of the network.

References

1. Agrawal, G.P.: Sistemas de Comunicação Por Fibra Óptica. Elsevier, São Paulo (2014)
2. Chiu, P.L., Lin, F.Y.S.: A simulated anneling algorithm to support the sensor placement for target location. In: Canadian Conference on Electrical and Computer Engineering 2004. IEEE, Piscataway (2004)
3. Floyd, R.W.: Algorithm 97: shortest path. Commun. ACM **5**(6), 345 (1962)

4. Gamvros, I., Raghavan, S., Golden, B.: An evolutionary approach to the multi-level capacitated minimum spanning tree problem. In: Telecommunications Network Design and Management, pp. 99–124. Springer, Berlin (2002)
5. Goldbarg, M.C., Luna, H.P.L.: Otimização Combinatória e Programação Linear: Modelos e Algoritmos. Campus, Rio de Janeiro (2000)
6. Goldberg, D.: Genetic Algorithms in Search, Optimization and Machine Learning. Adisson-Wesley, Reading (1989)
7. Grennsmith, J., Withbrook, A., Aickelin, U.: Handbook of Metaheuristics. Springer, Berlin (2010)
8. Gross, J.L., Yellen, J.: Graph Theory and Its Applications. Chapman & Hall, Boca Raton (2006)
9. Kampstra, P., van der Mei, R.D., Eiben, A.E.: Evolutionary computing in telecommunication network design: a survey. In: Vrije Universiteit, Faculty of Exact Sciences and CWI, Advanced Communication Networks. Amsterdam (2006)
10. Lee, J.H., Hancock, M.G., Hu, M.C.: Towards an effective framework for building smart cities: lessons from Seoul and San Francisco. Technol. Forecast. Soc. Change **89**, 80–99 (2013)
11. Maia, L.P.: Arquitetura de Redes de Computadores. LTC, Rio de Janeiro (2013)
12. Martins, E.A.: Um Sistema Computacional Para Apoio a Projetos de Redes de Comunicação em Sistemas Centralizados de Medição de Consumo e Tarifação de Energia elétrica: Desenvolvimento e Implementação Através de uma Abordagem Metaheurística. UNISINOS, São Leopoldo (2013) (in Portuguese)
13. Monteiro, M.S.R., Fontes, D.B.M.M., Fontes, F.A.C.C.: The hop-constrained minimum cost flow spanning tree problem with nonlinear costs: an ant colony optimization approach. Optim. Lett. **9**(3), 451–464 (2015)
14. Netto, P.O.B.: Grafos: Teoria, Modelos, Algoritmos. Blucher, São Paulo (2011)
15. Papadimitriou, C.H., Lewis, H.R.: Elementos de Teoria da Computação. Prentice Hall, São Paulo (1998)
16. Petrolo, R., Loscri, V., Mitton, N.: Towards a smart city based on cloud of things. In: Proceedings of the 2014 ACM Internacional Workshop on Wireless and Mobile Technologies for Smart Cities – WiMobCity 2014, pp. 61–66. ACM, New York (2014)
17. Piro, G., Cianci, I., Grieco, L.A., Boggia, G., Camarda, P.: Information centric services in Smart Cities. J. Syst. Softw. **88**(1), 169–188 (2014)
18. Segarra, J., Sales, V., Prat, J.: Planning and designing FTTH networks: elements, tools and practical issues. In: International Conference on Transparent Optical Networks, pp. 1–6. IEEE, Piscataway (2012)
19. Silva, H.A.D., De Souza Britto, A., Oliveira, L.E.S.D., Koerich, A.L.: Network infrastructure design with a multilevel algorithm. Expert Syst. Appl. **40**(9), 3471–3480 (2013)
20. Sukode, S., Gite, P.S., Agrawal, H.: Context aware framework in IoT: a survey. Int. J. **4**(1), 1–9 (2015)
21. Varvarigos, E.A., Christodoulopoulos, K.: Algorithmic aspects in planning fixed and flexible optical networks with emphasis on linear optimization and heuristic tecniques. J. Lightwave Technol. **32**(4), 681–693 (2014)
22. Watcharasitthiwat, K., Wardkein, P.: Reability optimization of topology communication network design using an improved ant colony optimization. Comput. Electr. Eng. **35**(5), 730–747 (2009)

Chapter 8
An Experimental Study of Regression Techniques for the Residential Energy Consumption Forecast in the Brazilian Scenario

Leonardo Vasconcelos, José Viterbo, Igor Machado Coelho, and João Marcos Meirelles da Silva

Abstract The world population has been rising on a large scale, and this directly reflects on the electricity consumption. In this scenario, techniques for accurately forecasting energy consumption are particularly important, since these data can be applied in decision-making and good planning aimed at providing constant and reliable energy. Forecasting energy consumption with the most accurate value possible is not a trivial task and depends on some factors. One of the most recent works dealing with the subject at the Brazil presented a fuzzy logic-based prediction model using consumption, Gross Domestic Product index (GDP), and population and obtained good results. This work aims to evaluate, through an experimental study, the performance of classical regression techniques—linear regression, multi-layer perceptron, and support vector regression—in energy consumption forecast in the Brazilian scenario. Also, we verified whether the inclusion of additional socioeconomic data could contribute to obtaining a more efficient model. When compared to the results available in the literature, our approach demonstrated superior performance in some situations.

L. Vasconcelos (✉) · J. Viterbo
Instituto de Computação, Universidade Federal Fluminense, Niterói, RJ, Brazil
e-mail: lvasconcelos@ic.uff.br; viterbo@ic.uff.br

I. M. Coelho
Instituto de Matemática e Estatística, Universidade do Estado do Rio de Janeiro, Rio de Janeiro, RJ, Brazil
e-mail: igor.machado@ime.uerj.br

J. M. M. da Silva
Escola de Engenharia, Universidade Federal Fluminense, Niterói, RJ, Brazil
e-mail: jmarcos@vm.uff.br

V. N. Coelho et al. (eds.), *Smart and Digital Cities*, Urban Computing,
https://doi.org/10.1007/978-3-030-12255-3_8

8.1 Introduction

The world population has been rising on a large scale. There was a jump from 3 billion people in 1960 to 7 billion in 2015 [1]. This result directly reflects the consumption of electricity. In the public, residential and agricultural sectors from 1971 to 2012, consumption rose from 4.65 to 18.6 billion GWh [2], which shows that world energy consumption has increased much more than population [3].

In Brazil, this growth followed the same proportion. According to the IBGE, the population rose from approximately 157 to 206 million in the period from 1996 to 2016 [4]. And, the consumption of electricity, according to data from Eletrobras—Centrais Elétricas Brasileiras SA—increased from 20 thousand GWh to 38 thousand GWh in the period from 1996 to 2016 [5]. As in the whole world, both in the economic space and in the political field, the electric sector is changing. This factor is due to the deregulation of the electricity sector and privatization. In Brazil, for example, this reform began in 1995 with the publication of the Concession Law in 1995 [6].

The hydroelectric plants are the ones that produce the most energy in Brazil, yet the country has faced significant problems related to energy. A large socioenvironmental is generated to build a hydroelectric plant, such as destruction of fauna and flora by creating dams with large areas of flooding and expropriation of rural or urban areas [7].

Residential energy consumption represents 30% of the total consumption of the entire planet, and in each country, the residential energy consumption varies from 16% to 50%. Therefore, it is interesting to understand the profile of this class of consumption, aiming at the implementation of more efficient technologies, population awareness, energy conservation, constant supply, and even substitution (if possible) for renewable energy sources [8].

Given this scenario, techniques for accurately predicting energy consumption are advantageous, since these data can be applied for decision-making and proper planning aimed at the uninterrupted and reliable supply of energy [9]. In developed and developing countries, the forecast of energy consumption is of high relevance to their governments, since a forecast below actual consumption would lead to possible interruptions of energy and increase the operating costs of energy suppliers, and these costs would be passed on to the consumer. On the other hand, a forecast above actual consumption could result in idle capacity making it unnecessarily wasteful of financial resources [10].

Predictions of short-, medium-, and long-term electricity consumption play an essential role for governments in developing countries as they are the basis for planning investment in the energy sector [10]. In addition, predictions are very important to sustain industrialization, growing demand for electricity, and stable long-term energy policies. A quality forecast is essential in designing tools that will fuel consumers when needed since it is impossible to store the alternating current [11].

According to Todesco [12], one of the primary goals of an energy concessionaire is to provide quality service at a fair price through efficient energy distribution management. In Turkey, for example, official consumption estimates are higher than consumption values, and because of this, the results are reviewed every 6 months by official government bodies [10]. To improve such cases, the distribution networks must do their utmost to allocate their investments and operational resources optimally. Therefore, the concessionaires continuously seek the improvement and optimization of energy distribution.

It is also worth noting that in some countries like Brazil, for example, some consumers do not have a conscious energy consumption, wasting energy with the use of electrical appliances without worrying about the consequences. Therefore, it is interesting to guide the population with the objective of increasing the conservation of electric energy [7].

Smart meters solutions [13–15] can help users to reduce energy consumption. Each consumer or supplier can consult information about electricity consumption. In addition to being accurate, this information is presented in detail, making it easier to define a consumption profile. This profile can be used to find ways to save energy and reduce costs. In addition, smart meters can help consumers compare the prices of competing suppliers [16], and based on forecast models anticipate their costs with energy bills [17].

Several techniques are usually used to predict energy consumption, among which regression techniques are being utilized as one of the most applied in the forecast of electricity consumption [10]. Regression is an approach that makes it possible to obtain a functional relationship between dependent variables (y) and explanatory variables (x) [18]. In other words, regression in a general way is to calculate values using a function obtaining a response (in our case a prediction) through explanatory variables. The best-known regression techniques are: linear regression [19], neural networks [20], and Support Vector Regression—SVR [10].

In the papers presented in the literature, several methodologies are found, with the objective of solving problems specific to particular countries or regions [10, 11, 21]. Classical techniques such as regression support, neural networks, and linear regression techniques, but also emerging techniques of optimization and nebulous logic, such as particle swarm optimization, are usually employed [22], gray model [11], Fuzzy Logic [21], and evolutionary strategies with fuzzy model [23, 24], among others. However, to obtain a model with good accuracy, some points must be analyzed [10]. Firstly, the parameters that affect the consumption of electricity must be well defined. Second, an appropriate methodology must be chosen to generate the model. And, thirdly, the model created should provide extensions for future improvements in performance.

This article aims to evaluate energy consumption forecasting techniques for the Brazilian scenario. Our objective was to answer two questions: (1) "Using the data of consumption, GDP, and population, as in [25], can one obtain better results by employing classical regression techniques?"; and (2) "Given additional socioeconomic data, can a better quality be obtained?". We considered linear regression, multilayer perceptron, and SVR techniques, which are the widely applied algorithms in prediction and regression problems [10, 26]. We compare the results of the forecast with the data presented in [25] which is one of the most recent work in the literature to perform residential energy consumption forecasting in Brazil.

This work is organized into seven sections, including this introduction. In the next section, we discuss the regression methods and evaluation parameters using in this work. In Sect. 8.3, we describe the prediction of energy consumption in the world and then present what has been developed in Brazil. In Sect. 8.4, we discuss the materials and methods involved in the prediction of regression models. In Sect. 8.5, we present our experimental analysis and compare it with the present results in the literature. In Sect. 8.6, we insert new explanatory variables and see if they can improve the quality of the forecast. Finally, in Sect. 8.7 we present our conclusions.

8.2 Regression Methods

Regression is an approach that allows obtaining a functional relationship between dependent variables and independent variables. Below, we will present a more detailed analysis of linear regression, multilayer perceptron, and support vector regression which are the methods used in our experiments.

Regression in a general way consists of calculating values through a function receiving a response (in our case a prediction) through explanatory variables. Regression, in complex systems such as energy consumption, can be observed as an iterative process, that is, where its output is used to identify, validate, criticize, and supposedly modify its entries [18]. A linear regression is called simple when it has only one dependent variable and multiple when it has more than one dependent variable. For example, when generating a prediction with linear regression using just a database of energy consumption, we are performing a simple linear regression. Similarly, by generating a linear regression using a basis containing energy consumption together with GDP, we are generating a multiple linear regression.

Multilayer Perceptron (MLP) neural networks are composed of neurons that interact with each other using weighted connections [27]. They have obtained good results in several problems among these; some considered difficult [28]. Mainly, an MLP consists of an input layer, a set of sensory units (source nodes), one or more hidden layer(s), and an output layer. The latter two layers have a computational capacity. The input signal propagates from layer to layer always forward. The input layer is where the input data resides and is composed of sensory units. Its function is to forward the data to the first hidden layer. The hidden layer(s) has the function of receiving input layer data, processing it, and routing it to the output layer. The output layer has the function of responding propagated by the hidden layer.

The support vector regression (SVR) is based on the inductive principle of structural risk minimization, which seeks to minimize an upper limit of the generalization of error consisting of the sum of the training error and a level of confidence. This principle makes many problems in machine learning have better generalization performances with SVR than with other techniques. The SVR training is equivalent to solving a restricted problem of quadratic programming linearly so that the solution of the algorithm is always globally optimal and unique, which avoids the risk of local minimums. The solution to the problem in the SVR is dependent only on a subset of training data points, called support vectors. According to the literature, SVR has been shown to be a very efficient algorithm for solving general regression problems [26, 29, 30].

The metrics used in our experimental analysis to compare the performance of the models are Mean Absolute Percentage Error (MAPE), Mean Absolute Deviation (MAD), and the Root-mean-square error (RMSE). They are calculated as follows:

$$\text{MAPE}(\%) = 100\left(\frac{1}{n}\sum_{t=1}^{n}\left|\frac{y_r - y_p}{y_r}\right|\right) \tag{8.1}$$

$$\text{MAD} = \frac{1}{n}\sum_{t=1}^{n}\left|y_r - y_p\right| \tag{8.2}$$

$$\text{RMSE} = \sqrt{\frac{\Sigma_{t=1}^{n}(y_r - y_p)^2}{n}} \tag{8.3}$$

where y_r = Real value, y_p = Estimated value, and n = number of points analyzed (in our case, the total of the analyzed years).

8.3 Energy Consumption Forecast

Growing energy consumption has raised concerns worldwide about the depletion of energy resources, the massive environmental impacts on energy production, and supply difficulties [3]. Finding the best way to provide energy without affecting the global economy and polluting as little as possible is an urgent problem for all countries [31].

The forecast of energy consumption is also very useful for planners and energy managers [9]. Through knowledge of price and income elasticities, utilities can create convenient demand policies and thus help policymakers develop policies to reform the electricity sector. Given that forecasting demand is important to potential investors in the electricity sector and to companies involved in the electricity business they need to plan their future strategies. In the next subsections, we will address the work that uses prediction of energy consumption at the national level and smart homes.

8.3.1 Forecast of Aggregate Energy Consumption at the National Level

Hamzacebi and Es [11] used an algorithm that performs predictions with incomplete data and little information called Optimized Gray model (OGM). It was applied to forecast the energy consumption in Turkey in the years 2011–2025. Two approaches were used for direct OGM and iterative OMG. The direct OMG only uses the base data to carry out the forecast. The iterative approach uses the baseline data but its own predictions to get a result. The database contains energy consumption for the period from 1945 to 2010. The years 1945–2005 were used for training and the years 2006–2010 for testing. The MAPE results of the direct OGM (3.28%) were higher than those of the iterative OGM (5.36%) and also those presented in the literature (3.43% and 4.6%).

Kaytez et al. [10] observed that technique least squares support vector machines (LS-SVM) had not been applied in the prediction of energy consumption. Based on this, he presented the LS-SVM for the consumption forecast and compared the data with the algorithms of multiple linear regression analysis and neural networks. The algorithms had as input a database of Turkey with the attributes of gross electricity generation (TWh), population (millions), installed capacity (GW), and total subscribers (millions) covering the period from 1970 to 2009. The LS-SVM obtained the best result as the other algorithms obtaining 0.876% training error and 1.004% error in the test using the MAPE metric. Its performance outperformed the neural networks having 4.86% training error and 1, 19% error in the test, and multiple linear regression with 4.01% training error and 3.34% error in the test.

Torrini et al. [21, 25] proposed a fuzzy logic methodology to predict annual energy demand in Brazil. According to them, the recent models presented in the literature on fuzzy logic only involve GDP or univariate variables. So, they offered a model containing the total population and GDP. The database covers the period 1980–2013 including the attributes: population, GDP, and energy consumption of the residential, industrial, commercial, and other sectors. The first 25 years were used for training and the last 10 years for testing. A forecast was made from 2014 to 2030. In the first step, the data of the total consumption of Brazil were used, the results compared with the Holt 2-parameter model using the MAPE metric. The fuzzy model obtained better results in the prediction of Brazil containing 1.46 of MAPE than the model Holt 2-parameter that obtained 5.81 of MAPE. In the second stage, consumption data were used in classes, divided into residential, commercial, industrial, and other sectors. The results showed that the incorporation of the population growth rate together with the GDP had a significant contribution to the quality of the forecast of residential consumption models and the total consumption of Brazil.

8.3.2 Use of Consumption Forecast in Smart Homes

In [32], Arora and Taylor studied the forecast energy consumption of smart meters using the conditional kernel density estimation. The data density present in the smart meters was modeled using distinct implementations of kernel density and conditional kernel density estimators. Then, density predictions were derived to predict the cost of energy consumption for six types of tariffs. They considered two points: the first was to compare the different costs that would have the potential to occur in the future. The second was, for each available rate, to choose which rate would generate a lower cost. Three different criteria were used for the choice of tariffs. The results showed that the exchange between tariffs generated a lower cost compared to cases where consumers had only one tariff over the whole period of time.

In [33], Laurinec and Lucká used clustering and the Seasonal naive, Multiple Linear Regression, Random Forests, and Triple Exponential Smoothing algorithms to predict the disaggregated electricity load of a final consumer in a smart grid. Two smart meter databases were used of the countries: Irish and Slovak. The Irish database is comprised of residential, small- and medium-sized, and other customers. Most of the data is residential consumption. The database, after removing the missing data, was left with 3639 instances. Each measurement was performed within 30 min, that is, 48 measurements per day. The Slovak database, after removing the missing data, was 3607 instances. Each measurement was performed within 15 min, that is, 96 measurements per day. The data were evaluated using the evaluation measures: Mean Absolute Error (MAE) and Root Mean Square Error (RMSE). The results showed that clustering reduced the prediction error in the MAE metric; however, they did not decrease the median error because of the type of the smart meter data. The results also showed that the proposed approach needs to train fewer models, which reduces the processing load of the training stage.

In [34], Ponoćko and Milanović used neural networks to predict the flexibility of the aggregate residential power load from a day-ahead. Data was collected from smart meters. Two methodologies were proposed. In the first one, the load was broken down considering the individual consumption of each item in the residence. In the second methodology, the artificial neural networks were used to disaggregate the total load of energy consumption. The proposed methodology was tested in statistics in the United Kingdom. Shortly there after, it was validated with data from a real dataset. The results showed that the coverage of the smart meters interfered more in the accuracy than the missing data. It was also identified that no meteorological data or a high number of historical data was required to train the model. Although the methodology has been applied in the residential sector, it can be used in other sectors such as commercial, industrial, and both together. The authors also argued that the accuracy of the model was greater when the data used for training were composed by users' neighbors.

8.4 Materials and Methods

For the computational experiments carried out in our case study, we used eight public databases. The instances and the explanatory variables were extracted in the respective periods of each case. We chose to use the data of residential energy consumption, Gross Domestic Product (GDP), and population because it is the same bases used by used by Torrini in [25], to compare our results.

In addition to the residential energy consumption, GDP, and population bases presented in [25], we believe that the minimum wage could interfere with consumption, since raising the minimum wage may increase the purchasing power of some citizens and consequently these citizens could invest in new electrical appliances contributing to the increase in energy consumption.

The decrease in the exchange rate for the purchase of the dollar could lead to a fall in the price of imported products and indirectly the purchase of electrical and electronic technological products, also contributing to the increase in consumption. The average tariff could contribute to an increase in consumption if its value is reduced or also to decrease consumption if its value is increased since the citizen would strive to save energy in order to reduce the value of his bill. The price of a barrel of oil can indirectly interfere with the economy of users, and a citizen could save his residential energy to travel by car, for example.

For the realization of our experiments, we used public databases available in open data portals in Brazil and worldwide. The Brazilian bases are of residential energy consumption [5], Gross Domestic Product (GDP) [35], population [36], minimum wage [5], commercial exchange rate to purchase US$ (US$) dollars in real (R$) [5], and average consumption tariff [5]. The foreign bases are the price per barrel of Brent crude oil [37], and the price per barrel of WTI crude oil [37]. Below is a description of each base used in our experiments.

In Table 8.1, we present a list of the open databases selected, with the sample periodicity and coverage period of each time series. The base energy consumption 1 only had data until the year 2012. To complete the remaining years, we selected

Table 8.1 List of the open databases selected, with the sample periodicity and coverage period

Base	Series	Period
Energy consumption 1	Annual	1963–2012
Energy consumption 2	Monthly	1976–2017
GDP	Annual	1962–2016
Population	Annual	1980–2050
Minimum wage	Monthly	1940–2017
Exchange rate	Annual	1942–2015
Average consumption tariff residential electricity	Annual	1942–2015
Price per barrel of Brent crude oil	Monthly	1987–2017
Price per barrel of WTI crude oil	Monthly	1986–2017

Table 8.2 Correlation matrix of energy consumption with other explanatory variables

Explanatory variable	Correlation
Energy consumption (consolidated)	99.45%
GDP	93.04%
Population	97.45%
Minimum wage	92.86%
Exchange rate	76.99%
Average consumption tariff	92.09%
Price per barrel of Brent crude oil	85.93%
Price per barrel of WTI crude oil	84.35%

the monthly base energy consumption 2. We added the monthly consumption of the years 2013, 2014, 2015, and 2016 to generate the annual series. Therefore, in all this work, when we refer to energy consumption, it will refer to the annual energy consumption (consolidated) base formed by the combination of the two series of residential energy consumption.

The monthly bases: minimum wage, barrel price of Brent crude oil, and barrel price of crude oil WTI were transformed into annual bases, for this was calculated the arithmetic average of the months of January to December of the current year. The bases: price per barrel of Brent oil and price per barrel of WTI oil contained missing data from 1980 to 1987 and 1980 to 1985, respectively. In place of these missing data was inserted the symbol '?' to be treated by the algorithms according to the standard form of each implementation used in WEKA. In SMOreg and linear regression, the missing data is filled by the mean. In multilayer perceptron neural networks, the missing data is ignored by setting the input value to 0. Table 8.2 shows a correlation matrix of the explanatory variable energy consumption with the others variables. We chose to use the data from 1980 because it is the same initial period used by Torrini in [25], to compare our results.

The computational experiments were performed using the WEKA tool in version 3.8.1 [38, 39], along with the Time Series Analysis and Forecasting plugin 1.1.25 [40]. The following Weka algorithms were employed: Linear Regression that implements simple and multiple linear regression [38]; Multilayer Perceptron that implements the multi-layer perceptron neural network [38]; and the *Sequential Minimal Optimization* (SMOreg) [41] which implements the regression support vector machine. The SMOreg algorithm was proposed by Shevade et al. in [41] in the year 2000. It offers an optimization of the SVR algorithm proposed by Smola and Scholkopf in [42].

Training of all models was performed using lags. In an experimental analysis, we achieved the best results when we used five lags. The first containing 1 year of lag, the second 2, and so on until the fifth that contained 5 years of lags. The linear regression model has no parameters for configuration. However, the multilayer perceptron neural network models and the support vector regression require their parameters to be calibrated. However, no technique in the literature determines the choice of parameters, and they are chosen according to the user experience.

Table 8.3 Parameters used in SVR calibration

Kernel	Valor (kernel)	C	ε
Polynomial	Exponent (1; 0.5; 0.25)	1; 0.1; 10	0.0001; 0.001; 0.01
Normalized polynomial	Exponent (2, 0.5, 0.25)	1; 0.1; 10	0.0001; 0.001; 0.01
RBF	Gamma (1; 0.5; 0.25)	1; 0.1; 10	0.0001; 0.001; 0.01
PUK	Omega (1; 0.5; 0.25) Sigma (1; 0.5; 0.25)	1; 0.1; 10	0.0001; 0.001; 0.01
Kernel matrix	–	–	–
String kernel	–	–	–

Thus, to select the set of parameters that would be tested, we did a preliminary empirical study, performing some experiments with the algorithms neural network multilayer perceptron, and the support vector regression. With the results, we verified the best calibration of the algorithm for our case study. The multilayer perceptron (MLP) were tested in the following configurations: learning rate (0.3 and 0.5), momentum: (0.2 and 0.4), training time (500, 250), and hidden nodes (1, 2, 7). In the execution of the experiments, we verified that the variation of the seed and of validation threshold did not present significant values, so they used the default setting value. Other tests were performed with different configurations; however, because they did not present significant results, they were removed. The configurations that obtained the best results were: learning rate (0.3), momentum (0.2), training time (500, 250), hidden nodes (2, 7).

The support vector regression (SVR) machine was tested in the configurations presented in Table 8.3. After running the experiments, we verified that the kernel PreComputedKernelMatrix and StringKernel were not applied to the type of data that we used in our experimental analysis. The polynomial kernel obtained the best results with the exponent (1), and the following combinations: the complexity parameter—C (1) and ε (0.01, 0.001, 0.0001), C (0.01) and ε (0.01, 0.001, 0.0001), and C(10) and ε (0.01).

8.5 Comparative Analysis of Regression Techniques

In this section, we generate the computational models for the Brazilian scenario and compare it with the model created by Torrini in [25], one of the most recent in the Brazilian scenario literature. Our approach was to generate models that would allow us to predict a year of the data period used to create the model. For example, considering a series from the period of 1980 to 2014, we will have the forecast for the year 2015. The tests presented here were carried out in two steps. The first step was to generate forecasts for the years 2004–2013. The second step consisted of generating forecasts from 2014 to 2016. These steps were defined to allow a comparison with the results described in [25], which used the fuzzy model to generate its predictions. This work was selected as a reference because it is one of the most recent in the Brazilian literature. The first step corresponds to the validation period in [25]. In the second step, we have the values of the projections of [25] and so we can calculate its result and compare with our forecast. The models were generated using the following configurations:

- **Configuration 1**: Energy consumption.
- **Configuration 2**: Energy consumption and GDP.
- **Configuration 3**: Energy consumption, GDP, and population.

8.5.1 Linear Regression (LR)

In this subsection, we present the results of our computational experiments generated by linear regression (LR) models. The experiments were performed in two steps as discussed above.

In the first step, forecasts were made for the years 2004–2013. Table 8.4 presents the MAPE (%) and MAD values of each approach. As can be observed, the approach using energy consumption (configuration 1) obtained the best approximations of MAPE (%) and MAD, followed by the approach using energy consumption together with GDP (configuration 2). The approach using energy consumption together with GDP and population (configuration 3) achieved the worst results. When the consumption was combined with GDP, the forecast worsened 2.14% in the MAPE metric and 1990.36 in the MAD metric. When consumption was combined with GDP and population, the forecast worsened by 2.60% in the MAPE metric and 2161.44 in the MAD metric.

In the second step, forecasts were made from 2014 to 2016, following the same techniques of the first step. In Table 8.4, the MAPE (%) and MAD values of each approach are presented. As can be seen, unlike the first step, the approach using energy consumption together with GDP and population (configuration 3) obtained the best approximations, followed by the approach using consumption together with GDP. The approach using only the energy consumption obtained the worst approximations of MAPE (%) and MAD. When the consumption was combined with GDP and population, the forecast improved by 3.15% in the MAPE metric

Table 8.4 Value of MAPE (%) and MAD for all combinations of Linear Regression

Linear Regression				
	First step (2004–2013)		Second step (2014–2016)	
Approach	MAPE (%)	MAD	MAPE (%)	MAD
Configuration 1	**17.2835**	**19,017.7661**	5.0848	6710.3617
Configuration 2	19.4284	21,008.1287	3.8055	5031.1198
Configuration 3	19.8908	21,179.2057	**1.9313**	**2549.3500**

The lowest error values are in bold

and 4161.01 in the MAD metric. When consumption was combined with GDP, the forecast improved by 1.28% in the MAPE metric and 1679.24 in the MAD metric.

Based on this information, we can conclude that neither GDP nor GDP together with population improves the accuracy of the linear regression model in previous years (2004–2013). However, for the most recent years (2014–2016), GDP and GDP together with the population contribute to the improvement of the approximation of the linear regression model presented in this section.

8.5.2 Neural Networks Multilayer Perceptron (MLP)

In this subsection, we present the results of our computational experiments generated by multilayer perceptron neural networks (MLP) models. The experiments were performed in two steps as discussed above. The first step was to generate forecasts for the years 2004–2013. For all the first stage approaches, the MLP obtained the best results in the configurations: learning rate (0.3), momentum (0.2), number of times (500), and hidden layer (s) (1). In Table 8.5, the MAPE (%) and MAD values of each approach are presented. As can be observed, the approach using energy consumption (configuration 1) obtained the best values of MAPE (%) and MAD, followed by the approach using energy consumption together with GDP (configuration 2). The approach using consumption together with GDP and population (configuration 3) obtained the worst values of MAPE (%) and MAD.

When the consumption was combined with the GDP, the forecast obtained a worsening of 0.69% of MAPE and 734.53 of MAD than the forecast using only the energy consumption. When consumption was combined with GDP and population, the forecast worsened by 1.04% in the MAPE metric and 1,214.10 in the MAD metric compared to the approach that used only the energy consumption.

In the second step, forecasts were made from 2014 to 2016, following the same techniques of the first stage. For all the approaches of the second step, MLPS obtained the best results in the configurations: learning rate (0.3), momentum (0.2), number of times (500), and hidden layer (s) (7). Table 8.5 displays the MAPE (%) and MAD values of each approach. As in the first step, the approach using only the energy consumption (configuration 1) obtained the best approximations, followed

Table 8.5 Value of MAPE (%) and MAD for all combinations of Multilayer Perceptron

Multilayer Perceptron				
	First step (2004–2013)		Second step (2014–2016)	
Approach	MAPE (%)	MAD	MAPE (%)	MAD
Configuration 1	**13.2716**	**14,602.4138**	**1.7556**	**2321.5181**
Configuration 2	13.9575	15,336.9395	2.9692	3910.6734
Configuration 3	14.3144	15,816.5176	3.0418	4016.7485

The lowest error values are in bold

by the approach using the energy consumption together with the GDP. The approach using consumption together with GDP and population obtained the worst values of MAPE (%) and MAD.

In the same way as in the first step, when consumption was combined with GDP, the forecast obtained a worsening of 1.21% of MAPE and 1589.16 of MAD compared to the forecast using only the energy consumption. When the forecast was made using consumption, GDP, and population, the result worsened, increasing by 1.29% the MAP rate and MAD 1695.23 compared to energy consumption.

Based on this information, we can conclude that neither the GDP nor the GDP together with the population improves the precision of the neural network multilayer perceptron model presented in this section.

8.5.3 Support Vector Regression (SVR)

In this subsection, we present the results of our computational experiments generated by the support vector regression (SVR) models. The experiments were performed in two steps as discussed above. The first step was to generate forecasts for the years 2004–2013. For all the approaches of the first step, the SVR obtained the best results in the configurations: exponent (1), C (0,1), ε (0,01). In Table 8.6, the values of MAPE (%) and MAD of each approach are presented. As can be seen, the approach using energy consumption together with GDP obtained the best approximations, followed by the approach using energy consumption together with GDP and population. The approach using only the energy consumption obtained the worst values of MAPE (%) and MAD.

When consumption was combined with GDP, the forecast improved by 0.96% in the MAPE metric and 828.40 in the MAD metric than the forecast using only the energy consumption. When consumption was combined with GDP and population, the forecast improved by 0.57% in the MAPE metric and 426.27 in the MAD metric compared to the approach that used only the energy consumption.

In the second stage, forecasts were made from 2014 to 2016, following the same techniques of the first stage. For the approach using only the explanatory variable energy consumption and the approach using the explanatory variables: energy consumption and GDP, the SVR obtained the best results in the configurations:

Table 8.6 Value of MAPE (%) and MAD for all combinations of Support Vector Regression

Support Vector Regression				
	First step (2004–2013)		Second step (2014–2016)	
Approach	MAPE (%)	MAD	MAPE (%)	MAD
Configuration 1	2.6886	2591.6944	3.0894	4086.4618
Configuration 2	**1.7298**	**1763.2968**	3.2863	4347.1674
Configuration 3	2.1177	2165.4251	**2.4485**	**3242.9483**

The lowest error values are in bold

exponent (1), C (0,1), ε (0.01). For the approach using the explanatory variables energy consumption, GDP, and population, the SVR obtained the best results in the configurations: exponent (1), C (1), ε (0.001). Table 8.6 displays the MAPE (%) and MAD values of each approach. Unlike the first step, the approach using energy consumption, GDP, and population achieved the best results, followed by the approach using only energy consumption. The approach using energy consumption together with GDP obtained the worst results from MAPE (%) and MAD.

Unlike the first one, when the consumption was combined with the GDP and the population, the forecast obtained a reduction of 0.64% of MAPE and 843,51 of MAD compared to the forecast using only the energy consumption. When the forecast was performed using consumption and GDP, the result worsened by 0.20% MAPE metric and 260.71 MAD metric compared to energy consumption.

Based on this information, we can conclude that for the years 2003–2013 the GDP was the factor that contributed most to the quality of the SVR model. However, for the most recent SVR models (2014–2016) the GDP together with the population were the factors that most contributed to the increase of the accuracy of the model presented in this section.

8.5.4 Comparison of Best Results

The computational experiments presented in this subsection were executed to answer the following research question:

- "Using the data of consumption, GDP, and population, as in [25], can one obtain better results employing classical regression techniques?"

In this subsection, we performed our computational experiments using the techniques of linear regression, multilayer perceptron neural networks, and support vector regression aiming to answer the above question.

In the first step (predictions from 2004 to 2013), the combinations of consumption, GDP, and consumption, GDP, and the population did not obtain better results than the predictions containing only input data. However, for the regression support vector (SVR) model, GDP contributed to the increase in the quality of the model, obtaining better approximations than the approaches using only energy consumption and consumption together with GDP and population. The SVR obtained the best result of all models generated in this step.

The work of Torrini [25] is one of the most recent in the literature on residential energy consumption forecasting in Brazil. In his thesis, he generated predictions using the fuzzy model using three approaches: the first used only energy consumption, the second used energy consumption and GDP, and the third used energy consumption, GDP, and population. The combined approach to consumption, GDP, and population achieved the best results.

In Table 8.7, we present the results of Torrini together with our best approaches. The results correspond to the forecasts from 2003 to 2013, the period used for

Table 8.7 Fuzzy model predictions and better approaches of LR, MLP, and SVR (2004–2013)

Approach	MAPE (%)	MAD
Fuzzy univariate	3.7590	3136.5020
Fuzzy with GDP	3.7235	3079.2810
Fuzzy with GDP + population	**1.6658**	**1366.9410**
LR—consumption	17.2835	19,017.7661
MLP—consumption	13.2716	14,602.4138
SVR—consumption + GDP	**1.7298**	**1763.2968**

LR Linear Regression, *MLP* Multilayer Perceptron, *SVR* Support Vector Regression
The lowest error values are in bold

Table 8.8 Projections of the fuzzy model, EPE, and better approaches of LR, MLP, and SVR (2014–2016)

Approach	MAPE (%)	MAD
Fuzzy + GDP + population	5.3750	85,680.3300
EPE	4.0289	5328.0000
LR—consumption + GDP + population	1.9313	2549.3500
MLP—consumption	**1.7556**	**2321.5181**
SVR—consumption + GDP + population	2.4485	3242.9483

EPE Empresa de Pesquisa Energética, *LR* Linear Regression, *MLP* Multilayer Perceptron, *SVR* Support Vector Regression
The lowest error values are in bold

validation in [25] and our first step. As we can see, the fuzzy model of [25] using consumption together with GDP and population obtained the best value of MAPE and MAD that all our models in this first stage.

In the second stage (predictions 2014–2016), for the linear regression (LR) and the support vector regression models the combination: consumption, GDP, and population obtained better results than the other approaches. For multilayer perceptron, the approach using only the energy consumption obtained better results than the other approaches. Among all our approaches, the neural networks using only the energy consumption obtained the best results of MAPE (%) and MAD in the predictions of the years 2014–2016.

Torrini [25] made projections using the fuzzy model using GDP and population up to the year 2030 and compared them with the projections of the Empresa de Pesquisa Energética (EPE). We calculate your MAPE (%) and MAD in the period from 2014 to 2016 which corresponds to our second stage. The results are presented together with our best approaches in Table 8.8.

According to Hamzacebi et al. [11], a prediction is considered successful when its MAPE is less than 10%, as can be seen in Table 8.8. All of our best approaches have achieved MAPE of less than 10%. Also, all of our approaches obtained better MAPE (%) and MAD values compared to the Torrini and Empresa de Pesquisa Energética (EPE) models from 2014 to 2016. Based on this, we can conclude that linear regression, multilayer perceptron, and support vector regression are interesting approaches to be applied in our country, considering that our recent forecasts are closer to the real value compared to recent projections in the literature.

8.6 Study of the Inclusion of New Explanatory Variables

In the previous section, we verified whether the approaches of linear regression, multilayer perceptron, and support vector regression would be better than the most recent approaches presented in the literature [25]. In this section, we will generate our models to verify if other socioeconomic factors can improve the quality of the forecasts of our models in the period from 2014 to 2016.

As can be seen in the Equation RMSE, the RMSE calculates the error and squares it before the average, because of this it gives considerable weight to the higher errors. For our experiments, we are more interested in knowing about the total of the periods analyzed, more than the weight of the errors of each year. So, in the occasions where we need to decide between the best values of MAPE (%), MAD, or RMSE we choose the lower values of MAPE (%) and MAD. The residential energy consumption approach is baseline, throughout this section it will be called configuration 1. Table 8.9 presents all approaches from configuration 2.

8.6.1 Linear Regression (LR)

In this subsection, we present the computational experiments generated by linear regression. The experiments were performed using all of the approaches described above. The best approximation of MAPE and MAD obtained was using the configuration approach 1. It used only the explanatory variable energy consumption obtaining 2.4255% of MAPE, 3204.7054 MAD, and 8905.7744 RMSE. The best approximation of RMSE obtained was using the configuration approach 18. It consisted of the combination of the explanatory variables: consumption, GDP, population and WTI crude obtaining 2.5831% MAPE, MAD 3406.4100, and 6001.4486 of RMSE. The five best results are shown in Table 8.10.

8.6.2 Neural Networks Multilayer Perceptron (MLP)

In this subsection, we present the computational experiments generated by multilayer perceptron. The experiments were performed using all of the approaches described above. The parameters that obtained the best results in each approach are shown in Table 8.11.

Table 8.9 Combination of all explanatory variables organized by configuration (approach)

Configuration	GDP	Population	Minimum wage	Exchange rate	Average consumption tariff	Price per barrelof Brent crude oil	Price per barrel of WTI crude oil
2	X						
3	X	X					
4			X				
5	X		X				
6	X	X	X				
7				X			
8	X			X			
9	X	X		X			
10					X		
11	X				X		
12	X	X			X		
13						X	
14	X					X	
15	X	X				X	
16							X
17	X						X
18	X	X					X
19	X	X	X	X			
20	X	X	X	X	X		
21	X	X	X	X	X	X	
22	X	X	X	X	X	X	X

Table 8.10 Forecast result using Linear Regression for the five best approaches

Linear Regression			
Approach	MAPE (%)	MAD	RMSE
Configuration 1	**2.4255**	**3204.7054**	8905.7744
Configuration 3	2.7540	3639.0265	7760.4100
Configuration 6	3.1708	4171.2667	9251.4151
Configuration 9	3.5055	4639.9933	8978.3134
Configuration 18	2.5831	3406.4100	**6001.4486**

The lowest error values are in bold

Table 8.11 Parameters of Multilayer Perceptron

Multilayer Perceptron				
Approach	Parameters			
Configuration	Learning rate	Momentum	Training time	Hidden layers
1	0.3	0.2	500	7
2	0.3	0.2	500	2
3	0.3	0.2	250	2
4	0.3	0.2	250	2
5	0.3	0.2	500	2
6	0.3	0.2	500	2
7	0.3	0.2	250	2
8	0.3	0.2	500	7
9	0.3	0.2	250	2
10	0.3	0.2	500	7
11	0.3	0.2	250	2
12	0.3	0.2	500	2
13	0.3	0.2	500	2
14	0.3	0.2	500	2
15	0.3	0.2	500	2
16	0.3	0.2	250	2
17	0.3	0.2	500	2
18	0.3	0.2	500	7
19	0.3	0.2	250	2
20	0.3	0.2	250	2
21	0.3	0.2	500	2
22	0.3	0.2	500	7

The best approximation of MAPE, MAD, and RMSE obtained was using the configuration approach 2. It used the explanatory variables: energy consumption and GDP obtaining 0.7646% of MAPE, 1005.8067 MAD, and 2152.7581 RMSE. The five best results are shown in Table 8.12.

Table 8.12 Forecast result using Multilayer Perceptron for the five best approaches

Multilayer Perceptron			
Approach	MAPE (%)	MAD	RMSE
Configuration 1	1.0429	1374.6433	2703.2750
Configuration 2	**0.7646**	**1005.8067**	**2152.7581**
Configuration 3	1.3934	1838.5767	4620.8096
Configuration 4	1.4488	1919.0833	3886.1274
Configuration 10	1.3962	1851.3933	4479.2248

The lowest error values are in bold

Table 8.13 Forecast result using Support Vector Regression for the five best approaches

Support Vector Regression			
Approach	MAPE (%)	MAD	RMSE
Configuration 3	1.5287	2008.3833	4750.4118
Configuration 6	**1.3046**	**1714.0700**	**4246.2019**
Configuration 9	1.8328	2420.8500	4252.2392
Configuration 18	2.0614	2713.1900	5679.3386
Configuration 19	1.8460	2436.2933	4534.5008

The lowest error values are in bold

8.6.3 Support Vector Regression (SVR)

In this subsection, we present the computational experiments generated by support vector regression (Table 8.13). The experiments were performed using all of the approaches described above. The parameters that obtained the best results in each approach are shown in Table 8.14.

The best approximation of MAPE, MAD, and RMSE obtained was using the configuration approach 6. It consisted of a combination of the explanatory variables: consumption, GDP, population, and minimum wage obtaining 1.3046% of MAPE, 1714.0700 of MAD, and 4246.2019 of RMSE. The five best results are shown in Table 8.13.

8.6.4 Comparison of Best Results

The computational experiments presented in this section were performed to answer the following research question:

- "Given additional socioeconomic data in the classical regression techniques, such as the minimum wage, the exchange rate, the average tariff of energy consumption, and the price of a barrel of oil, can one obtain better results?"

In the previous subsections, we execute our computational experiments to answer the above question. First, we add the explanatory variables in the linear regression model, then we add in the multilayer perceptron neural network models, and to finalize, in the support vector regression model.

Table 8.14 Parameters of Support Vector Regression

Support Vector Regression				
Approach	Parameters			
Configuration	Kernel	Exponent	C	ε
1	Polynomial	1	10	0.01
2	Polynomial	1	1	0.01
3	Polynomial	1	1	0.01
4	Polynomial	1	1	0.01
5	Polynomial	1	10	0.01
6	Polynomial	1	10	0.01
7	Polynomial	1	1	0.01
8	Polynomial	1	10	0.01
9	Polynomial	1	1	0.01
10	Polynomial	1	10	0.0001
11	Polynomial	1	1	0.01
12	Polynomial	1	1	0.0001
13	Polynomial	1	10	0.01
14	Polynomial	1	10	0.01
15	Polynomial	1	10	0.01
16	Polynomial	1	1	0.0001
17	Polynomial	1	10	0.01
18	Polynomial	1	10	0.01
19	Polynomial	1	1	0.01
20	Polynomial	1	1	0.01
21	Polynomial	1	1	0.01
22	Polynomial	1	1	0.01

Table 8.15 Better results of all approaches

Approach	MAPE (%)	MAD	RMSE
Configuration 1	2.4255	3204.7054	8905.77440
Configuration 2	**0.7646**	**1005.8067**	**2152.76**
Configuration 6	1.3046	1714.0700	4246.2019

The lowest error values are in bold

When comparing the best approaches of all models, we can observe that the linear regression model obtained the best results in the approach that used only the explanatory variable energy consumption (configuration 1). The multi-layered perceptron neural network models obtained better results using energy consumption together with GDP (configuration 2). And, to finalize the regression support vector machine models, we obtained the best results using the consumption approach along with GDP, population, and minimum wage (configuration 6).

In Table 8.15, we present the results of the best approaches of each algorithm shown in this section. As we can observe, the inclusion of explanatory variables did not increase the quality of the linear regression model. We can also conclude that there are correlations of all explanatory variables as shown in Table 8.2 but they are not linear.

When analyzing the multi-layered perceptron neural networks, we can observe that the inclusion of the explanatory variable PIB increased the approximation of the model. And, to finalize the support vector regression model, the best results were obtained when the explanatory variables were added: consumption, GDP, population, and minimum wage.

The model of multilayer perceptron networks using consumption together with GDP obtained the best result of all approaches presented in our experimental analysis.

8.7 Conclusion

Predicting energy consumption accurately is quite essential for the government, since a forecast below actual consumption can lead to potential power outages and raise the supplier's operating costs. In contrast, a forecast above actual consumption may result in idle capacity generating waste of financial resources unnecessarily. Therefore, a prediction model with good accuracy is fundamental to avoid costly errors.

In addition, in smart homes, the use of smart meters associated with consumption forecasting can help consumers to understand their consumption habits and eventually make adjustments to reduce their energy expenditures. The results presented in this chapter can be of great value to the Brazilian government, assisting it in the planning and investment in the country's energy sector.

The aim of this study was to evaluate the performance of classical regression techniques in the prediction of residential energy consumption in the Brazilian scenario and then to verify if any socioeconomic factors contribute to the generation of a more efficient model.

First, the models were generated using as input data the residential energy consumption, the residential consumption together with the GDP, and the residential energy consumption together with the GDP and the population of Brazil. The results of our approaches have demonstrated superior performance in some situations when compared to the fuzzy model approach, which is one of the most recent works in the literature.

Then, we verify if any socioeconomic factors contribute to the quality of our forecasts. The results showed that the linear regression approach has a decrease in quality when other socioeconomic factors are added. For the multilayered perceptron neural network models, GDP contributes to the increase of model precision in the most recent forecasts. And, for support vector regression models, the combination of socioeconomic factors: GDP, population, and minimum wage contribute to the generation of a more accurate model.

Although we used Torrini [25] as a reference to compare the results, we did not have access to its source code to reproduce forecasts or prediction values. We only had access to the MAP and MAD values of the model in the period from 2003 to 2013 in a window of 10 years, and the values of the projections in the period from

2014 to 2030. In practice, our comparison favors our model, considering that we are making medium-term forecasts and Torrini [25] made long-term forecasts.

In our research, we had some difficulties, such as the lack of part of the data of the barrel price of Brent crude oil and WTI. We could not find weather data to see if they would improve the accuracy of our approaches.

In a next step, we intend to find optimal values of training lags in larger parameter variations. We will also look to see if it is possible to improve our approaches taking into account shorter periods of time in order to check if some old data does not contribute to the quality of the forecasts. In addition, we intend to apply these algorithms in specific regions and then in each city with the objective of verifying the increase in the quality of the forecasting in order to offer suggestions for population awareness, conservation of energy, and to verify the possibility of substitution by sources of renewable energies.

References

1. CIA: Central Intelligence Agency - The World Factbook (2016). Available: https://www.cia.gov/library/publications/the-world-factbook/geos/xx.html. Accessed 17 Feb 2017
2. International Energy Agency (IEA): Key World Energy Statistics. IEA, Paris. https://doi.org/10.1787/key_energ_stat-2015-en (2015)
3. Pérez-Lombard, L., Ortiz, J., Pout, C.: A review on buildings energy consumption information. Energy Build. **40**(3), 394–398 (2008)
4. IBGE: IBGE - Instituto Brasileiro de Geografia e Estatística (2017). Available: http://www.ibge.gov.br/home/. Accessed 21 Feb 2016
5. IPEADATA: IPEADATA: Instituto de Pesquisa Econômica Aplicada. Available http://www.ipeadata.gov.br (2016). Accessed 28 Nov 2016
6. Campos, R.J.: Previsão de séries temporais com aplicações a séries de consumo de energia elétrica, Ph.D. dissertation, Universidade Federal de Minas Gerais (2008)
7. Tidre, P.V.V., Biase, N.G.G., de Sousa Silva, M.I.: Utilização dos modelos de séries temporais na previsão do consumo mensal de energia elétrica da região norte do brasil. Matemática e Estatística em Foco **1**(1), 57–66 (2013)
8. Swan, L.G., Ugursal, V.I.: Modeling of end-use energy consumption in the residential sector: a review of modeling techniques. Renew. Sustain. Energy Rev. **13**(8), 1819–1835 (2009)
9. Bianco, V., Manca, O., Nardini, S., Minea, A.A.: Analysis and forecasting of nonresidential electricity consumption in Romania. Appl. Energy **87**(11), 3584–3590 (2010)
10. Kaytez, F., Taplamacioglu, M.C., Cam, E., Hardalac, F.: Forecasting electricity consumption: a comparison of regression analysis, neural networks and least squares support vector machines. Int. J. Electr. Power Energy Syst. **67**, 431–438 (2015)
11. Hamzacebi, C., Es, H.A.: Forecasting the annual electricity consumption of Turkey using an optimized grey model. Energy **70**, 165–171 (2014)
12. Todesco, J.L., Pimentel, F.J., Bettiol, A.L.: O uso de famílias de circuitos e rede neural artificial para previsão de demanda de energia elétrica. Revista Produção Online, **4**(4), 1–8 (2004)
13. Mammen, P.M., Kumar, H., Ramamritham, K., Rashid, H.: Want to reduce energy consumption, whom should we call? In: Proceedings of the Ninth International Conference on Future Energy Systems, pp. 12–20. ACM, New York (2018)
14. Zhou, D.P., Balandat, M., Tomlin, C.J.: Estimating treatment effects of a residential demand response program using non-experimental data. In: 2017 IEEE International Conference on Data Mining Workshops (ICDMW), pp. 95–102. IEEE, Piscataway (2017)

15. Siddiqui, I.F., Lee, S.U.-J., Abbas, A., Bashir, A.K.: Optimizing lifespan and energy consumption by smart meters in green-cloud-based smart grids. IEEE Access **5**, 20934–20945 (2017)
16. Bedingfield, S., Alahakoon, D., Genegedera, H., Chilamkurti, N.: Multi-granular electricity consumer load profiling for smart homes using a scalable big data algorithm. Sustain. Cities Soc. **40**, 611–624 (2018)
17. Yan, X., Ozturk, Y., Hu, Z., Song, Y.: A review on price-driven residential demand response. Renew. Sustain. Energy Rev. **96**, 411–419 (2018)
18. Fumo, N., Biswas, M.R.: Regression analysis for prediction of residential energy consumption. Renew. Sustain. Energy Rev. **47**, 332–343 (2015)
19. Bianco, V., Manca, O., Nardini, S.: Electricity consumption forecasting in Italy using linear regression models. Energy **34**(9), 1413–1421 (2009)
20. Günay, M.E.: Forecasting annual gross electricity demand by artificial neural networks using predicted values of socio-economic indicators and climatic conditions: case of Turkey. Energy Policy **90**, 92–101 (2016)
21. Torrini, F.C., Souza, R.C., Oliveira, F.L.C., Pessanha, J.F.M.: Long term electricity consumption forecast in Brazil: a fuzzy logic approach. Socio Econ. Plan. Sci. **54**, 18–27 (2016)
22. Askarzadeh, A.: Comparison of particle swarm optimization and other metaheuristics on electricity demand estimation: a case study of Iran. Energy **72**, 484–491 (2014)
23. Coelho, V.N., Coelho, I.M., Coelho, B.N., Reis, A.J., Enayatifar, R., Souza, M.J., Guimarães, F.G.: A self-adaptive evolutionary fuzzy model for load forecasting problems on smart grid environment. Appl. Energy **169**, 567–584 (2016). Available: http://www.sciencedirect.com/science/article/pii/S0306261916301684
24. Coelho, I.M., Coelho, V.N., da S. Luz, E.J., Ochi, L.S., Guimarães, F.G., Rios, E.: A GPU deep learning metaheuristic based model for time series forecasting. Appl. Energy (2017). Available: http://www.sciencedirect.com/science/article/pii/S0306261917300041
25. Torrini, F.C.: Modelos de lógica fuzzy para a previsão de longo prazo de consumo de energia, Ph.D. dissertation, PUC-Rio (2014)
26. Dong, B., Cao, C., Lee, S.E.: Applying support vector machines to predict building energy consumption in tropical region. Energy Build. **37**(5), 545–553 (2005)
27. Pal, S.K., Mitra, S.: Multilayer perceptron, fuzzy sets, and classification. IEEE Trans. Neural Netw. **3**(5), 683–697 (1992)
28. Haykin, S.S.: Redes Neurais–Princípios e Prática (2001)
29. Li, Q., Meng, Q., Cai, J., Yoshino, H., Mochida, A.: Applying support vector machine to predict hourly cooling load in the building. Appl. Energy **86**(10), 2249–2256 (2009)
30. Zhao, H., Magoulès, F.: Parallel support vector machines applied to the prediction of multiple buildings energy consumption. J. Algorithms Comput. Technol. **4**(2), 231–249 (2010)
31. Hua, Y., Oliphant, M., Hu, E.J.: Development of renewable energy in Australia and China: a comparison of policies and status. Renew. Energy **85**, 1044–1051 (2016)
32. Arora, S., Taylor, J.W.: Forecasting electricity smart meter data using conditional kernel density estimation. Omega **59**, 47–59 (2016)
33. Laurinec, P., Lucká, M.: New clustering-based forecasting method for disaggregated end-consumer electricity load using smart grid data. In: 2017 IEEE 14th International Scientific Conference on Informatics, pp. 210–215. IEEE, Piscataway (2017)
34. Ponocko, J., Milanovic, J.V.: Forecasting demand flexibility of aggregated residential load using smart meter data. IEEE Trans. Power Syst. **33**(5), 5446–5455 (2018)
35. SGS: Banco Central do Brasil - SGS - Sistema Gerenciador de Séries Temporais - Produto interno bruto em R$ correntes (2017). Avaliable: https://www3.bcb.gov.br/sgspub/localizarseries/localizarSeries.do?method=prepararTelaLocalizarSeries. Accessed 20 Feb 2017
36. IBGE: IBGE - Instituto Brasileiro de Geografia e Estatística - Séries Históricas e Estatísticas (2016). Avaliable: https://seriesestatisticas.ibge.gov.br/. Accessed 21 Feb 2016
37. EIA: U.S. Energy Information Administration. Available: https://www.eia.gov/. Accessed 30 June 2016

38. Witten, I.H., Frank, E., Hall, M.A., Pal, C.J.: Data Mining: Practical Machine Learning Tools and Techniques. Morgan Kaufmann, Cambridge (2016)
39. Hall, M., Frank, E., Holmes, G., Pfahringer, B., Reutemann, P., Witten, I.H.: The WEKA data mining software: an update. ACM SIGKDD Explorations Newsl. **11**(1), 10–18 (2009)
40. Hall, M.: Time series analysis and forecasting with weka-pentaho data mining. Pentaho.com. http://wiki.pentaho.com/display/DATAMINING/Time+Series+Analysis+and+Forecasting+with +Weka. Accessed 28 Feb 2017
41. Shevade, S.K., Keerthi, S.S., Bhattacharyya, C., Murthy, K.R.K.: Improvements to the SMO algorithm for SVM regression. IEEE Trans. Neural Netw. **11**(5), 1188–1193 (2000)
42. Smola, A.J., Schölkopf, B.: A tutorial on support vector regression. Stat. Comput. **14**(3), 199–222 (2004)

Part II
Simulating the Possibilities and Getting Ready for Real Applications

Chapter 9
Simulation and Evaluation of a Model for Assistive Smart City

Marcelo Josué Telles, Jorge Luis Victória Barbosa, Rodrigo da Rosa Righi, José Vicente Canto dos Santos, Márcio Joel Barth, and Leandro Mengue

Abstract This chapter presents the evaluation of the Model for Assistive Smart Cities (MASC), which is intended for ubiquitous accessibility. The model was evaluated with data obtained from a contextual simulator (SIAFU), applied in the central region of São Leopoldo city, Brazil. Unlike other approaches, the evaluation considers multiple accesses asynchronously, indicating that the model meets massive applications with response time within the standards indicated for this type of application. The evaluation considered requests from three different groups of users, characterized as: people with disabilities (PwDs), health professionals involved in the care of PwDs, and managers of public administration. The results of the evaluation indicated the feasibility of implementing the model in Smart Cities. In addition to collaborating with accessibility, the model favors the decision-making in the management of services in the cities.

9.1 Introduction

The technologies for location-based systems (LBS) [16] provide services for various tasks and are applied in smart cities. Such technologies are increasingly present in environments, either by the low cost of their implementation or by the advantages of their services [21]. Among the challenges of this segment, it is possible to quote the large-scale systems, the coverage of great geographic regions, and the use of diverse sources of information [15].

Large urban centers create demands and need to meet a public increasingly more diversified. Researches conducted in recent years [9, 44] indicated that 15% of people live with some disability and are considered people with disabilities (PwDs). There was an increase of more than 27% in the number of people with limited

M. J. Telles (✉) · J. L. V. Barbosa · R. da Rosa Righi · J. V. C. dos Santos
M. J. Barth · L. Mengue
University of Vale do Rio dos Sinos—UNISINOS, São Leopoldo, RS, Brazil
e-mail: jbarbosa@unisinos.br; rrrighi@unisinos.br; jvcanto@unisinos.br

V. N. Coelho et al. (eds.), *Smart and Digital Cities*, Urban Computing,
https://doi.org/10.1007/978-3-030-12255-3_9

locomotion. This segment of the population can be supported through LBS, cloud computing, systems based on user profile, and internet of things (IoT) [3, 8, 38].

The advancement in information technology and communication systems (ICT) has provided the emergence of cities that adopt technological concepts in various fields. Among the concepts, it is possible to mention wireless technologies, ubiquitous computing, artificial intelligence, and IoT [17, 33, 34, 37]. The adopted technologies aim to offer new services, automate actions, and promote improvements in the daily lives of individuals and in the management of cities. A segment that deserves attention in this scenario is accessibility. The term accessibility includes people with disabilities (PwDs) or people with limited locomotion conditions that may be elderly, injured, or pregnant. For this group of citizens, ICT can offer services for access to resources and for monitoring and recommendations.

Typically, applications [2, 14, 19, 28, 39] with the focus on accessibility do not exchange information with other systems and were not evaluated with a large number of users. Another limitation of the applications is their coverage area, which requires prior mapping. In this scenario, simulation is proposed to evaluate a Model for Assistive Smart Cities (MASC). The authors suggest that smart cities helping PwDs can be called Assistive Smart Cities (ASC).

Providing support for accessibility in any environment can be an impracticable task, but it is viable the support in environments equipped with technologies such as sensor network, resource mapping, geolocation, and internet signal. An important aspect to be considered in this scenario is the use of the information generated during the use of the resources, that is, information generated by the interactions of the PwDs, may support other PwDs.

This chapter is structured in five sections. Section 9.2 discusses basic concepts about smart cities and their applications. MASC model is presented in Sect. 9.3. Section 9.4 describes the proposed simulation. Section 9.5 deals with the evaluation of the model with simulation data. Section 9.6 presents the conclusions and future work.

9.2 Smart Cities and Their Applications

Information technology and communication systems (ICT), especially ubiquitous computing, have a tooling for implementation aid and maintenance of the resources in cities demand [32]. The technology makes viable, not only the provision of new resources, but mainly the obtainment of detailed information about them. In addition, ICT offer the ease of monitoring and contact between PwDs and people who support them. In this scenario, ubiquitous computing is an important tool to be used for the monitoring of PwDs and the provision of contextual information to support them by family and health professionals.

The infrastructure available in urban areas, such as connectivity, sensor network, and ubiquitous computing, along with collaborative resources, that is, information provided by the citizen, can offer subsidies for the implementation of services for the PwDs. The ubiquitous computing involves Cloud Computing, Big Data and

IoT. Another important foundation for urban support is the set of technologies discussed by Costa et al. [11]. The work proposes a model composed by a framework and a middleware. Such model aims to meet the challenges of ubiquitous computing, among them the interpretation of information collected by different sensors, generation of automated actions and promotion of interoperability between systems.

The cities classification with regard to technology level receives four nomenclatures [25]. These take into consideration both the level of technology and its abundacy in cities. Hereafter, the four classifications are detailed.

The first classification is Digital City [24, 45] (or digital community, information city, or e-city). It refers to a connected community that combines broadband communications infrastructure, service-oriented computing based on open standards and innovative services that meet the needs of governments and their citizens. The use of open standards is considered an important issue for interoperability between the various information systems and computation systems.

The second classification is Intelligent City [22]. Cities are defined as territories that bring systems of innovation and ICT within the same locality, combining creativity of individuals composing the population of the city, institutions that improve learning and spaces of innovation, usually virtual, that facilitate the management of knowledge. The combination of people's creativity involves the collective intelligence strategy, where trends are identified and standardized, using people's experiences in order to collaborate collectively.

The third classification is Smart City [42]. In this paradigm, the use of ICT aims to make the components of infrastructure and services essential for a more intelligent, interconnected and efficient city [35]. This concept was implemented in some cities [3], such as Brisbane, Malta, Dubai, and Kochi. One of the main objectives of these cities is to improve the quality of life of people from different points of view, for example, the level of access to information, available resources, as well as the current state of such resources.

The fourth classification is Ubiquitous City [18]. In this scenario, the city is fully equipped with networks through which city authorities can monitor what is happening in the city, for example traffic monitoring, crime prevention, and fire prevention. The users can access any service on the network regardless of the place they are, although their positions are relevant. In addition to distinct systems sharing the same information, the number of devices is significantly higher than in the other classifications. This classification provokes different opinions between specialists and users, some are completely in favor, others argue that these systems violate users' privacy, and make the systems vulnerable.

In cities classified as Smart City, sensors are designed for the collection of information [31]. The functionalities of smart cities range from systems that store information for future queries to systems based on data collected in real time, through traffic monitoring using sensors [1, 4]. Having access to the detailed information decisions can be compared with decisions taken before making it possible to arrive more certainly in the best action to be taken. The MASC model can be implemented in smart cities, due to the level of technology available in this type of city.

9.3 MASC Model

MASC proposes the concept of assistive smart city (ASC), where support for PwDs is offered through ubiquitous computing. This concept consists of using information generated by the PwDs to build a historical database, offering full-time cloud computing, Big Data and IoT support, as well as attending the PwDs in their daily activities. MASC fits into the concept of assistive smart city, and is intended for users who present some type of disability and related segments, such as health professionals and public administrations. In ubiquitous computing, the Cloud Computing, Big Data, and IoT are strategic because they provide the technological basis of the model.

The MASC supports different types of disabilities and manages contexts, resources, and profiles [41] of PwDs. In addition, it meets massive applications and stores history [27] forming tracks [5]. The tracks are composed of actions and sequences of visited places. Tracks contain information related to a person, place, or object that occurred in an order [43].

Ubiquitous accessibility refers to the use of ubiquitous computing in accessibility [40]. A city that adopts ubiquitous computing becomes intelligent; already the concept assistive city is resulting from accessibility in cities. The integration of ubiquitous computing and accessibility in the smart city constitutes an Assistive Smart City. The MASC collaborates to the constitution of this concept as it adopts ubiquitous computing technologies, along with other resources described in this chapter. The MASC model provides PwDs support, so that they may use the resources in their daily life and perform their activities with more autonomy, performing the tasks more conveniently.

The concept of assistive city involves a mapping of accessible paths. The accessible paths should be mapped in order to provide detailed information for PwDs. Information such as height and width of the sidewalk, light post, or other objects are relevant information for PwDs. In addition, the places that have tactile floor also are relevant, in the case of PwD being visually impaired. Places that support PwDs also have particularities and they should be informed. All these concepts are part of an assistive city and were treated by the MASC.

The formalization of knowledge is an important step in consolidating intelligent cities. A proposal for formal representation of knowledge in intelligent cities was described by Komninos et al. [23] which presents an ontology for this domain. MASC has an ontology for representation of the disabilities and classification on particularities of each one. The ontology was extended from the work of Tavares et al. [39] which proposes to establish representation of the environments and tracks, besides the entities already covered by the original ontology (people, resources, and disability).

Next section presents the simulation for generating evaluation data of the MASC. The use of real data can also be performed, but this would involve a longer work and the participation of a large number of volunteers.

9.4 Simulation for Generate Data

The simulation generated contextual user information. The scenario for assessing the model was in the city of São Leopoldo, in Brazil, including places residential, commercial, leisure, and the city center. We adopted the SIAFU simulator [26]. In order to guarantee a real simulation, we used maps obtained from the Open Street Map OSM platform [30]. This platform allows to obtain maps of real locations. Figure 9.1 shows the place where the simulation was performed.

Figure 9.1 shows two maps, the real map (a) and the map adopted by the simulation in black and white (b). The simulation considered the black regions of the map, as areas for pedestrians. In addition to defining the physical location of the

(a)

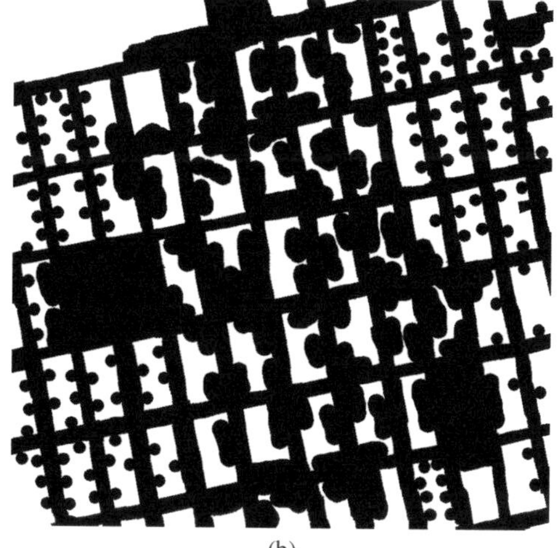

(b)

Fig. 9.1 Simulation maps. (**a**) Map indicating the simulation location. (**b**) Boundaries of the simulation location

Fig. 9.2 Places of residences, work, and leisure places of the simulation. (a) Residences. (b) Work places. (c) Leisure places

simulation, it was necessary to define places of residence, work, and leisure. The places of residence, work, and leisure can be checked separately in Fig. 9.2.

Figure 9.3 presents the WiFi network access locations. In these places, simulation agents make small stops to receive the internet signal. These places also were considered preferences in daily pedestrian traffic.

The simulation was parameterized with number of people, regions of the simulation (latitude and longitude), places of residences, work, and leisure. Also, it is possible to set the percentage of people using cars. In the test, this percentage was set to 0 (zero). Other configurations are possible, such as work schedule, sleep period, heat zones, assault index, and so on. As a result, the simulation generated the interactions of people in the city center on the basis of latitude and longitude. We considered 125 agents (PwDs) positioned in the residential places, on the periphery of the map.

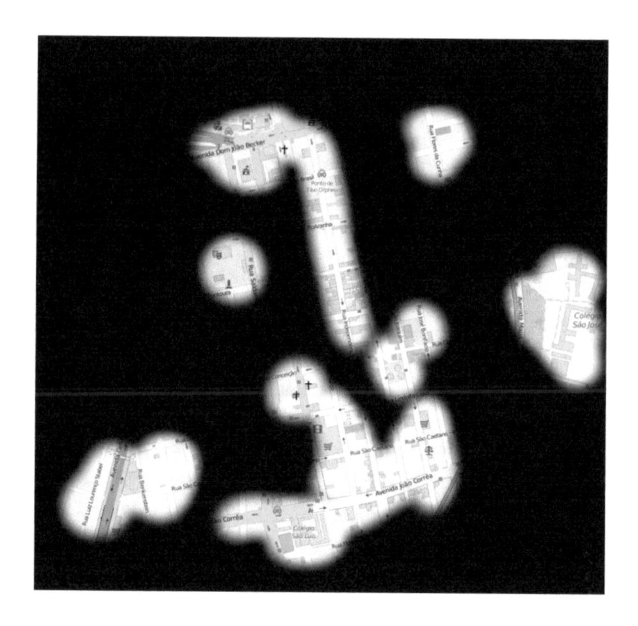

Fig. 9.3 WiFi signal coverage locations

9.4.1 Simulation for Smart Assistive Cities

The simulation included performance and functionality tests. The performance tests consisted in sending data to the model and carrying out queries, in order to verify if the MASC serves massive requests for both recording information and for performing queries.

The functionality tests aimed at verifying if the model meets needs of accessibility services for PwDs, health professionals, and public administration. At the beginning of the day, the simulation agents moved to the workplaces created in the center of the map. In this simulation, 43 places of work were defined.

At the end of the day, some agents remained in the workplace for some time, some moved to their residences, others moved to places of leisure. It were defined 37 places of leisure. The data of the simulation were exported to a Comma Separated Value (CSV) file, and then sent to the MASC. The simulation period was 7 days, generated 270,000 record histories, forming the iterations of the PwDs at the simulation place.

In a real-world situation, data on accessibility should be obtained in open data source such as data on urban mobility [36] and accessibility [12] and [13]. The initiatives proposed by the Semantic Web [6] are also important mechanisms to be used by the MASC. The public information known as open data sources and linked data [7] should also be used in our work.

9.5 MASC Assessment

The performance evaluation was applied with the aid of algorithms using parallel processes (Threads) to simulate PwDs by sending displacement data and query requests. The metric to test performance was the response time.

In the performance test, the contextual data generated in the SIAFU tool were sent, with the objective of simulating PwD interactions in the map. During sending of the data for the MASC and the respective insertion into the database, the response time was monitored. All tests were asynchronous, concurrent, and the response time was considered from the submission of the request until respective result [20]. The response time is composed of client, network, and server latency.

The response time is presented as follows. Twelve processes were run in parallel, each one was responsible for sending about 2160 records of the CSV file. The observed average for the response time was 286 ms (milliseconds), the shortest time was 164 ms, the longest response time was 33,006 ms, and the standard deviation was 858 ms. The total number of requests with a response time greater than or equal to 2000 ms was 836. The majority of the requests were answered at around 200 ms and 400 ms. Bypassing the requests that took 2000 ms or more, one can consider that the MASC met 99.69% of the requisitions in less than 2 s, which is the acceptable value for this type of application [29].

This percentage represents the guarantee that the information for the construction of the queries are stored, since the loss of 0.3096% of the requests does not compromise the construction of the base. The requests that took 2000 ms or more were dispersed, not concentrated in a single moment, indicating that the unanswered requests are only one or two of each PwD. In order to test the performance, in addition to the test for history generation, a test using algorithm in Java®, for requests of query to the MASC was conducted. This test processed the query in the database. The routine compares the distance between the current PwD position and resource positions.

The algorithm responsible for this test generated Threads to simulate PwDs. As the threads were started, the response times of the requests were stored for analysis. It was defined that every 100 ms, one more PwD would initiate request to the MASC. Each PwD performed 1 request every 5 s. The test was designed to identify the maximum number of PwDs supported, thus identifying the first request with no return the test was completed. At the end of the test, 415 Threads were generated. The total number of requests received was 1925 in 42 s. In Fig. 9.4, the response times for resource queries are presented.

9.5.1 Evaluation of Services

The tests consisted of attending accessibility services for PwDs, health care, and public administration professionals. For PwDs, a service was done that makes the composition of tracks. The services for health professionals and public administration are shown in Table 9.1.

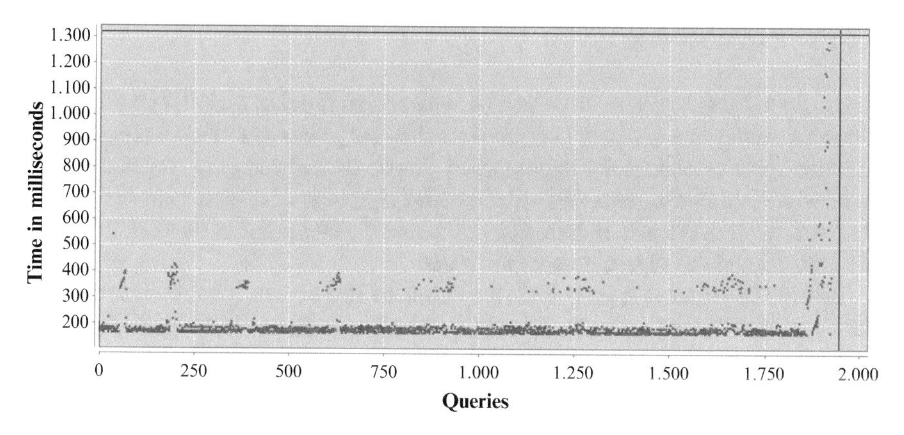

Fig. 9.4 Response time for resource queries

Table 9.1 Services for health professionals and public administration

URI	Example	Parameter	Return example
/health	e-mail of PwD	0; mark@server.com;	mark@server.com.br;2017-10-27 22:15:05;−29.767, −51.1439;
/health	For disability	1; Lower members;	blanch@server.com.br;2017-10-31 06:42:10;−29.768;−51.144; washi@server.com.br;2017-10-26 23:55:00;−29.761;−51.143; agnes@server.com.br;2017-10-26 23:55:00;−29.763;−51.151; wukong@server.com.br;2017-10-26 23:55:00;−29.761;−51.150;
/health	e-mail list	2; rusa@server.com.br; sophie@server.com.br; hikaa@server.com.br; zoraida@server.com.br; tapah@server.com.br;	rusa@server.com.br;2017-10-26 23:55:00;−29.763;−51.151; sophie@server.com.br;2017-10-26 23:55:00;−29.764;−51.142; hikaa@server.com.br;2017-10-26 23:55:00;−29.767;−51.144; zoraida@server.com.br;2017-10-26 23:55:00;−29.764;−51.151; tapah@server.com.br;2017-10-26 23:55:00;−29.765;−51.151;
/adm1	PwD trail	blanch@server.com.br; 2017-10-31;	1;−29.761;−51.15065; 2;−29.769;−51.15054; 3;−29.762;−51.15049; 4;−29.765;−51.15041; 5;−29.762;−51.15018; ...
/adm2	Area of interest	−29.76348; −51.1483; 12;	n sets of points, each set is formed by a sequence with identifier, latitude, and longitude in the same format of the service /adm1

9.5.1.1 PwDs Services

The service for this test receives as a parameter a start point, e-mail, and end point. Upon receiving this entry, the service searches for points near the start point, and then the route of each point is composed. It was considered only routes of PwDs with profile similar to requesting PwD, comparing type of disability of the person. An identification is made if there is a point near the end point, if there is, this route is indicated and the MASC returns the route.

Figure 9.5 presents a map with the result of this search. In this example, the indicated route (continuous line) is not the shortest path, but is the path accessible to the requesting PwD. Figure 9.5 shows the shortest route along the line dashed.

Figure 9.6 presents a map with the result of the same search, but with distinct start and end points. In the test, the query presented the route with more resources that passes through the more central region of the map. Other alternatives would be

Fig. 9.5 PwDs service, map with accessible trail

Fig. 9.6 PwD Service, second trail

possible but the routes are ordered by the amount of resources being indicated the route that passed by the street with more resources.

9.5.1.2 Services for Health Professionals

This service receives an identifier followed by a text, namely. Identifier 0 and e-mail of a PwD or identifier 1 and a type of disability or identifier 2 and an e-mail list of PwDs. The result consists of points with the last recorded position of each PwD. This service is intended for monitoring PwDs, allowing family members to monitor real-time location or health professionals to monitor several PwDs simultaneously. It possible to present a map with the last positions of all PwDs with limitations in the lower limbs or other desired classification. Table 9.1 presents three examples of this service.

In Fig. 9.7, another map of the same region of the city is presented, but with the position of a set of PwDs with disabilities in the lower limbs.

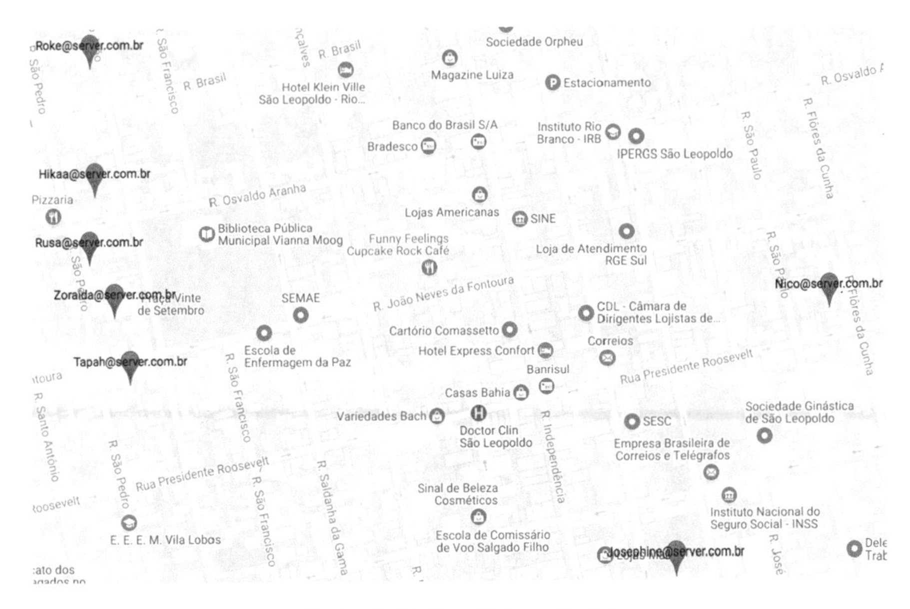

Fig. 9.7 Service for health professionals (PwDs with lower limbs)

9.5.1.3 Services for Public Administration

The functionality tests consisted of servicing situations for three segments. Tests were carried out for services for PwDs, services for health professionals, and services for the public administration.

Two services were developed for this test. The first receives the e-mail of a PwD and a date and returns a list with the points that the PwD has shifted on the informed date. Table 9.1, line /adm1, shows an example of this service. The number of points was limited by 5.

The second service receives a point and a distance that is used as radius. Such information delineates an area of interest in tracking the routes that pass through in this area. For each point found this area, a route is build. In the last line of Table 9.1, an example of this service is presented. With this service, city hall can identify the flow of PwDs for installation of accessibility equipment and resources in public transport. It also can provide installation of safety equipment in places that concentrate more PwDs, allocating guards, and security agents in strategic locations, identifying locations for access ramp installation and predicting the flow of PwDs for decision-making. Figure 9.8 shows a map with the area of interest around the location of a PwD.

Another feature for public administration was the automatic identification of places with concentration of PwDs not only in real time but over time. For this purpose, a service was developed that receives as parameter a point, a distance, and one or two dates. When receiving a point, a distance, and a date, the service departs from the point and finds places with route concentration on the date entered.

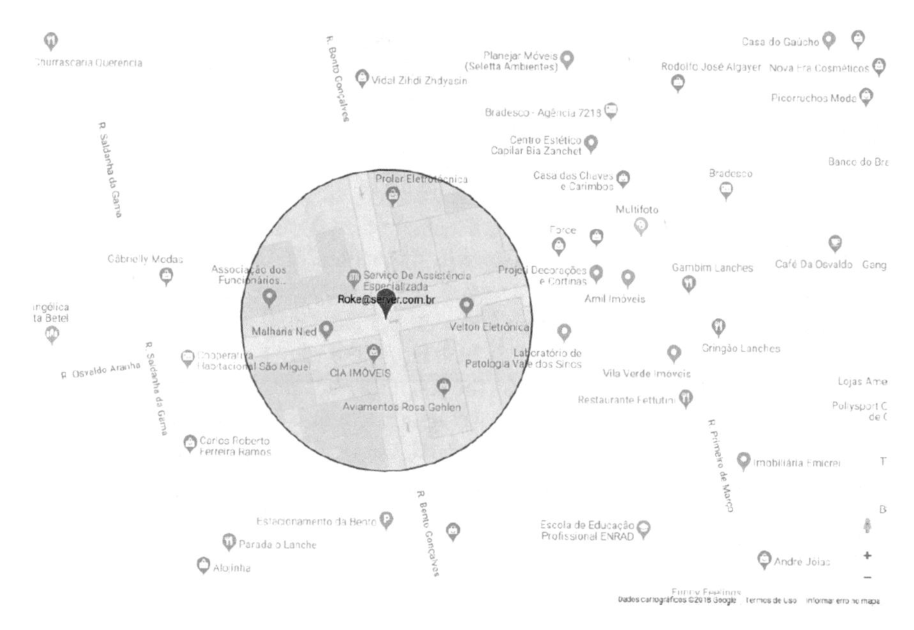

Fig. 9.8 City hall service, point, and area of interest

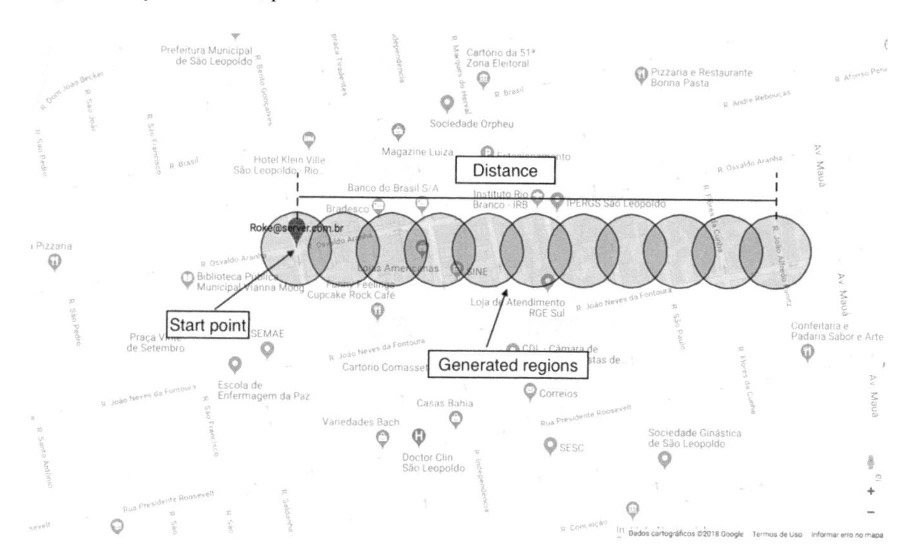

Fig. 9.9 Regions of interest being generated to the east

Figure 9.9 shows regions east of the starting point. The regions are generated until the distance reported by parameter is reached. In the test, regions of interest were generated obeying the average of the dimensions of the block of the simulation region, which is 110 m. The areas of interest are generated with 55 m radius until reaching the distance reported by parameter.

The service algorithm continues to generate areas of interest to the west, north, and south. For each region is counted the routes that pass through it. Finally, it is counted which region contains the most routes, thus the service returns a list with the points of the routes that pass through the region, as well as the central point of the generated region.

The same is done if two dates are received, the single difference is that in the query a period is taken into account, not a single date. With this service, the public administration can exploit the regions with high flow of PwDs in large areas of automated form. It is possible to suggest a new approach for the treatment of regions, in circle format of different sizes. In this way, regions are overlapped by proportionally sized circles. With this technique, the routes and resources can be treated optimally. Figure 9.10 presents the proposal that adopts the theory of central places [10]. This technique considers hexagons, which were generated east of the point of interest. Figure 9.10 has been adapted for circular regions. The treatment of routes can also be carried out using classifiers.

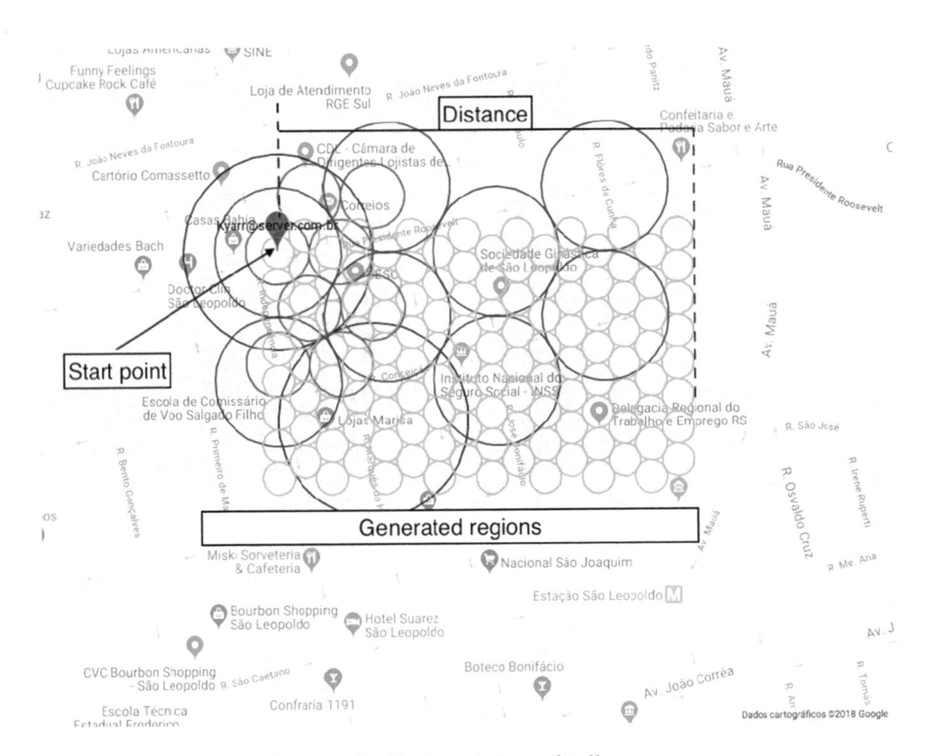

Fig. 9.10 Central places theory applied in the technique of trails

9.6 Conclusion and Future Work

The services described in the previous section perform queries in the databases to attend PwDs, health professionals, and public administration. The last described service motivates the development of new functionalities such as automatic search for places that have resources but do not receive the presence of PwDs in specific days or times, allowing the reallocation of resources.

The performance tests allowed to conclude that the MASC fulfills requests for the recording of the information of the routes and requests for consultations. The MASC supported the sending of information on routes generated by the 125 agents of the simulation and answered requests for consultations for 415 agents. The tests allowed to realize that new functionalities can be developed without the need of new components in the model. Using routes as a service, it is possible to make the information stored in the model available in different ways, either in specific applications for PwDs or for professionals related to this segment.

The application of classifiers can also be considered in a future evaluation of the model. At the present time, we seek the development of a mobile client to make available to the PwD and to initiate the records of information of day-to-day location. This will allow a new step in research and future services, in addition to collaborating with users, since accessible locations will be shared according to profile similarity. Research is also under way to implement contextual recommendations and context prediction.

Acknowledgements The authors wish to acknowledge that this work was financed by CNPq/Brazil (National Council for Scientific and Technological Development—http://www.cnpq.br) and Capes/Brazil (Coordination for the Improvement of Higher Education Personnel—http://www.capes.gov.br). We are also grateful to Unisinos (http://www.unisinos.br) for embracing this research.

References

1. Albino, V., Berardi, U., Dangelico, R.M.: Smart cities: definitions, dimensions, performance, and initiatives. J. Urban Technol. **22**(1), 3–21 (2015). https://doi.org/10.1080/10630732.2014.942092
2. Aly, W.H.F.: Mnd wsn for helping people with different disabilities. Int. J. Distrib. Sens. Netw. 2014(Article ID 489289) (2014). https://doi.org/10.1155/2014/489289
3. Anthopoulos, L., Fitsilis, P.: From digital to ubiquitous cities: defining a common architecture for urban development. In: Proceedings of the 2010 Sixth International Conference on Intelligent Environments, IE '10, pp. 301–306. IEEE Computer Society, Washington, DC (2010). http://dx.doi.org/10.1109/IE.2010.61
4. Barba, C., Mateos, M., Soto, P., Mezher, A., Igartua, M.: Smart city for vanets using warning messages, traffic statistics and intelligent traffic lights. In: Intelligent Vehicles Symposium (IV), 2012 IEEE, pp. 902–907. IEEE, Piscataway (2012). https://doi.org/10.1109/IVS.2012.6232229

5. Barbosa, J., Tavares, J., Cardoso, I., Mota, B., Martini, B.: Trailcare: an indoor and outdoor context-aware system to assist wheelchair users. Int. J. Hum. Comput. Stud. **116**, 1–14 (2018). https://doi.org/10.1016/j.ijhcs.2018.04.001
6. Berners-Lee, T., Hendler, J., Lassila, O.: The semantic web. Sci. Am. **284**(5), 34–43 (2001)
7. Bizer, C., Heath, T., Berners-Lee, T.: Linked data - the story so far. Int. J. Semant. Web Inf. Syst. **5**(3), 1–22 (2009). https://doi.org/10.4018/jswis.2009081901
8. Borgia, E.: The internet of things vision: key features, applications and open issues (2014). http://dx.doi.org/10.1016/j.comcom.2014.09.008. http://www.sciencedirect.com/science/article/pii/S0140366414003168
9. Brazilian Institute of Geography and Statistics. Available at: http://www.ibge.gov.br. Accessed 13 Jan 2018 (2010). Censo 2010
10. Christaller, W., Baskin, C.W.: Central Places in Southern Germany. Prentice-Hall, Englewood Cliffs (1966)
11. da Costa, C.A., Yamin, A.C., Geyer, C.F.R.: Toward a general software infrastructure for ubiquitous computing. IEEE Pervasive Comput. **7**(1), 64–73 (2008). http://dx.doi.org/10.1109/MPRV.2008.21
12. Ding, C., Wald, M., Wills, G.: A survey of open accessibility data. In: Proceedings of the 11th Web for All Conference, W4A '14, pp. 37:1–37:4. ACM, New York (2014). http://doi.acm.org/10.1145/2596695.2596708
13. Ding, C., Wald, M., Wills, G.: Linked data-driven decision support for accessible travelling. In: Proceedings of the 12th Web for All Conference, W4A '15, pp. 39:1–39:2. ACM, New York (2015). http://doi.acm.org/10.1145/2745555.2746681
14. Fernandes, H., Filipe, V., Costa, P., Barroso, J.: Location based services for the blind supported by {RFID} technology. Procedia Comput. Sci. **27**, 2–8 (2014). http://dx.doi.org/10.1016/j.procs.2014.02.002. http://www.sciencedirect.com/science/article/pii/S1877050914000040. 5th International Conference on Software Development and Technologies for Enhancing Accessibility and Fighting Info-exclusion, {DSAI} 2013
15. Hashem, I.A.T., Yaqoob, I., Anuar, N.B., Mokhtar, S., Gani, A., Khan, S.U.: The rise of big data on cloud computing: Review and open research issues. Inf. Syst. **47**, 98–115 (2015). http://dx.doi.org/10.1016/j.is.2014.07.006. http://www.sciencedirect.com/science/article/pii/S0306437914001288
16. Hightower, J., Borriello, G.: Location systems for ubiquitous computing. Computer **34**(8), 57–66 (2001). https://doi.org/10.1109/2.940014
17. Isotani, S., Ibert Bittencourt, I., Francine Barbosa, E., Dermeval, D., Oscar Araujo Paiva, R.: Ontology driven software engineering: a review of challenges and opportunities. IEEE Lat. Am. Trans. **13**(3), 863–869 (2015)
18. Jang, M., Suh, S.T.: U-city: new trends of urban planning in Korea based on pervasive and ubiquitous geotechnology and geoinformation. In: Computational Science and Its Applications–ICCSA 2010, pp. 262–270. Springer, Fukuoka (2010)
19. Kbar, G., Aly, S., Elsharawy, I., Bhatia, A., Alhasan, N., Enriquez, R.: Smart help at the workplace for persons with disabilities (shw-pwd). Int. J. Comput. Control Quantum Inf. Eng. **9**(1), 84–90 (2015). http://waset.org/Publications?p=97
20. Kim, E.: Oasis advancing open standards for the information society. Available at: http://docs.oasis-open.org/wsqm/WS-Quality-Factors/v1.0/WS-Quality-Factors-v1.0.html. Accessed 15 Jan 2016 (2012). Web Services Quality Factors Version 1.0. 31 Oct
21. Kitchin, R.: The real-time city? Big data and smart urbanism. GeoJournal **79**(1), 1–14 (2014)
22. Komninos, N.: The architecture of intelligent cities: integrating human, collective and artificial intelligence to enhance knowledge and innovation. In: 2nd IET International Conference on Intelligent Environments, 2006. IE 06, vol. 1, pp. 13–20. IET, Athens (2006)
23. Komninos, N., Bratsas, C., Kakderi, C., Tsarchopoulos, P.: Smart city ontologies: improving the effectiveness of smart city applications. J. Smart Cities **1** (2015). https://www.researchgate.net/publication/281740518_Smart_city_ontologies_Improving_the_effectiveness_of_smart_city_applications

24. Lee, J.H., Hancock, M.G., Hu, M.C.: Towards an effective framework for building smart cities: lessons from Seoul and San Francisco. Technol. Forecast. Soc. Chang. **89**, 80–99 (2014). https://doi.org/10.1016/j.techfore.2013.08.033. http://www.sciencedirect.com/science/article/pii/S0040162513002187
25. Leem, C.S., Kim, B.G.: Taxonomy of ubiquitous computing service for city development. Pers. Ubiquit. Comput. **17**(7), 1475–1483 (2013). http://dx.doi.org/10.1007/s00779-012-0583-5
26. Martin, M., Nurmi, P.: A generic large scale simulator for ubiquitous computing. In: Third Annual International Conference on Mobile and Ubiquitous Systems: Networking & Services, 2006 (MobiQuitous 2006). IEEE Computer Society, San Jose (2006). https://doi.org/10.1109/MOBIQ.2006.340388
27. Martins, M.V.L.: Frametrail: um framework para o desenvolvimento de aplicações orientadas a trilhas. Dissertação (mestrado em ciência da computação), Universidade do Vale do Rio dos Sinos, Programa de Pós-Graduação em Computação Aplicada (PIPCA), São Leopoldo (2011)
28. Mirri, S., Prandi, C., Salomoni, P., Callegati, F., Campi, A.: On combining crowdsourcing, sensing and open data for an accessible smart city. In: Next Generation Mobile Apps, Services and Technologies (NGMAST), 2014 Eighth International Conference on, pp. 294–299. IEEE, Oxford (2014). https://doi.org/10.1109/NGMAST.2014.59
29. Molyneaux, I.: The art of application performance testing. from strategy to tools. SciTech Book News (2014)
30. OSM. Available at: http://www.openstreetmap.org/. Accessed 10 Nov 2017 (2015). Open Street Map
31. Perboli, G., Marco, A.D., Perfetti, F., Marone, M.: A new taxonomy of smart city projects. Transp. Res. Procedia **3**, 470–478 (2014). https://doi.org/10.1016/j.trpro.2014.10.028. http://www.sciencedirect.com/science/article/pii/S2352146514001914. 17th Meeting of the EURO Working Group on Transportation, EWGT2014, 2–4 July 2014, Sevilla, Spain
32. Petrolo, R., Loscrí, V., Mitton, N.: Towards a smart city based on cloud of things. In: Proceedings of the 2014 ACM International Workshop on Wireless and Mobile Technologies for Smart Cities, WiMobCity '14, pp. 61–66. ACM, New York (2014). https://doi.org/10.1145/2633661.2633667
33. Piro, G., Cianci, I., Grieco, L., Boggia, G., Camarda, P.: Information centric services in smart cities. J. Syst. Softw. **88**, 169–188 (2014). https://doi.org/10.1016/j.jss.2013.10.029. http://www.sciencedirect.com/science/article/pii/S0164121213002586
34. Shin, D.H.: Ubiquitous city: urban technologies, urban infrastructure and urban informatics. J. Inf. Sci. **35**(5), 515–526 (2009). http://dx.doi.org/10.1177/0165551509100832
35. Shin, D.H.: Ubiquitous computing acceptance model: end user concern about security, privacy and risk. Int. J. Mobile Commun. **8**(2), 169–186 (2010). http://dx.doi.org/10.1504/IJMC.2010.031446
36. Sozialhelden, E.V.: Wheelmap open and free online map (2015). Available at: http://wheelmap.org/en/map. Accessed 22 Jan 2015. Wheelchair accessible places
37. Sukode, S., Gite, S., Agrawal, H.: Context aware framework in IoT: a survey. Int. J. Adv. Trends Comput. Sci. Eng. **4**(1), 1–9 (2015). http://warse.org/pdfs/2015/ijatcse01412015.pdf
38. Tang, S., Lee, B.S., He, B.: Towards economic fairness for big data processing in pay-as-you-go cloud computing. In: Cloud Computing Technology and Science (CloudCom), 2014 IEEE 6th International Conference on, pp. 638–643. Singapore (2014). http://dx.doi.org/10.1109/CloudCom.2014.120
39. Tavares, J., Barbosa, J., Cardoso, I., Costa, C., Yamin, A., Real, R.: Hefestos: an intelligent system applied to ubiquitous accessibility. Universal Access in the Information Society pp. 1–19 (2015). https://doi.org/10.1007/s10209-015-0423-2
40. Vanderheiden, G.C.: Ubiquitous accessibility: building access features directly into the network to allow anyone, anywhere access to ubiquitous computing environments. In: C. Stephanidis (ed.) Universal Access in Human-Computer Interaction. Intelligent and Ubiquitous Interaction Environments. Lecture Notes in Computer Science, vol. 5615, pp. 432–437. Springer, Berlin (2009). https://doi.org/10.1007/978-3-642-02710-9_47

41. Wagner, A., Barbosa, J.L., Barbosa, D.N.: A model for profile management applied to ubiquitous learning environments. Expert Syst. Appl. **41**(4, Part 2), 2023–2034 (2014). https://doi.org/10.1016/j.eswa.2013.08.098. http://www.sciencedirect.com/science/article/pii/S0957417413007203
42. Washburn, D., Sindhu, U., Balaouras, S., Dines, R., Hayes, N., Nelson, L.: Helping CIOs understand smart city initiatives: defining the smart city, its drivers, and the role of the CIO. Forrester Research (2010)
43. Wiedemann, T., Barbosa, J.L.V., Barbosa, D.N.F., Rigo, S.: Recsim: a model for learning objects recommendation using similarity of sessions. J. Univ. Comput. Sci. **22**(8), 1175–1200 (2016). http://www.jucs.org/jucs_22_8/recsim_a_model_for
44. World Report on Disability. Available at: http://www.who.int. Accessed 12 Jan 2017 (2011). World Health Organization
45. Yovanof, G., Hazapis, G.: An architectural framework and enabling wireless technologies for digital cities & intelligent urban environments. Wirel. Pers. Commun. **49**(3), 445–463 (2009). http://dx.doi.org/10.1007/s11277-009-9693-4

Chapter 10
Realistic Vehicular Networks Simulations

Tiago do Vale Saraiva and Carlos Alberto Vieira Campos

Abstract Intelligent transportation systems are one of the components to make smart cities through which they have sought to improve levels of safety, comfort, and efficiency of transportation systems. Vehicular networks support the exchange of messages by vehicles with the information necessary for proper functioning of these systems. Because of complexity of these communication networks, a technique widely used to evaluate their performance is simulation. However, a simulation problem involves choosing appropriate parameters to achieve realistic results. This work deals with the problem of realistic simulation in vehicular networks, through the simulation of a message dissemination application, where various simulation and application parameters are varied. The main contribution consists in analysis of results obtained according to the chosen parameters and the finding that these parameters must be adjusted properly to obtain results consistent with reality.

10.1 Fundamentals

10.1.1 Vehicular Networks

Vehicular networks, also known as vehicular ad hoc networks (VANETs), are networks formed by moving vehicles equipped with wireless communication devices. These networks are a special case of mobile ad hoc networks (MANETs), with the difference that mobility of nodes (vehicles) in VANETs is restricted by roads and traffic characteristics (traffic jam, speed limit, signaling, etc.) in each region.

Given the movement of vehicles, the topology in a vehicular network is generally quite dynamic, with frequent disconnections and mobility limited by transit routes. The communication can be of type V2V (between vehicles), V2I (between vehicles

T. do Vale Saraiva (✉) · C. A. Vieira Campos
Universidade Federal do Estado do Rio de Janeiro - UNIRIO, Rio de Janeiro, RJ, Brazil
e-mail: tiago.saraiva@uniriotec.br; beto@uniriotec.br

V. N. Coelho et al. (eds.), *Smart and Digital Cities*, Urban Computing,
https://doi.org/10.1007/978-3-030-12255-3_10

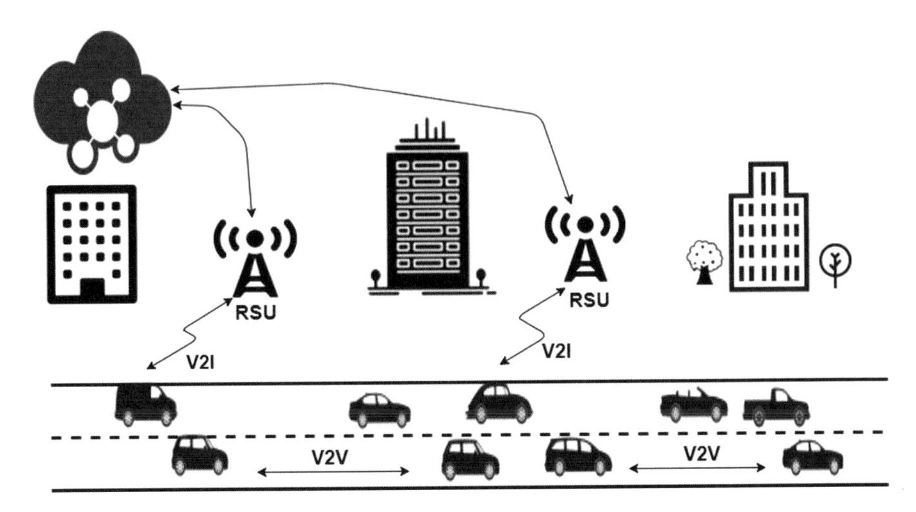

Fig. 10.1 A VANET and its components in V2I and V2V approaches

and infrastructure), or hybrid (combination of V2V and V2I). Other general characteristics are low bandwidth and wireless transmission over short distances [8].

Each vehicle in a VANET has an onboard unit (OBU) and when the vehicles are communicating with each other, the transmission/reception happens between these OBUs. In the infrastructured case, the communication takes place between the OBUs in each vehicle and the road side units (RSUs) of city.

Figure 10.1 shows a VANET in a smart city. The reference architecture of VANETs is called WAVE (wireless access in vehicular environment) and is defined by the IEEE 1609 standards [26]. IEEE 1609.4 provides for multi-channel operations using IEEE 802.11p, which is part of the dedicated short range communications (DSRC), in the 5.9 GHz frequency band, where the spectrum is divided into seven channels of 10 MHz, being three of control, and data rate between 3 and 27 Mbps [21]. An inter-vehicle communication architecture for Europe is described in ETSI EN 302 665 [25].

In urban scenarios, the topology of a VANET may have hundreds of vehicles in a relatively small region. In this case, it is necessary to design protocols to control access to the medium that deals especially with collisions, given the high probability of occurrence. On the other hand, in expressway scenarios, the topology is sparser and the connectivity is more intermittent. This second scenario then suggests the need for disconnection tolerant protocols. In addition, vehicles traveling in both scenarios need to adapt their behavior to variations in network density to provide good data transfer [8].

10.1.2 Smart Cities and Intelligent Transportation System

In the last decade, there has been a growing demand for more attractive and efficient cities, in an attempt to reduce the negative effects of urbanization. The high concentration of people raised different challenges for the government, such as uncontrolled growth, traffic congestion, crime, waste resource management, and others [2]. Furthermore, owing to globalization, cities started to compete with each other to attract the best professionals, by providing them attractive environments where they could live [2]. These challenges have led governments to adopt technologically based approaches and handle the negative effects of urbanization via broadband interconnected cities, known as smart cities [2].

A concept closely linked to intelligent cities is that of ubiquitous or pervasive computing. According to [29], it is a scenario in which day-to-day objects have communication, storage and processing skills, providing services to users anytime, anywhere. Future smart cities will generate and process a lot of data. As vehicles are continuously improving in terms of processing power and networking capacities, they are one of the most promising building blocks for these smart cities [10].

Intelligent transportation system (ITS) represents a new concept that emerges in the context of intelligent cities, with the objective of providing better levels of safety, comfort, and efficiency to all those involved in a transportation system.

VANETs can be used to boost the use of ITS [23], and to provide communication and automation services infrastructure to make cities smarter [10]. According to [8], advancements in mobile communications and protocols for VANETs will allow the emergence of architectural solutions for vehicular networks, in both road and urban environments, to support applications with different requirements. This will allow the provision of new efficiency, traffic road and monitoring services, and new comfort/entertainment options.

As an example that VANETs can serve to support ITS-related applications and also provide communication infrastructure for smart cities, [4] evaluates the taxi mobility in the city of Rome (Italy), in order to verify the efficiency of these vehicles as "mules" that can provide communication for the city's automation.

So far, the major practical focus on ITS applications has been in the development of protocols to support road traffic safety through the transmission of messages with information such as speed, position, and direction of vehicles [25]. The official name of the protocol that sends these position messages in Europe is CAM (cooperative awareness messages). In the USA the equivalent are the BSM (basic safety messages). In [24] the applications of BSM messages dissemination are classified in single hop and multiple hops. The difference being that in the second case we have the routing of messages by intermediate nodes. The implementation of a multi-hop approach is more complex because it requires a routing algorithm for messages. In contrast, the single-hop strategy is less efficient because of smaller range.

There are also DENM (decentralized environmental notification messages) messages, which are event triggered messages that are transmitted in case of specific events (e.g., accidents). While the event is valid, the DENMs will be transmitted alongside CAMs [25]. Different applications in VANETs have varying requirements, forward collision warning (FCWS) applications, for example, require fast message dispersal for nearby vehicles, while other messages, such as bottleneck alerts, support greater latency [5].

10.1.3 Vehicular Networks Simulation

For proper adoption of VANETs, studies of the various components of these networks are required. One of the major challenges for research in this area is the logistical complexity to achieve realistic results. Experiments with real vehicles are expensive and difficult to implement. Even the access to communication radios in the IEEE 802.11p standard is not easy for all research groups. In such complex scenarios, a widely used solution is simulation. So, researchers are increasingly using simulation strategies.

There are several tools that can be used to simulate the behavior of communication networks. However, the majority of the existing simulation software (i.e., simulators) is designed for traditional wireless networks and technologies, not for VANETs. In summary, there are two main points to consider when choosing a simulation strategy in VANETs: the network simulator and the mobility simulator. Despite the advantages in using the simulators, these software need to be properly adjusted, so that the simulated scenarios are realistic.

The main aspects that need to be modeled and simulated are: motion restrictions (start and end, topologies, traffic lanes, signaling), traffic dynamics (vehicle density, direction, speed, change of lane, pedestrians, stops), scenarios (accidents, bottleneck, signal violation, emergency situations), and communication channels (channel strength, duration, reception level, packet loss, obstructions) [23].

Two widely network simulators used in VANETs research are NS-3 [17] and OMNET++ [18], since they allow the use of real mobility traces, implement the IEEE 802.11p standard, and impart realism, including propagation characteristics of electromagnetic signal in communication through models like two ray ground and Nakagami. Another widely used network simulator is NS-2, which has a long history in academia. However, compared to NS-2, NS-3 models the network more accurately and has more functionality, achieving even better levels of efficiency in the use of computational resources [22].

As for mobility, the traces available in the literature can be classified as synthetic or realistic [7]. Synthetic traces are constructed by tools that consider the characteristics of cities, such as population, type of area (e.g., residential, industrial), among other aspects.

It is worth noting that synthetic traces may have a greater capacity for representation than real traces (collected from vehicle positions, usually via GPS), since, depending on how the actual trace is generated, gaps can lead to erroneous results [7]. Some of the best-known synthetic mobility traces are the ones from Cologne

(Germany) and Zurich (Switzerland). Due to their high granularity, these traces in terms of space and time provide great realism [7]. In the experiments described in this chapter, we have decided to utilize both Zurich and Cologne traces.

Synthetic models can still be classified in macroscopic and microscopic. The macroscopic models deal with the traffic density, the flows, and the initial distribution of vehicles. Microscopic models deal with the movement of each vehicle, its location, speed, acceleration, and other attributes of its context, such as the changes of track and the vehicles around. These are considered to be more realistic models than the macroscopic ones [22]. In [24] some mobility models are presented, such as the random waypoint (RWP) model that has been used in the past because of its simplicity, but is not realistic. Other models cited are: constant speed and uniform speed, Manhattan model, Krauss mobility model, and CA-based mobility model.

Another aspect that can be considered in a performance evaluation concerns vehicle applications. Saini et al. [23] classifies the applications in VANETs between those focused on safety and non-safety. In this second category are the applications of efficiency, comfort, and entertainment. Each application has its various parameters that can impact the performance of the network and must be validated for study and deployment in an appropriate way.

10.1.4 Contributions of This Chapter

The objective of this chapter is to deal with the problem of realistic representation of VANETs in a simulated environment. In general the simulators have a fundamental role, however, if the applied parameters are not properly selected, the results obtained may lead to conclusions that are not in agreement with reality or even contradictory. We have used realistic simulation parameters to verify the performance metric behavior as a function of changes in these elements. We also considered two traces of mobility with different characteristics.

The present work differs from previous ones by addressing two fundamental aspects for the adoption and study of VANETs, simultaneously: simulation and vehicular applications. Related works treat these subjects separately, without evaluating the impacts of both on results due to variation in the choice of parameters adopted in simulation. In this study, we present the reader with an integrated view of relationship between themes. Thus, main contribution of this chapter is the analysis of results generated in realistic simulations, as a function of parameters modified in the VANET simulations supporting a message dissemination application.

The remainder of the chapter is organized as follows. Section 10.2 describes the related work . In Sect. 10.3, the evaluated scenarios are listed, with the implementation details. In Sect. 10.4 are the results obtained in the simulations with the corresponding analysis. Finally, conclusion and possibilities for future work are presented in Sect. 10.5.

10.2 Related Work

There are several academic papers dealing with the applications and the simulation question in VANETs. These two issues are being widely discussed in the academic community and will soon allow widespread adoption of VANETs, initially in the USA, Europe, and Japan, where bands of the electromagnetic spectrum have already been reserved. In [25], authors suggest that commercial adoption in Europe can start in 2018.

In [1] the authors evaluated the impact on the average end-to-end delay and the packet delivery rate of the network, according to the Friis, TRG, log-normal, and Nakagami propagation models, through the OMNET++ network simulator. The mobility was generated from the SUMO urban mobility simulator [12], with a map of the Curitiba city (Brazil) with 1800 m $\times$ 2800 m and 5 different demands: 10, 25, 50, 75, and 100 vehicles with a maximum speed of around 40 km/h. This way, they evaluated only the question of propagation models and did not use realistic mobility traces. In [11] real data of Shanghai city is used to evaluate the proposal of delay reduction for data dissemination in VANETs. The trace used is admittedly valid; however, it represents only the taxi mobility. In [4], authors used a similar strategy, with the generation of real traces, where mobility data of 320 taxis was collected during 6 months in the city of Rome (Italy).

In [3], the authors proposed a realistic channel model based on empirical measurements for use in simulation. Tests were performed on the OMNET++ with the traffic generated through the mobility simulator SUMO and the results analyzed. The mobility used was not realistic, since they used a unique scenario, composed of two roads of 4 km with a total of 1043 vehicles. In [27] the authors evaluated the impact of MANETs simulations and proposed a more realistic physical layer model. For evaluation, the simulation results were compared with field measurements in the center of the German city of Stuttgart (1.5 km $\times$ 1.5 km). This work focused only on MANETs, in scenarios between 10 and 300 pedestrians walking randomly and used the IEEE 802.11b protocol. In [31] an analytical model was proposed to evaluate the MAC and application layers in multi-channel communications following the IEEE 1609.4 standard in VANETs, and the results validated via simulation in NS-2. This work did not consider the mobility of nodes.

In [32] authors performed an analysis of vehicular mobility in express routes with free flow. For that, they used mathematical modeling and simulation through MATLAB software. In [21] is demonstrated the importance of the packet inter-reception (PIR) time metric in addition to packet delivery rate in evaluation of VANETs. For this, real experiments were carried out with vehicles equipped with IEEE 802.11p radios generating GPS location data on journeys through Italy.

In this article, we evaluate the impact on communication of VANETs in function of both inherent simulation parameters (propagation models and mobility trace) and application parameters such as the interval between message transmissions and data packets size.

10.3 Evaluation Methodology

In this section, we will present and describe in detail all the components used in our simulations, such as propagation models, mobility traces, and the application of message dissemination alerts.

Although both OMNET++ and NS-3 are modern simulators compatible with VANET simulations, the simulations performed in this work were based on [6], which presents a script for the NS-3 simulator (available at [16]), which allows the evaluation of various aspects of VANETs in realistic scenarios. We have made changes in the script to fit the scope of this work. The routing part and the computations related to the network throughput were removed, besides rewriting some functions. The metric chosen to verify the behavior of VANET in the proposed scenarios was the PDR, which is the percentage of delivered messages in relation to what was transmitted in the network.

In this work, packet delivery rate (PDR) ranges are values that determine a distance for the calculation of the PDR. For example, if the PDR in a given scenario was 0.8 for a range of 200 m, this means that 80% of messages received by the vehicles were received in a distance of up to 200 m from the transmitters.

In Eq. (10.1) the PDR formula is shown, where QP_r is the amount of packets received in a given area and QP_t is the number of packets transmitted in that area.

$$\mathrm{PDR} = \frac{QP_r}{QP_t} \tag{10.1}$$

For the presentation of the results in the next section, some graphs were generated with several lines of different colors, each one referring to a PDR range value, being the lowest PDR 50 m (dark green) and the largest 500 m (golden). For these graphs, the accumulated PDR was calculated with each second of simulation. According to Eq. (10.2), the PDR at any moment t_1 during the simulation period will be the sum of all packets received at a given distance, divided by the sum of all transmitted packets.

$$\mathrm{PDR}(t_1) = \frac{\sum_{t=0}^{t_1} QP_r}{\sum_{t=0}^{t_1} QP_t} \tag{10.2}$$

We consolidated the simulation data for each of the 10 PDR ranges into a relative frequency plot of the PDRs with the PDR values accumulated every second (300 values) with the configuration of each specific scenario. According to [23], VANETs were standardized by establishing a coverage area for 30 m communication in the North America, 15–20 m in Europe and 1 km in Japan. Thus, in order to compare the results, we choose intermediate PDR distance values.

The script in the NS-3 implements in each evaluated scenario an application running on all vehicles sending BSM messages. According to theory, these messages are periodically in broadcast (for all vehicles within its range) on the intervals corresponding to the control channels in an IEEE 1609 application. As already

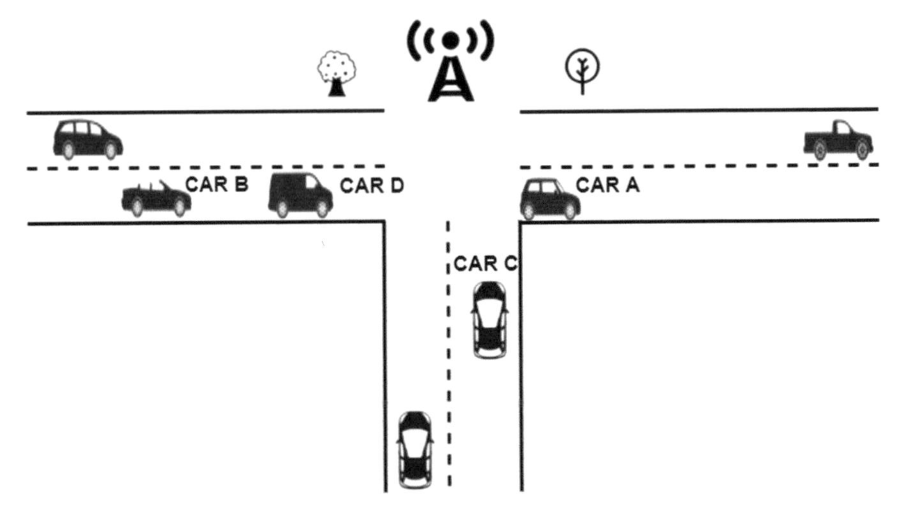

Fig. 10.2 Crossing of urban roads in which one can make use of a VANET, with the exchange of BSM messages, to avoid accidents

mentioned, these messages generally contain kinematic data (e.g., position, speed, and direction) in order to support security-related services [31] and other more complex applications besides those designed to detect and prevent accidents.

We have in Fig. 10.2 a representation of vehicles in a VANET. These vehicles could send periodic BSM messages in broadcast through its IEEE 802.11p radio equipment connected in their OBUs. Based on these messages, vehicles identified as "Car A" and "Car D" could avoid an accident. Even if they collided, messages could reach vehicles *B* and *C*, causing they don't contribute to congestion or even get involved in the accident. The application brings this result in function of the data collected in the vehicle itself and in the external context, through the BSM messages.

One of the two mobility traces used in the results presented in this chapter was obtained through a multi-agent microscopic traffic simulator, able to simulate traffic on map regions of Switzerland with great realism and based on people and their movements in a 24-h period. We use a part that has the traffic of 99 vehicles, in the region of Unterstrass (Zurich), in a time frame of 300 s, in an approximate area of 4650 m $\times$ 3000 m. The complete trace was presented in [15] and covers an area of approximately 50 km $\times$ 260 km and contains 260,000 vehicles. The trace has already been generated in a NS-3 compatible format.

The second trace was derived from the city of Cologne (Germany), covering a region of 400 km^2 in a period of 24 h in a typical work day, involving more than 700,000 individual vehicle trips. It was generated from OpenStreetMap [19] data in conjunction with the SUMO [19] mobility simulator and survey data. The complete and detailed trace is in [28]. For this chapter, we have used a slice of this trace with 300 s and 1547 vehicles. In Fig. 10.3 there is an illustration of regions corresponding to two traces used.

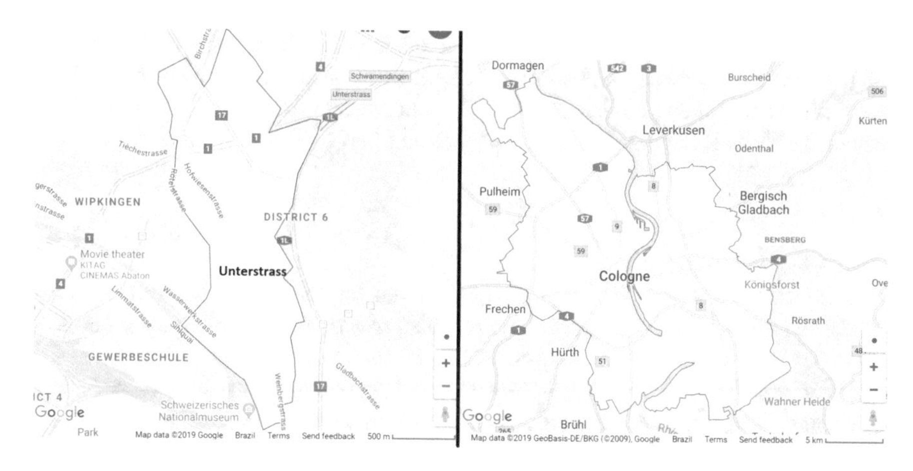

Fig. 10.3 Regions from where the mobility traces used have been derived

As for signal propagation models, Stepanov and Rothermel [27] demonstrates the importance of the adoption of a modeling that can adequately represent the effects of the electromagnetic propagation waves when simulating mobile ad hoc networks.

Mean signal strength at the receiver as a function of distance from transmitter can be estimated by large-scale propagation models (e.g., two ray ground and Friis), while rapid fluctuations of the signal at the wave-length scale are better represented by small-scale fading models (e.g., Nakagami) [22]. In the present work we will evaluate three models used in outdoor communications: Friis, known as free space propagation model [9], and two ray ground (TRG) model [20], which considers part of the negative effects of electromagnetic signals reflections, both combined with Nakagami probabilistic model [14], in order to give even more realism within the simulations. Yaqub et al. [30] have used a similar approach, with the combination of TRG and Nakagami.

We know the limitations of Nakagami model. Since it is based on probabilistic distributions that do not use information about the specific scenario, it does not consider the difference between line of signal (LOS) and non-line of signal (NLOS) [13]. Even so, we use it because it provides a more comprehensive characterization of signal propagation in urban environments [22], and has been shown to fit the amplitude envelope of empirical data for 802.11p channel well [31].

We have in Eq. (10.3) the calculation of the Friis model, where P_r is the power level reaching the receiver (in Watts), P_t is the transmitted power level, and G_r and G_t are respectively the gains in the receiving and transmitting antennas, while λ is the wavelength, L is the loss in the system, and d is the distance between transmitter and receiver.

$$P_r = \frac{P_t G_t G_r \lambda^2}{(4\pi)^2 d^2 L} \tag{10.3}$$

We see that this model considers only the distance between source and destination, the frequency (inversely proportional to the wavelength) and the transmission and reception gains.

Equation (10.4) is the formula for the TRG model, which considers the height of the antennas between transmitter and receiver, respectively, h_t and h_r. This additional component of antenna height is necessary for the calculation of the negative effects on signal strength as a function of ground reflection function.

$$P_r = P_t G_t G_r \frac{h_t^2 h_r^2}{d^4 L} \tag{10.4}$$

TRG model presents better prediction results for long distances and the Friis model is more suitable for short distances. The two models have the same level of losses up to a distance $dc = \frac{4\pi h t h r}{\lambda}$ [22]. With the operation of the vehicles at the frequency of 5.9 GHz and the height of 1.5 m of the antennas, we have $dc \approx 556$ m. Thus, until 556 m the perceived effects are the same, depending on the model used.

In addition to these deterministic models, there are probabilistic models that can be used in the simulation to add more realism and precision, provided they are properly adjusted, preceded preferably by empirical measures. Among the probabilistic models we can mention the log-normal shadowing (LNS) and the Nakagami (NAK).

With the Nakagami model, we have shown more realism in the obtained data, through the fading effects of the multiple paths traveled by the electromagnetic waves. We leave the default NS-3 Nakagami model parameters, where the parameter of form m for distance $d < 80$ m was in m $= 1.5$ and for greater distance $m = 0.75$. As quoted in [31], the form parameter m of the Nakagami model is approximately 3 for values where the distance d between transmitter and receiver is less than 50 m, 1.5 for distances between 50 and 150 m, and 1 for distances greater than 150 m.

Since different applications may require different packet sizes to transmit the information, we vary the size of the packets transmitted by the BSM application in order to see how this change impacts the PDR. Another parameter of the application that was varied in the simulations was the transmission interval between the successive packages generated and sent in broadcast by the BSM application in vehicles. There are several reasons for which the packet transmission interval can be varied in an application. One of these reasons is saving energy, since more time transmitting causes higher energy consumption.

Two sets of configurations were used and are listed in Tables 10.1 and 10.2. One set (in Table 10.1) was used to evaluate the impact on the evaluation metric (PDR) according to messages transmission interval and packet size. The second set (in Table 10.2) was used to evaluate the impact in function of the mobility traces, in this case we use different packet sizes and transmissions interval and 10 rounds simulations for each graph. In this way we can obtain a greater number of situations in our experiments. These parameters can be subdivided into BSM application parameters and inherent simulation parameters. Changes in both will generate different results in the metrics (PDR in our case), but are independent of each other. The application parameters used between the simulations were the packet size and the transmission interval, while the inherent parameters of simulation used were the propagation model and the mobility traces.

Table 10.1 Configurations used to evaluate packet size and transmission interval

Parameters	Values
Evaluated distances	50, 100, 150, 200, 250, 300, 350, 400, 450, and 500 m
Number of vehicles	99 (Unterstrass trace), 1547 (Cologne trace)
Mobility	2 realistic traces (Unterstrass and Cologne)
Propagation models	Two ray ground and Friis (+ Nakagami)
Frequency	5.9 GHz (OFDM modulation—10 MHz channel)
Data rate	6 Mbps
Packets size	100, 200, and 400
Interval transmission	10 ms, 1, 10 s, and 10 s
Number of rounds	1 and 10 (time interval)
Maximum transmission delay	10 ms
GPS accuracy	40 ns

Table 10.2 Configurations used to evaluate mobility traces

Parameters	Values
Evaluated distances	50, 100, 150, 200, 250, 300, 350, 400, 450, and 500 m
Number of vehicles	99 (Unterstrass trace), 1547 (Cologne trace)
Mobility	2 realistic traces (Unterstrass and Cologne)
Propagation models	Two ray ground and Friis (+ Nakagami)
Frequency	5.9 GHz (OFDM modulation—10 MHz channel)
Data rate	6 Mbps
Packets size	(150, 300 and 600 bytes)
Interval transmission	10 ms, 1 s and 10 s
Number of rounds	10
Maximum transmission delay	10 ms
GPS accuracy	40 ns

10.4 Results Obtained

The first results are presented in Fig. 10.4 and concerns for the Unterstrass region trace with respect to the impact of the propagation models. Although the theoretical separation distance between the TRG and Friis models is approximately 556 m considering the range we are operating (5.9 GHz) and the height of the antennas of the vehicles (1.5 m), we can see the models behave as if they were beyond this limit, with the losses resulting from the TRG model being much more pronounced, considering the effects of signal reflection. It is worth noting that in both cases we have the fading effect added by the Nakagami model (NGK).

In relation to packets size, it can be seen from Fig. 10.5 that the larger the number of bytes being transmitted, the smaller the PDRs. This result is consistent with the theory, since larger packets require more time transmitting, which can lead to a greater interference probability.

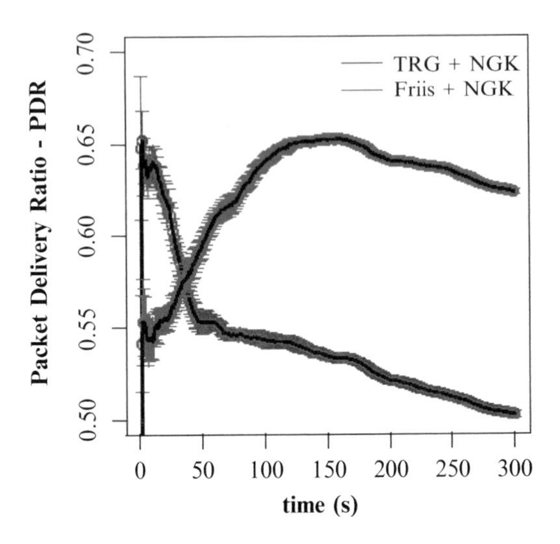

Fig. 10.4 TRG and Friis propagation models with Unterstrass trace

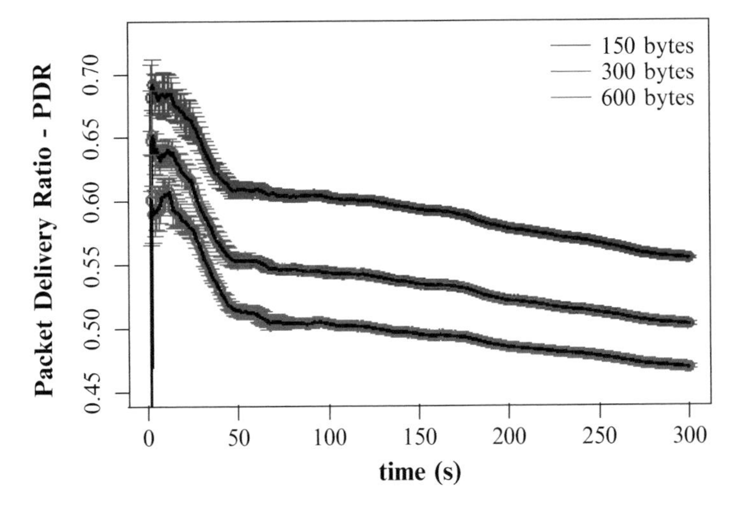

Fig. 10.5 Different sizes of transmitted packets with Unterstrass trace

We also have combined the results in function of the data packets size (values of 100, 200, and 400 bytes) with the TRG and Friis propagation models in simulations with fixed 1 s transmission interval between packets. The results are shown in Figs. 10.6 and 10.7. Although each simulation round has 300 s of duration, the graphs of Fig. 10.6 go only up to 150 s, considering that this time is adequate to show the results obtained for these size of packages, since from that moment on we do not have significant changes in PDRs. We can observe that for both propagation models, the bigger size of packages the smaller are values of PDR. This is consistent with theory, since larger packets increase the likelihood of collisions between signals

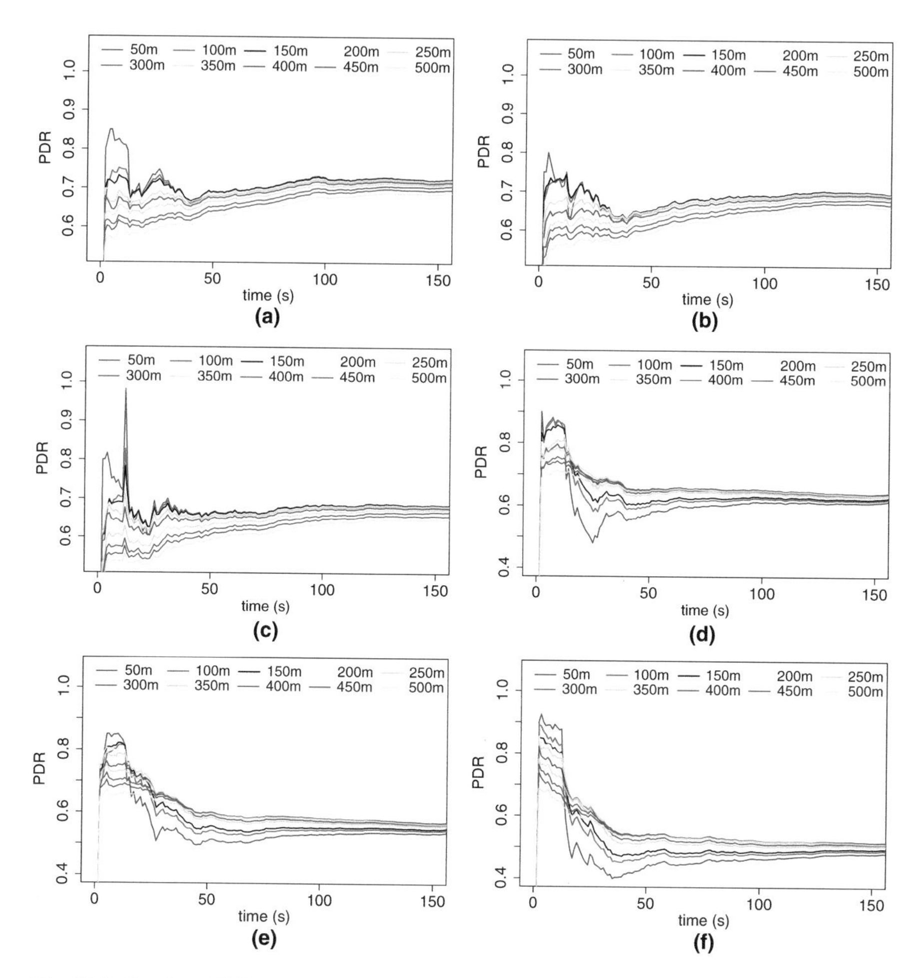

Fig. 10.6 Results of PDR measured in rages of 50 m to 500 m to each vehicle, varying the packets size between Friis and TRG models with Unterstrass mobility trace. (**a**) 100 bytes—Friis. (**b**) 200 bytes—Friis. (**c**) 400 bytes—Friis. (**d**) 100 bytes—TRG. (**e**) 200 bytes—TRG. (**f**) 400 bytes—TRG

transmitted by vehicles and, consequently, result in smaller PDRs. The effects of PDR fall as a function of packages size is much more evident when using TRG model also here.

From Fig. 10.7, which shows the relative frequencies, it is possible to reach the same conclusions. It is observed that the calculated accumulated PDR values for packet sizes between 100 and 400 bytes are in their significant majority between 60% and 80% for the Friis propagation model and between 40% and 70% of PDR for the TRG propagation model. These results are also consistent with the theory, since the lower values for the TRG model are due to the fact that this model considers the negative effects of signal reflections.

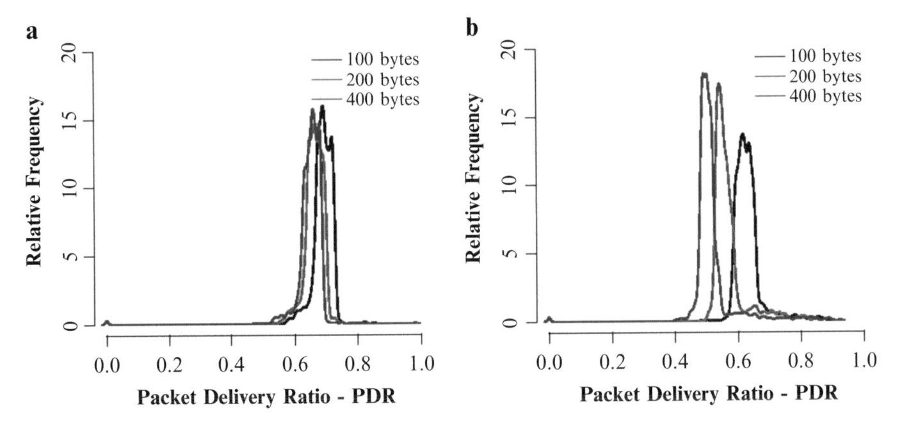

Fig. 10.7 Relative frequencies based on packet size and propagation models with Unterstrass trace. (**a**) Friis. (**b**) TRG

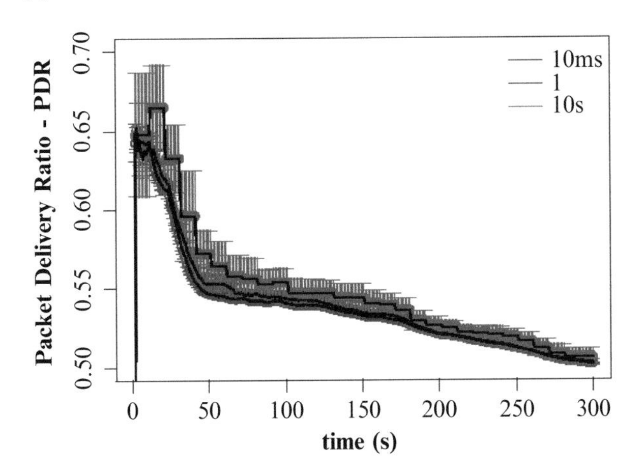

Fig. 10.8 Different transmission intervals with Unterstrass trace

Thus, as an example of how we might arrive at a contradictory conclusion for this type of scenario, considering the other fixed parameters, if it is established that the PDR cannot be less than 50% for the proper operation of some application transmitting packets of 400 bytes, the results would be positive, considering the Friis model, or negative, considering the TRG.

The last evaluated parameter of the application was the transmission interval between packets. Figure 10.8 shows the results for PDRs over time. PDRs are slightly larger for the 10 s value, which is consistent with the theory, since larger transmission intervals reduce the probability of interference. It is worth noting that as PDR is calculated every second, smaller values in the transmission interval between packages have little influence.

We also have combined the results in function of the transmission interval between the packets (values of 100 ms, 1 s, and 10 s) with the TRG and Friis

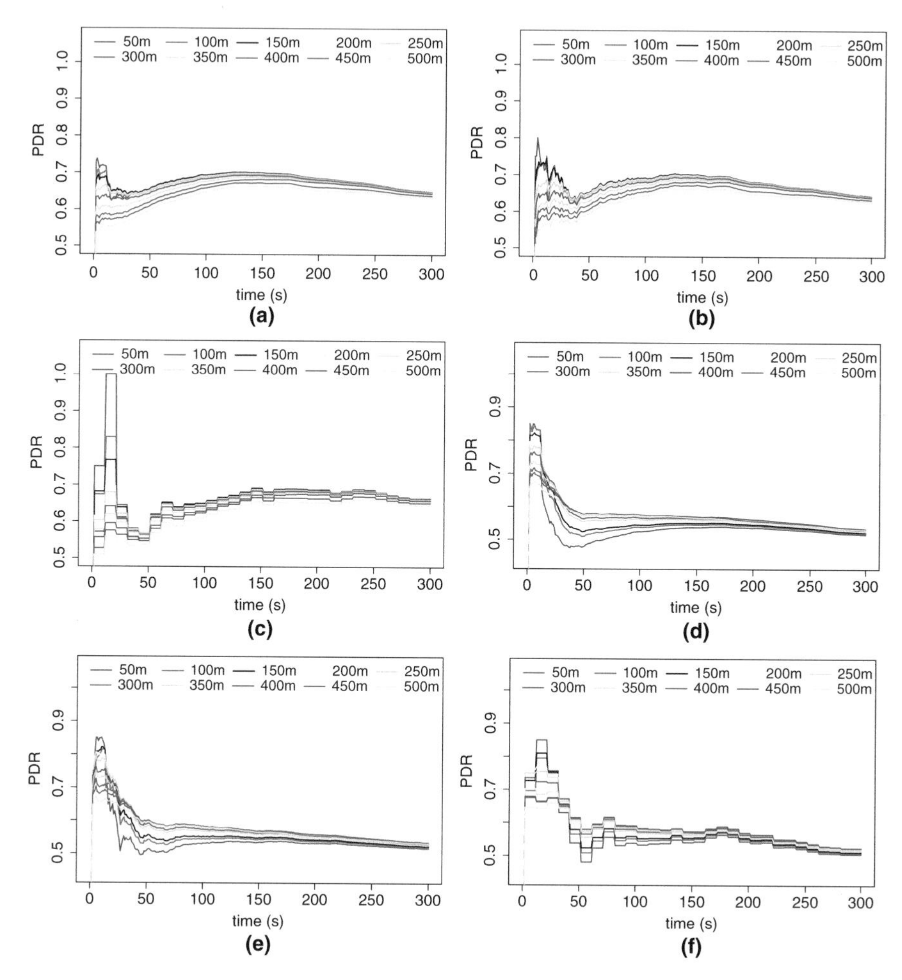

Fig. 10.9 Results of PDR measured in rages of 50 m to 500 m of each vehicle, varying the transmission interval between 100 ms, 1 s, and 10 s for the Friis and TRG models with Unterstrass mobility trace. (**a**) 100 ms—Friis. (**b**) 1 s—Friis. (**c**) 10 s—Friis. (**d**) 100 ms—TRG. (**e**) 1 s—TRG. (**f**) 10 s—TRG

propagation models in simulations with fixed 200 bytes packet transmission size. The results are shown in Figs. 10.9 and 10.10. It is noted that the changes did not have a significant impact on the PDR. The graph of Fig. 10.10 with the relative frequencies reinforces this finding. Lower values for the transmission interval increase the time that the vehicles are with their radios occupying the electromagnetic spectrum, transmitting messages, and, consequently, the increase in the probability of collision occurs, resulting in smaller PDRs. However, since the cumulative PDR is calculated every second, and the speed of the vehicles is compatible with urban environments, changes in the transmission interval of less than 1 s do not cause significant change in the metric. It is worth noting that

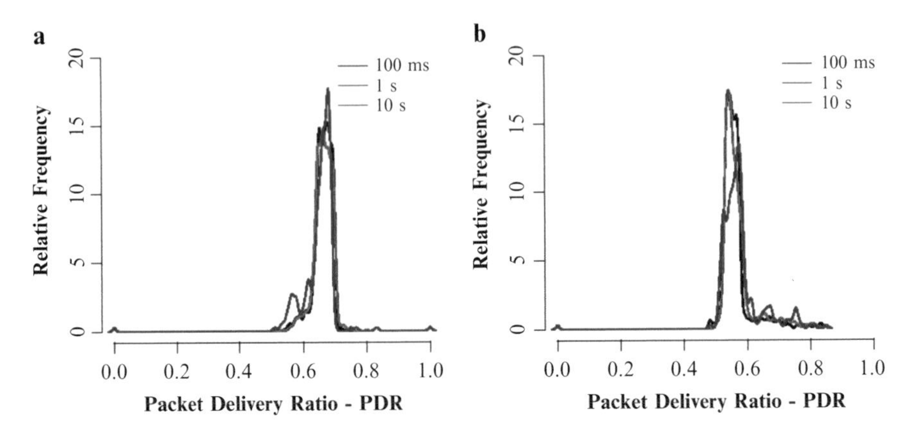

Fig. 10.10 Relative frequencies in function of packets transmission interval and propagation loss models only with Unterstrass trace. (**a**) Friis. (**b**) TRG

$\forall t_1 = (t_0 + dt) \rightarrow \mathrm{PDR}(t_1) \neq \mathrm{PDR}(t_2)$ with $0 < t_2 < 300$ and $t_2 \neq t_1$, where t_0 is the beginning of BSM message transmission between vehicles, t_1 a moment in the future during the simulation period dt seconds later, PDR (t_1) a PDR at time t_1 and t_2 another time in the simulation other than t_1. This is because the probability of collision at any point in the simulation tends to be always different since vehicles are moving. So, slight difference in the graphic format.

It is important to emphasize that due to the random components inherent to the simulation, such as delay for radio transmission or the accuracy in the synchronism of each transmitter, although the simulation happens with the vehicles transmitting with the same frequency and all begin at the same moment, the transmissions are not exactly simultaneous. A maximum broadcast delay has been set for 10 ms radios and GPS accuracy of 40 ns. These are important parameters that must be taken into account for the correct data interpretation.

We can also observe from the graph of Fig. 10.10 that for a 10 s transmission interval there is a slightly higher probability of the PDR being above 60%. In relation to the other intervals for the TRG model, the absolute majority of the PDRs is between 50% and 60%. For the Friis model, the values were between 60% and 70%.

Another result that contributes to the major differential of present work is the comparative analysis of two realistic traces of vehicular mobility. We have in Fig. 10.11 the results for 10 different PDRs in a round of simulation with vehicles transmitting 300 bytes packets every 1 s, with a modeling propagation signal loss equivalent to Nakagami + TRG. Although each scenario has a simulation time of 300 s, Fig. 10.11 shows the results up to the time of 150 s, since that interval was chosen because it is sufficient to adequately represent the effects on the PDRs. It is clear that although the simulations are running the same application, the results are quite different. By analyzing only the number of vehicles, it could be assumed that the Cologne scenario had smaller PDRs, since the level of interference would be

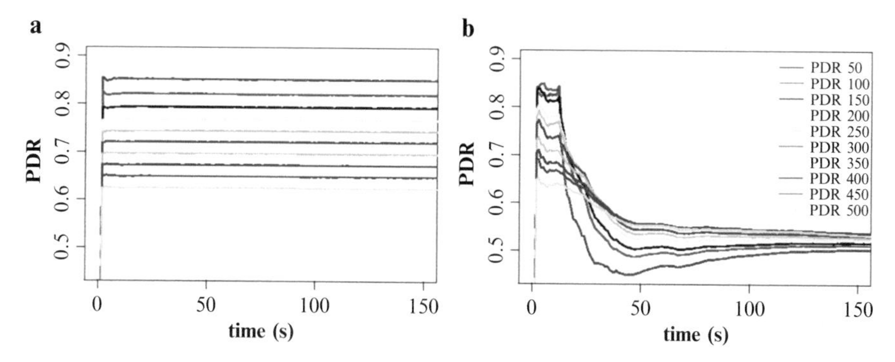

Fig. 10.11 Different results of PDR measured in ranges of 50 m to 500 m of each vehicle, using Cologne (**a**) and Unterstrass (**b**) mobility traces

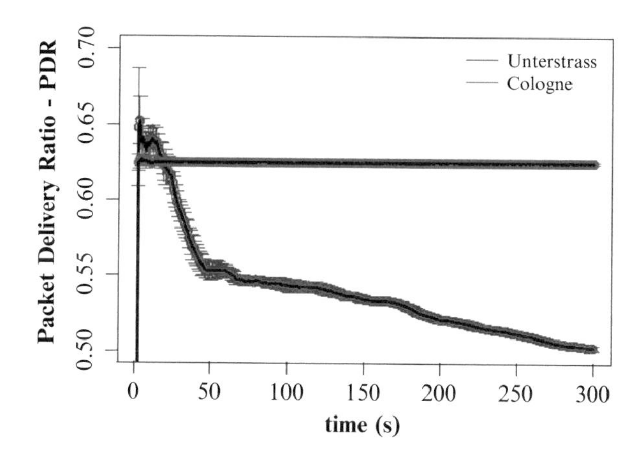

Fig. 10.12 Analysis of PDR equal to 500 m for Unterstrass and Cologne traces

higher and, as mentioned in [22], the signal-to-noise ratio is given by $\text{SINR} = \frac{S}{I+N}$, where N is the noise level, S is the signal level, and I is the accumulated power of the interfering signals.

However, when we analyze the vehicle density (dv) in each trace, which equals the number of vehicles per km^2, we have in the Cologne trace $dv = \frac{1547\,\text{vehicles}}{400\,\text{km}^2} \rightarrow dv = 3.87\ \text{vehicles/km}^2$. In the Unterstrass trace the density equals to $dv = \frac{99\,\text{vehicles}}{4.65\times 3} \rightarrow dv = 7.1\ \text{vehicles/km}^2$. Thus, the result is consistent with the theory, since, despite having the smallest number of vehicles, the Unterstrass trace has almost twice the density, resulting in a higher probability of interference and, consequently, lower PDRs.

In Fig. 10.12 we present the 500 m PDR calculated over the simulation time for the two traces, along with the error bars. The discrepancy remains evident, since we have a PDR practically continuous for the Cologne trace and, for Unterstrass, a PDR that starts close to 70% and decreases up to 50% at the end of simulation.

10.5 Conclusion and Future Work

In this work we verified the impact on results of vehicular network communications simulations in realistic scenarios, obtained as a function of inherent application parameters (transmission interval and message size) and inherent simulation parameters (propagation model and mobility traces). With this, it was evidenced that such parameters must be properly adjusted so that results obtained can be consistent with reality.

It was observed that the chosen signal propagation model affects the probability of packet delivery. Although the distance evaluated was at the operating threshold between the TRG and Friis models, we obtained results consistent with the related theory, being possible to observe different results depending on each model. In addition, the size of messages sent by application influences the PDR inversely proportional, due to the increase in interference probability.

It was mentioned that although vehicles start the message transmission simultaneously, the transmissions do not occur at exactly same time for all, as depending on the random components of simulation (e.g., synchronism and delay). Due to vehicle speed and the way of calculating the accumulated PDR (every second), changes of up to 1 s in transmission interval between the messages had little influence on the metric. The small influence observed was coherent with related theory, proving the level of experiment realistic representation.

Finally, we verified the need to evaluate several traces of vehicular mobility. BSM applications, for example, are related to traffic safety, so they should be verified in all possible scenarios. We found that, in relation to the two mobility traces used, the same application (with the same parameters) generated very different results for each mobility trace.

Thus, it is concluded that through the results of presented simulations, the parameters inherent to the simulation or application can alter the metrics defined for vehicular network analysis, making it clear that in order to obtain realistic results the researchers cannot ignore these parameters or evaluate them separately or even leaving them in some inaccurate default setting, different from what will occur in the real world.

In some results obtained, we can see that after the first 50 s of simulation, PDR values tend to approach/stabilize. As a future work, an analytical modeling based on these results can be a future research object, as well as the evaluation of multi-hop BSM applications using routing algorithms for VANETs and considering additional metrics such as latency, throughput, and inter-packet reception time.

References

1. Angeles, W., Borin, V.P., Munaretto, A., Fonseca, M.: The impact of propagation models in the performance of ad hoc routing protocols for urban VANET. In: 84th Vehicular Technology Conference (VTC-Fall), pp. 1–5. IEEE, Piscataway (2016)

2. Avelar, E., Marques, L., dos Passos, D., Macedo, R., Dias, K., Nogueira, M.: Interoperability issues on heterogeneous wireless communication for smart cities. Comput. Commun. **58**, 4–15 (2015)
3. Bastani, S., Ozalla, D.T., Karaca, M.: On the performance of vehicular communications with a measurement-based radio propagation model. In: 21st International Workshop on Computer Aided Modelling and Design of Communication Links and Networks (CAMAD), pp. 6–11. IEEE, Piscataway (2016)
4. Bonola, M., Bracciale, L., Loreti, P., Amici, R., Rabuffi, A., Bianchi, G.: Opportunistic communication in smart city: experimental insight with small-scale taxi fleets as data carriers. Ad Hoc Netw. **43**, 43–55 (2016)
5. Carpenter, S.E.: Obstacle shadowing influences in VANET safety. In: 22nd International Conference on Network Protocols, pp. 480–482. IEEE, Piscataway (2014)
6. Carpenter, S.E., Sichitiu, M.L., Underwood, D.A., Patwardhan, M., Starr, S.: Evaluating VANET performance using ns-3. In: WNS3 Workshop on NS-3, pp. 3–4 (2014)
7. Celes, C., Silva, F.A., Boukerche, A., Andrade, R.M. de C., Loureiro, A.A.F.: Improving VANET simulation with calibrated vehicular mobility traces. IEEE Trans. Mob. Comput. **16**, 3376–3389 (2017)
8. Cunha, F., Villas, L., Boukerche, A., Maia, G., Viana, A., Mini, R.A.F., Loureiro, A.A.F.: Data communication in VANETs: protocols, applications and challenges. Ad Hoc Netw. **44**, 90–103 (2016)
9. Friis, H.T.: A note on a simple transmission formula. In: Proceedings of the Institute of Radio Engineers, vol. 34, no. 5, pp. 254–256. IEEE, Piscataway (1946)
10. Hagenauer, F., Sommer, C., Onishi, R., Wilhelm, M., Dressler, F., Altintas, O.: Interconnecting smart cities by vehicles: how feasible is it? In: IEEE Conference on Computer Communications Workshops (INFOCOM WKSHPS), pp. 788–793. IEEE, Piscataway (2016)
11. He, J., Cai, L., Cheng, P., Pan, J.: Delay minimization for data dissemination in large-scale VANETs with buses and taxis. IEEE Trans. Mob. Comput. **15**, 1939–1950 (2016)
12. Krajzewicz, D., Erdmann, J., Behrisch, M., Bieker, L.: Recent development and applications of SUMO - simulation of urban mobility. Int. J. Adv. Syst. Meas. **5**, 128–138 (2012)
13. Martinez, F.J., Fogue, M., Toh, C.K., Cano, J., Calafate, C.T., Manzoni, P.: Computer simulations of VANETs using realistic city topologies. Wirel. Pers. Commun. **69**, 639–663 (2013)
14. Nakagami, M.: The m-distribution, a general formula of intensity of rapid fading. In: Statistical Methods in Radio Wave Propagation, pp. 3–36. Pergamon Press, New York (1960)
15. Naumov, V., Baumann, R., Gross, T.: An evaluation of inter-vehicle ad hoc networks based on realistic vehicular traces. In: Proceedings of the Seventh ACM International Symposium on Mobile Ad Hoc Networking and Computing - MobiHoc 06, p. 108. ACM Press, New York (2006)
16. NS-3 Consortium. vanet-routing-compare.cc. https://www.nsnam.org/doxygen/vanet-routing-compare_8cc_source.html. Accessed 20 May 2018
17. NS-3 Consortium. NS-3 Discrete Event Network Simulator. https://www.nsnam.org. Accessed 20 May 2018
18. OpenSim Ltd. OMNET++ Discrete Event Simulator. https://www.omnetpp.org. Accessed 20 May 2018
19. OSM Foundation. OpenStreetMap. https://www.openstreetmap.org. Accessed 20 May 2018
20. Rappaport, T.S.: Wireless Communications: Principles and Practice, vol. 2. Prentice Hall, Upper Saddle River (1996)
21. Renda, M.E., Resta, G., Santi, P., Martelli, F., Franchini, A.: IEEE 802.11p VANets: experimental evaluation of packet inter-reception time. Comput. Commun. **75**, 26–38 (2016)
22. Ros, F.J., Martinez, J.A., Ruiz, P.M.: A survey on modeling and simulation of vehicular networks: communications, mobility, and tools. Comput. Commun. **43**, 1–15 (2014)
23. Saini, M., Alelaiwi, A., Saddik, A. El: How close are we to realizing a pragmatic VANET solution? A meta-survey. ACM Comput. Surv. **48**, 1–40 (2015)

24. Sanguesa, J.A., Fogue, M., Garrido, P., Martinez, F.J., Cano, J.-C., Calafate, C.T.: A survey and comparative study of broadcast warning message dissemination schemes for VANETs. Mob. Inf. Syst. **2016**, 1–18 (2016)
25. Sjoberg, K., Andres, P., Buburuzan, T., Brakemeier, A.: Cooperative intelligent transport systems in Europe: current deployment status and outlook. IEEE Veh. Technol. Mag. **12**, 89–97 (2017)
26. Standards Development Working Group. 1609 - Dedicated Short Range Communication Working Group. IEEE, Piscataway. https://standards.ieee.org/develop/wg/1609.html (2017). Accessed 20 May 2018
27. Stepanov, I., Rothermel, K.: On the impact of a more realistic physical layer on MANET simulations results. Ad Hoc Netw. **6**, 61–78 (2008)
28. Uppoor, S., Trullols-Cruces, O., Fiore, M., Barcelo-Ordinas, J.M.: Generation and analysis of a large-scale urban vehicular mobility dataset. IEEE Trans. Mob. Comput. **13**, 1061–1075 (2014)
29. Vahdat-Nejad, H., Ramazani, A., Mohammadi, T., Mansoor, W.: A survey on context-aware vehicular network applications. Veh. Commun. **3**, 43–57 (2016)
30. Yaqub, M.A., Ahmed, S.H., Bouk, S.H., Kim, D.: FBR: fleet based video retrieval in 3G and 4G enabled vehicular ad hoc networks. In: IEEE International Conference on Communications (ICC), pp. 1–6. IEEE, Piscataway (2016)
31. Yin, X., Ma, X., Trivedi, K.S., Vinel, A.: Performance and reliability evaluation of BSM broadcasting in DSRC with multi-channel schemes. IEEE Trans. Comput. **63**, 3101–3113 (2014)
32. Zarei, M., Rahmani, A.M.: Analysis of vehicular mobility in a dynamic free-flow highway. Veh. Commun. **7**, 51–57 (2017)

Chapter 11
LoRaWan: Low Cost Solution for Smart Cities

Fernando Vinícios Manchini de Souza and Roberto Dos Santos Rabello

Abstract Communication and easy access to information are the main factors in reducing waste of time and resources. Communication systems are heavily dependent on telecommunications companies, which provide high speed services, but charge a high monthly fee per device, thus making some applications unaffordable. A LoRaWan is an emerging network technology that stands out for having characteristics such as long range capability, low power consumption and low cost, and focuses on sensing, thus proving to be a good alternative as a communication solution in applications for smart cities. This article highlights projects that use LoRaWan in simple applications with the potential to improve the quality of life of citizens by implementing the concept of a smart city in a distributed manner. In this study, tests are carried out on libraries, and frameworks are developed by open software communities on how they play a key role in the popularization of technology around the world.

11.1 Introduction

In various scenarios, communication and facilitated access to information are the key to reducing waste of time and resources. For instance, receiving prior notification that traffic will be blocked and suggestions for detour, making medical appointments via Internet, providing essential services such as location of banks throughout the city, transportation optimization through sharing, production and distribution of food in urban farming. However, the communication system is heavily dependent on telecommunications companies, which provide high speed services, but charge high monthly fees per device, thus making some applications unaffordable. The telecommunications market is governed by complex regulations. As a result, few companies offer these services, and competition is discouraged.

F. V. M. de Souza (✉) · R. D. S. Rabello
Universidade de Passo Fundo, São José, Passo Fundo, Brazil
e-mail: 168730@upf.br; rabello@upf.br

V. N. Coelho et al. (eds.), *Smart and Digital Cities*, Urban Computing,
https://doi.org/10.1007/978-3-030-12255-3_11

For instance, in Brazil, prices increase considerably due to high taxes. As a result, investments in small towns and rural areas are insufficient, and many places have no network coverage. These obstacles make it difficult to implement the Internet of Things (IoT) on a large scale in rural areas or even in the urban area of small cities as the cost of each device increases because it requires reasonable processing power to support network protocols, such as TCP/IP, in addition to the maintenance fee to keep each device connected.

Solutions for smart cities are very dependent on information and communication technology (ICT) companies for the collection, communication, and analysis of data. Platforms such as IBM have a strong appeal to their products and services. However, it is impossible that they have a generic model for smart cities [25], even with high technological standards and interoperability of devices and systems, since each city has its own problems and peculiarities. Therefore, strategies for national smart cities fail for not being able to capitalize local resources and for ignoring their regional needs and priorities [3].

As a local solution, this study presents the use of LoRaWan technology which is easy and inexpensive, and provides fast results, even though most of the projects are still prototypes or proofs of concept, but promise future solutions as large-scale customizable projects.

11.2 Smart Cities

The concept "smart city" has become popular in the political field in recent years. The main focus is on ICT infrastructure, but there is also much research on social well-being, education, environmental sustainability, and mobility as drivers of urban growth [5]. Yet, there is not a definitive concept of a "smart city" as research groups and companies approach the subject based on their own interests. The major technology companies, Cisco in 2015, IBM in 2009, and Siemens in 2004, were the first ones to define and adopt the term "smart city," linking it to complex systems that integrate urban infrastructure and services of building, transportation, water distribution, energy, and public safety [10].

The definition of IBM starts with the optimization of interconnected information to better understand, control, and optimize the use of limited resources. A smart city balances its social, commercial, and environmental needs optimizing available resources. IBM's mission is to provide smart cities with solutions to facilitate the development and sustainability of a city for the benefit of its population. The smartness of cities is measured by the improvements in quality of life and economic well-being that are achieved by the application of ICT to plan, design, and build the city's infrastructure [13].

With the beginning of the fourth industrial revolution, new paradigms have been created, which were strongly supported by technology and integration between the physical, virtual, and biological world promoted by IoT and artificial intelligence (AI). A paradigm example is the local distributed production that meets specific demands with customized products, such as 3D printing [26]. The concept of IoT arose in the late 1990s by ASHTHON in his lecture on RFID to P&G [4]. However, since this first explanation, the concept has been evolved. Presently, it can be defined as global infrastructure for the interconnection of physical and virtual objects through communications systems for information gathering and performance in the environment [12, 31]. In order to read the real world, a large number of devices and sensors are used, which are distributed over long distances and often at difficult places to reach. Yet it is possible that they are in a position that power supply is not available and a long battery life is required.

The popularization of IoT solutions has been increasing due to several factors such as evolution of miniaturization, reduction of sensor costs, connection speed, unique ID to each device, power to process and store information in the cloud at low cost, and development of cognitive computing and machine learning [7]. The convergence of these aspects together with environmental concerns and social welfare, when balanced, meets the needs and desires of the population with a sustainable functional capacity using technology to transform infrastructure and optimize resources, thus creating an environment based on the concept of a smart city [9].

11.3 LoRaWan

By definition, LoRa is the radio frequency (RF) technology whereas LoRaWan is the network protocol, which is a low power wide area network (LPWAN), consists of a network architecture focused on low power consumption, low data rate and long range, key points for IoT, and promotes bidirectional safe communication, mobility, and location service. It uses star topology with a gateway as a bridge between end-devices and the network server, which in turn performs the routing of the packets from each device to the application server [15].

Another relevant point is that LoRaWan is open, which allows any company to produce devices. However, they must be certified by LoRa Alliance [15] as LoRa transmitter owns Semtech's patents. The specification also predicts the join server to manage the authentication of the devices in the network. The network reference model can be observed in Fig. 11.1.

End-devices can be sensors and actuators that are connected wirelessly to the LoRaWan network through the radio gateway. The transmission uses the concept of chirp spread spectrum (CSS) as a modulation technique that provides strength and robustness against interference and the Doppler effect, thus being this the great differential of LoRa technology [1]. The other communication parameters are:

- Bandwidth (BW): it can vary from 125 to 500 kHz in the American continent [1].
- *Spreading Factor* (SF): it can vary from 7 to 12, and defines the time and speed that the information will be conveyed. The smaller the SF, the lower the transmission time and the greater the speed, thus covering smaller range. The

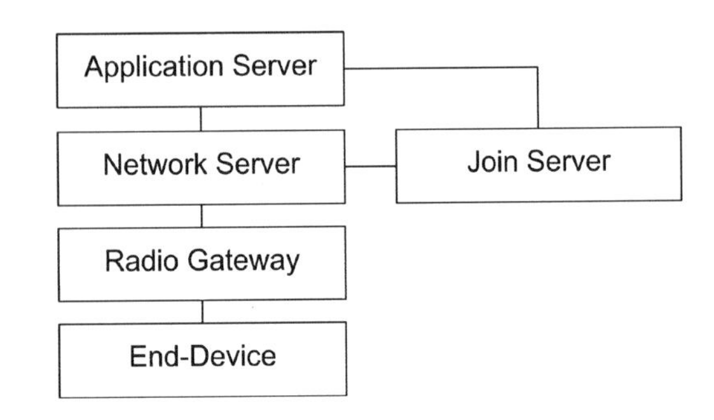

Fig. 11.1 Network reference model according to Lora Alliance [17]

higher the SF, the longer the transmission time, and the slower the speed, but it will have longer range capability [1].
- TXPower: it indicates the signal strength. In the American standard it varies from 10 to 30 dBm [16].

The combination of bandwidth and spreading factors provides transmission speeds between 0.3 and 27 kb/s [1], and adaptive data rate parameterization (ADR) which allows SF and the end-device of TXPower to set to optimize power consumption. This, in practice, provides the same coverage range of a gateway to have different data rates, as the devices located near the gateway can use less SF, thus saving energy.

LoRa devices are categorized into classes for appropriate use in each project, and they are always focused on the lowest energy consumption:

- Class A: bidirectional devices, but with the possibility to receive restricted messages (downlink) immediately after sending (uplink) them. This kind of operation reduces energy consumption as the rest of the time LoRa module will be off.
- Class B: bidirectional devices, but able to receive messages with synchronized tasks from time to time. When sending a message, a task can be created for synchronization, so the server can know when the device is listening.
- Class C: bidirectional device, but with the possibility of receiving messages at almost any moment.

Table 11.1 summarizes the difference of LoRaWan device classes.

Table 11.1 Comparative table between LoRa device class

Class	A	B	C
Multicast messages	No	Yes	Yes
Latency	High	Low	No
Device initiates communication	Yes	Yes	Yes
Server initiates communication	No	Yes, with intervals	Yes, constantly
Energy consumption	Low	Moderate	High

The use of radio frequency bands is limited by local governments that define rules and organization, and avoid interference problems between equipment of critical use. Therefore, there is an international agreement which defines some frequencies without the need for regulation to encourage industrial, scientific, and medical development. This agreement is called ISM (industrial, scientific, and medical) band. In Brazil, the telecommunications regulator body is the National Telecommunications Agency (ANATEL) and its responsibilities are to define the frequency bands that need to be registered for operation. However, according to Anatel Resolution 506, Chapter II, it is defined that the following restricted frequency bands do not require registration with the Federal Government: 902–907.5 MHz, 915–928 MHz, 2400–2483.5 MHz, 5725–5875 MHz, and 24.00–24.25 GHz [2].

The LoRa Alliance instruction for the American continent is to use the frequency range of 902–928 MHz. As can be observed, the bands between 907.5 and 915 MHz cannot be used. Therefore, when using American standard devices, one should be careful to disable these channels via software in the application. However, this limitation already exists in other countries which have standardized the distribution of frequency bands in a way that channels such as the Australian standard would not be disabled. However, currently there is no defined standard to Brazil [16].

Packets that reach gateway are forwarded with no processing or interpretation to the network server, which is the center of the star topology and has the following functions:

- End-device address verification;
- Authentication of frames;
- Manage adaptive data rate, ADR;
- Respond to all end-device requests;
- Forward all uplink messages to application servers;
- Schedule and forward all downlink messages from application servers to the end-device;
- Forward the end-device authentication requests to the join server.

Security is a critical point in IoT infrastructure. LoRaWan protocol has a security mechanism based on the AES encryption standard and has two authentication methods for the end-device to connect to the network and to send information to the application server. In the activation by personalization (ABP) method, the end-device receives the identifiers in manufacturing or encoding process, so it is not possible to transfer this end-device to another network. The parameters defined in this method are:

- *Device Address* (DevAddr): Unique identifier in 32 bit device. Present in every data frame.
- *Network Session Key* (NwkSKey): Unique identifier on the network layer with 128 bit with AES encryption, ensuring message integrity between the end node and the network server.
- *Application Session Key* (AppSKey): Unique identifier at the application layer with 128 bit AES encryption. Used for encryption of messages in the application.

In the over-the-air (OTA) method, the end-device has a global unique DevEUI identifier, through which it sends an unencrypted join request. In case it is accepted on the network, DevAddr is generated and sent to the device that will use it to exchange the next encrypted messages [15]. The join server manages authentication only in this method.

In the application server, messages of the devices are treated (payloads) to the end user. These messages are decrypted through the use of AppSKey, taking into account that it is only in this layer that the information generated by the devices is actually accessed.

11.4 LoRaWan in Smart Cities

This session analyzes studies that address LoRaWan in applications of smart cities. Some studies are still in the initial stages. However, they demonstrate how simple solutions can contribute to the improvement of the structure of public services to the population.

11.4.1 Evaluation of LoRa LPWAN Technology for Indoor Remote Health and Well-being Monitoring

The study of Petäjäjärvi et al. [21] reports experiments that use commercial LoRaWan devices at the University of Oulu in Finland such as different SF, bandwidth, and TXPower. In this study, a transmission power of 14 dBm and SF = 12 was used, being able to cover the entire campus area of 570 × 320 m, with a mean of 96.7% streaming success.

There are reports of tests both outside and inside of the buildings between 55 and 370 m distance using spreading factors from 7 to 10. A table and a graph show the percentage of packets with error and RSSI (received signal strength) assessed in the form of a heat map. At the end, a brief account of energy consumption with this variation of spreading factors was conducted. It was concluded that consumption is greater due to the time the transmitter remains active in larger SF.

As it is one of the first practical studies on the use of LoRa technology in indoor environments, Petäjäjärvi et al. conclude that it is possible to have indoor transmission within 300 m even though it presents losses and delays; it is feasible and has potential to use technology to monitor health and to control non-critical signals, such as monitoring of physical activities, location, pets and personal management, among others.

11.4.2 IoT-Based Health Monitoring via LoRaWan

Mdhaffar et al. of the University of Sousse, Tunisia [19] present a new low cost collector for medical data using LoRaWan for transmission of blood pressure, glucose, and temperature data. The so-called IOT4HC is based on higher cost devices that use other transmission technologies. However, this prototype simplifies the device allowing people with specific health problems who live in rural areas or have no internet access, to not need to go and collect preventive data.

This project involves common users and medical doctors from the sensor to the web access. However, the study covered in this article is limited to the device and is divided into two subgroups. The first one involved the coverage of the area of the LoRaWan gateway, which in the urban area obtained results of up to 1.89 km in one direction and 0.73 km in a region of high urban density whereas in rural areas it reached 33 km^2 using a 3 dBi gain antenna.

The second group focused on energy consumption of the device using LoRaWan, when compared to general packet radio service (GPRS) technology that, even in idle mode, the device equipped with GPRS consumes approximately 20 mA. However, with LoRa, it consumes approximately 4.7 mA and, using an estimate of consumption of a power bank of 2200 mA, it would have autonomy of up to 10,875 days.

This equipment proves to be of interest for remote areas as it has a range of 33 km^2 in rural areas and, in experiments, it showed 10 times less energy consumption than GPRS transmission. However, the discontinuous sending of data does not allow critical patients to use this solution. Therefore, an alternative solution should be implemented for such cases.

11.4.3 Smart Cities: A Case Study in Waste Monitoring and Management

The study carried out by Castro Lundin et al. [6] reports the construction of a waste monitoring system at the Technical University of Denmark. This study addresses management and optimization of waste control as one of the main characteristics of smart cities. However, commercial solutions make large-scale applications unaffordable due to their very high cost.

The construction of the whole system is described in a summarized way: the devices have ultrasonic sensors to determine the waste levels, the equipment is used for the assembly of the gateway, and technologies and protocols are applied for the backend and frontend implementation.

While the data collected during the test phase was being analyzed, it was verified that some dumps were filled before the day of the expected collection as they were located near a snack bar, while others remained at a low level. This scenario was already predicted by garbage collectors as a result of their experience and knowledge of the site, so no predictions were applied to the results obtained. However, the author argues that in case of unavailability of current collectors, substitutes would not have such knowledge, and the environment would be at the expense of garbage accumulation, and any other unexpected event would also cause problems. Therefore, the implementation of a smart monitoring system is crucial as it allows the application of a proactive and non-reactive approach.

In conclusion, in spite of being simple and inexpensive, the smart system proves to improve services to citizens. In spite of replacing the knowledge of the collectors, it improves feedback in some atypical situations. The cost per trash can was also respected during the study. It did not exceed 100 dollars, and was one of the main factors for the implementation of this system in large scale.

11.4.4 Smart LED Street Light Systems: A Bruneian Case Study

A Suhaili's article [27] reports the LED lighting project in Bruneian Malaysia. This article explains its current situation where about 2% of the population has little access to LED lamps. Consumption was reduced by 40% as a result of a simple exchange from high-pressure sodium (HPS) lamps to LEDs. Besides, with the implantation of this smart system, optimization occurred even more either by reducing power or switching off the lights according to real-time sensor data.

This project proposes the use of light dependent resistor (LDR), which relies on the intensity of natural illumination in conjunction with the passive infrared (PIR) sensor to detect the presence of people or vehicles by increasing or reducing its brightness intensity. When there is no movement, it turns off. As a result, energy was saved between 67% and 71%.

LoRaWan is defined in architecture as a means of communication due to its simplicity, energy consumption, and star topology, which is contrary to other standards that use mesh topology, and consume more resources as they require devices to perform the mediation in the communication.

This article demonstrates the benefits of a smart system applied in lightning of public areas, adding greater smartness, reducing human intervention, saving energy, and maximizing comfort and security of the areas covered by the system.

11.5 Preliminary Results

LoRaWan is still considered an emerging standard in LPWan, with an increasing popularity in several regions of the world, as can be observed in the studies cited above, mainly by the open source community that develops and shares libraries and frameworks to accelerate development. In spite of that, companies start to produce

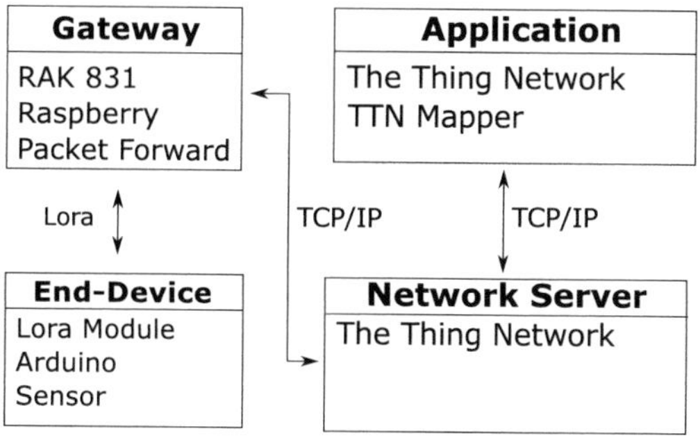

Gateway
RAK 831
Raspberry
Packet Forward

Application
The Thing Network
TTN Mapper

Lora

TCP/IP

TCP/IP

End-Device
Lora Module
Arduino
Sensor

Network Server
The Thing Network

Fig. 11.2 Overview of the prototype developed according to the simplified LoRaWan architecture

Fig. 11.3 The device constructed using the LoRaWan HOPERF RFM95W module with a helical antenna, board Arduino Pro Mini Atmega328 and a DHT 22 temperature sensor

hardware to target this public, thus boosting the ecosystem. This chapter aims to describe the end-device and gateway built with the lowest possible hardware cost, using libraries and applications created by the community members as facilitators to capture and measure data as well as outdoor coverage of the LoRa signal. Figure 11.2 shows an overview of the prototype of the hardware and software used in each layer of LoRaWan architecture. Because it is using the ABP authentication method, it is not necessary for the join server to transport the data to the application layer.

In the construction of an end-device, the initial attractiveness is the cost of LoRaWan modules as it is possible to purchase simple modules for less than 7 dollars. However, in this case, in addition to sensors, it will be necessary for the microcontroller and antenna to have a functional device, as can be observed in Fig. 11.3. There are already several solutions encapsulated with the LoRa transmitter and microcontroller in a single board, and there are also alternatives like shields in the Arduino standard, which simplify the development of prototypes [8].

In the development of the end-device, the biggest challenge is to use libraries appropriately for the LoRaWan module and the frequency of the region. In this case, the main libraries are developed in Europe and operate in other frequency bands. As a result, one should be careful to make adjustments to the correct ranges for the American standard. The LMiC library was initially developed by IBM Zurich for LoRa Semtech SX1272/SX1276 and HopeRF RFM92/95 modules. However, it has now been modified by the community members in fork repositories of the original, making it possible in a relatively simple way to integrate them into the Arduino and other frameworks [11].

For the implementation of a gateway, the initial challenge was to find suppliers of LoRaWan equipment at retail affordable prices, as the market until early 2017 had only professional or industrial gateways which cost more than 300 dollars [18]. However, with the dissemination of this technology, new solutions have emerged, such as the RAK 831 board, based on the Semtech SX1301, as well as other more expensive concentrators on the market, being able to operate in the frequency of 433, 868, and 915 MHz, with reception sensitivity of −138 dBm, support to 8 channels of reception and 1 to send information [22]. Thus, solutions aimed at prototyping that has limited resources are half the cost, as it is possible to mount a gateway for indoor installation with approximately 165 dollars. The gateway is still composed of a Raspberry board and runs the Raspbian operating system [23]. The physical interconnection between the RAK modules was performed using the GPIO ports of Raspberry and following the official documentation [24] and community members [14]. The packet forwarder software installed on the gateway is responsible for forwarding packets received from end-devices to the network server [28].

The things network (TTN) is a project that created an open LoRaWan network and in 2015, in just 6 weeks, covered the city of Amsterdam with 10 antennas. With the expansion to other areas, mainly in Europe, this community is very active. It distributes public gateways around the world with the help of its members. Thus, any developer inside the covered area can simply create devices and an application, without the concern of having infrastructure [29]. Public gateways built by the authors of this study are available on TTN platform in the city of Ijuí, in the state of Rio Grande do Sul, Brazil.

In general, TTN network operates as a network server in LoRaWan architecture, so the packet information does not become public, not even for the gateway owner. All the information is routed to the servers of TTN network. Only the user who owns the device can access the payload at the application layer through various integration methods that the platform provides [29].

TTN mapper application is a community project derived from the things network. It consists of a system that maps and publicizes coverage of gateways, thus encouraging other developers to use them. One of the mapping methods is through Android application available on the website where the user links a device registered in TTN network and moves around the city. The received packets will have the analyzed information as the received signal strength indicator (RSSI), and a heat map will be compiled with the gateway coverage on the platform [30], as can be seen in Fig. 11.4.

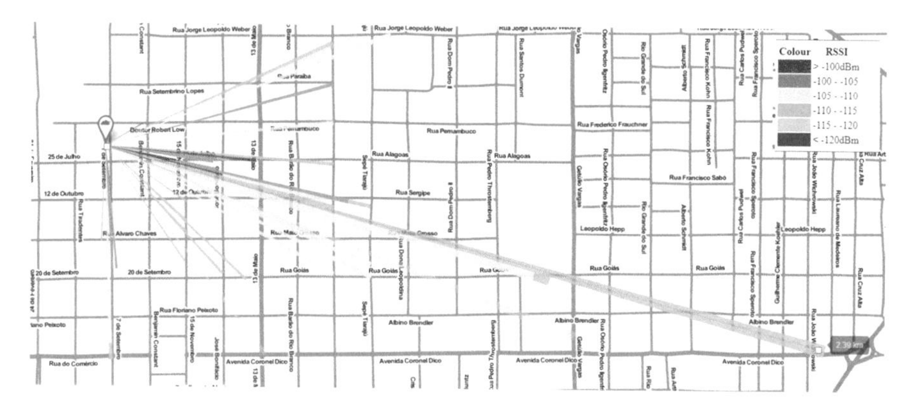

Fig. 11.4 A view of TTN mapper application of the gateway described above. A 2.39 km coverage can be observed in the gateways built by the authors in the city of Ijuí, in the state of Rio Grande do Sul, Brazil

Figure 11.4 shows the coverage of the prototype obtained by TTN mapper. The "red-tone" tasks indicate greater power of the signal, yellow ones represent medium, and blue ones represent low. In the initial tests, the largest distance obtained was 2.39 km due to the geography of the city of Ijuí [20] which has moderate ripples. However, more distant places, but higher, have better coverage than nearer places which are lower due as during the movement the line of sight to the gateway was occasionally obtained.

11.6 Remarks

This article discusses how LoRaWan technology meets the expectations in obtaining communication solutions for low cost and long distance smart cities, since there is no need to use a data plan for an end-device. It uses ISM frequencies and can be implemented almost anywhere. LoRa modules are affordable to the market, and the 2.39 km urban signal coverage obtained in the experiments demonstrates that technology has promising results. The open software community platforms assist in the dissemination of technology as they allow users to cooperate directly with new solutions, or to customize applications implemented in other regions using the local reality.

The initial tests and other related studies demonstrate that even with low cost equipment, they provide a very considerable outdoor coverage in urban environment and much more than expected in rural environment [19]. However, for indoor monitoring, effectiveness is limited [21]. If LoRaWan is chosen as communication technology, the criticality of the data when the focus is on low consumption should be taken into account. It may not be very effective when maintaining intermittent connections due to the packet loss. Therefore, for real-time applications, LoRaWan is not recommended [19, 21].

Following this study, local reality problems that can be solved or mitigated using simple sensors in conjunction with LoRaWan, such as an application for counting number of vehicles in traffic, finding parking spaces for people with special needs, and monitoring public lighting, in order to validate the limitations of a larger number of simultaneous devices and the impediments of a network with heterogeneous devices should be addressed.

References

1. Adelantado, F., Vilajosana, X., Tuset-Peiro, P., Martinez, B., Melia-Segui, J., Watteyne, T.: Understanding the limits of LoRaWAN. IEEE Commun. Mag. **55**(9), 34–40 (2017)
2. Anatel. Resolução n 506, de 1° de julho de 2008, Agência Nacional de Telecomunicações. http://www.anatel.gov.br/legislacao/resolucoes/23-2008/104-resolucao-506 (2018)
3. Angelidou, M.: Smart city policies: a spatial approach. Cities **41**, S3–S11 (2014)
4. Ashton, K.: That 'internet of things' thing. RFID J. **22**(7), 97–114 (2009)
5. Caragliu, A., Del Bo, C., Nijkamp, P.: Smart cities in Europe. J. Urban Technol. **18**(2), 65–82 (2011)
6. Castro Lundin, A., Ozkil, A.G., Schuldt-Jensen, J.: Smart cities: a case study in waste monitoring and management. In: Proceedings of the 50th Hawaii International Conference on System Sciences (2017)
7. Dias, R.R.F.: Internet das Coisas Sem Mistérios: Uma nova inteligência para os negócios. Netpress Books, São Paulo (2016)
8. Dragino Technology Co: Arduino Shield Featuring LoRa® Technology. http://www.dragino.com/products/module/item/102-lora-shield.html (2017). Accessed 17 Jan 2018
9. Farias, J.E.P., Alencar, M.S., Lima, Í.A., Alencar, R.T.: Cidades inteligentes e Comunicações. Rev. Tecnol. Inf. Comun. **1**(1), 28–32 (2011)
10. Harrison, C., Donnelly, I.A.: A theory of smart cities. In: Proceedings of the 55th Annual Meeting of the ISSS-2011, Hull, vol. 55, no. 1 (2011)
11. IBM Zurich: IBM lmic. https://github.com/mcci-catena/ibm-lmic (2017). Accessed 4 Jan 2018
12. ITU: Overview of the Internet of Thing - International Telecommunication Union. https://www.itu.int/rec/T-REC-Y.2060-201206-I (2012). Accessed 20 Sept 2017
13. Kehoe, M., Cosgrove, M., Gennaro, S.D., Harrison, C., Harthoorn, W., Hogan, J., Meegan, J., Nesbitt, P., Peters, C.: Smarter cities series: a foundation for understanding IBM smarter cities. Redguides for Business Leaders, IBM (2011)
14. Krish, N.: Getting Started with the RAK 831 Lora Gateway and RPi3. https://www.hackster.io/naresh-krish/getting-started-with-the-rak-831-lora-gateway-and-rpi3-e3351d (2017). Accessed 5 Jan 2018
15. Lora Alliance: LoRaWAN™ 1.1 Specification. https://www.lora-alliance.org/technology (2017). Accessed 15 Jan 2018
16. Lora Alliance: LoRaWAN™ 1.1 Regional Parameters. https://www.lora-alliance.org/for-developers (2017). Accessed 15 Jan 2018
17. Lora Alliance: LoRaWAN™ Backend Interfaces 1.0 Specification. https://www.lora-alliance.org/technology (2017). Accessed 15 Jan 2018
18. LORIOT: Lora Gateways and Concentrators. https://www.loriot.io/lora-gateways.html (2017). Accessed 15 Jan 2018

19. Mdhaffar, A., Chaari, T., Larbi, K., Jmaiel, M., Freisleben, B.: IoT-based health monitoring via LoRaWAN. In: 17th International Conference on Smart Technologies, IEEE EUROCON 2017, pp. 519–524. IEEE, Piscataway (2017)
20. Município de Ijuí: Geografia de Ijuí. http://www.ijui.rs.gov.br/paginas/geografia (2017). Accessed 28 Dec 2017
21. Petäjäjärvi, J., Mikhaylov, K., Yasmin, R., Hämäläinen, M., Iinatti, J.: Evaluation of LoRa LPWAN technology for indoor remote health and wellbeing monitoring. Int. J. Wireless Inf. Netw. **24**(2), 153–165 (2017)
22. Rakwireless Technology Co: RAK831 - LoRaWan Gateway Module. http://www.rakwireless.com/en/WisKeyOSH/RAK831 (2017). Accessed 6 Jan 2018
23. Raspberry Pi Foundation: Raspbian. https://www.raspberrypi.org/downloads/raspbian/ (2017). Accessed 12 Jan 2018
24. Raspberry Pi Foundation: GPIO: Raspberry PI Models A and B. https://www.raspberrypi.org/documentation/usage/gpio/ (2017). Accessed 12 Jan 2018
25. Rede Brasileira de Cidades Inteligentes e Humanas: Brasil 2030: Indicadores Brasileiros DE Cidades Inteligentes e Humanas. http://redebrasileira.org/indicadores (2017). Accessed 15 Jan 2018
26. Schwab, K.: The Fourth Industrial Revolution. Crown Business, New York (2017)
27. Suhaili, W.S.: Smart LED street light systems: a Bruneian case study. In: Proceedings of Multi-Disciplinary Trends in Artificial Intelligence: 11th International Workshop, MIWAI 2017, Gadong, Brunei, November 20–22, 2017, vol. 10,607, p. 370. Springer, Cham (2017)
28. The Things Network: The Things Network - Packet Forwarder. https://github.com/TheThingsNetwork/packet_forwarder (2017). Accessed 5 Jan 2018
29. The Things Network: https://www.thethingsnetwork.org/ (2017). Accessed 4 Jan 2018
30. TTN Mapper: https://ttnmapper.org/ (2016). Accessed 4 Jan 2018
31. Vermesan, O., Friess, P. (eds.): Internet of Things-from Research and Innovation to Market Deployment, vol. 29. River Publishers, Aalborg (2014)

Part III
The Social Science Behind the Cities of the Future

Chapter 12
Digital Cities and Emerging Technologies

Thays A. Oliveira, Vitor N. Coelho, Helena Ramalhinho, and Miquel Oliver

Abstract Cities are undergoing transformations in several respects, but, mainly, regarding novel technologies. In this sense, the aim of this study is to understand the relation between the cities and the use of emerging technologies such as digital democracy, blockchain and, in particular, smart contracts. To discuss and analyze the possibilities of this evolution and transformation, a bibliometric search is carried out in this chapter. A search of keywords in some refereed world databases was made: Springer, SCOPUS, IEEExplore, Science Direct, Google Trends, Web of Science, and Taylor & Francis. This combination of databases was selected in order to become possible to reach a high coverage of numbers of works about digital city and related technologies.

12.1 Introduction

Cities have been evolving along with the humankind. From modern ecovillages, smart condominiums, digital urban, and sustainable rural areas, society has been using information and communication technologies for our human benefits [4, 28]. The evolution is now being driven by a sea of shared information, sometimes called big data. Fully distributed cities can now reach agreements and evolve in accordance with the wishes and goals of those who lives there. According to Batty and Marshall [8], cities are "a collection of elements that act independently of one another but nevertheless manage to act in concert." [p. 567].

T. A. Oliveira (✉) · H. Ramalhinho · M. Oliver
Universitat Pompeu Fabra, Barcelona, Spain
e-mail: thaysaparecida.deoliveira01@estudiant.upf.edu; helena.ramalhinho@upf.edu; miquel.oliver@upf.edu

V. N. Coelho
Institute of Computer Science, Universidade Federal Fluminense, Niterói, RJ, Brazil
e-mail: satoru@ic.uff.br; http://vncoelho.github.io/

V. N. Coelho et al. (eds.), *Smart and Digital Cities*, Urban Computing,
https://doi.org/10.1007/978-3-030-12255-3_12

The computer science has a "(...) a progressive approach to urban studies started during the last years of the twentieth century when the digital revolution began to transform urban areas" [20, p. 4] in a "constellation of computers" [7, p. 155]. The ICT (information and communication technologies) also have an important position in this digital revolution, because of its accomplishment for innovations and support to the new science of cities [20].

This chapter is developing the works that have been done in the researches [6, 12, 13, 22–24], these articles approach smart cities, challenges to connect SC and citizens, as well as new cities' technologies that came to facilitate citizens life, marketing, digital cities, among others. The team of researchers has been participating on congress and conferences, also organizing special sessions and round table, sharing knowledge and learning with other researchers who work with different themes in the field of smart and digital cities. In these studies they pointed ways for the citizens to be more participatory in cities' decisions together with the government and new forms of judicial and legislative decision making in digital cities. Groundbreaking technologies such as blockchain, digital democracy and smart contracts have the potential to reconfigure the way we storage data and manage information [29].

The concept around digital cities is broad, involving technology and infrastructures that aims to build good interactions between humankind and technologies, becoming a reference and indicator to develop the urban innovation and receiving more attention from universities and researchers around the world [20]. According to Nam and Pardo [21], the concept of digital city is around the "integration of infrastructures and technology-mediated services, social learning for strengthening human infrastructure, and governance for institutional improvement and citizen engagement." (p. 282).

This chapter explores the digital city (DC), connecting sustainable and emerging technologies, such as: blockchain, smart contracts, and digital democracy.

Through a bibliographic research this chapter analyzes the relations between these technologies with the aim of discussing what are the impacts of digital democracy, blockchain, and smart contract for digital city and how these concepts differ and correlate among them.

In this sense, the contributions of this work involve:

- Analyze studies between the DC with blockchain, digital democracy, and smart contract;
- Comparing the publications in each area;
- Analyze the relevance of these terms to cities' evolution;
- Consider the impact of these terms in the context of smart cities and DC.

The remainder of this chapter is organized as follows: Sect. 12.2 presents a background surrounding the aforementioned themes, a bibliographic search is conducted in Sect. 12.3, and the final considerations and future works are presented in Sect. 12.4.

12.2 Background

In this section, we presented a general view about cities in Sect. 12.2.1, in particular, the blockchain technology and smart contracts are presented in Sects. 12.2.2 and 12.2.3, respectively, while digital democracy is approached in Sect. 12.2.4.

12.2.1 Cities' Evolution: Connecting Real, Sustainable, Smart, and Digital Cities

Technologies are surely the basins of human development. Since plenty of years ago, the humankind has been using its skills for achieving better life quality. The evolution of the species [14], undoubtedly, guided the permanence of those who are able to refine their arts from those technological tools. The word "Techne" means art, skill, craft, or the way, manner, or means by which a thing is gained. The Roman philosopher Cicero commented [10] that humans had an ability of transforming the environment and, consequently, creating a "second nature." The evolution is an environment that evolves in accordance with those who manage and govern that space. Technology, in the deep sense of view, is not only related to digital and computational intelligence technologies, but can reach different aspects according to the desired evolution that a group of individuals are working on.

Cities, ubiquitous and smart cities are usually advocated as: "(...) a model of city where information systems are sharing data, like the cloud computing." [27]. The International Organization for Standardization complements that "Developing Smart Cities can benefit synchronized development, industrialization, informatization, urbanization and agricultural modernization and sustainability of cities development." ISO2014 [25, p. 02]. Around the definition of term for an intelligent city, different authors defend distinct points of view. For instance, Ahvenniemi et al. [1] mention that a city that is not sustainable is not really "smart." Furthermore, the authors add that "Sustainability assessment should be part of the SC development and therefore we find it important to integrate sustainability and smart city frameworks, so that both views are accounted for in performance measurement systems (...)" [1, p. 235]. Finally, Ahvenniemi et al. [1] recommended the use of the term: "smart sustainable cities" instead of "smart cities." On the other hand, smart city or digital city can be seen as a cyber-physical integration in the urban space, along this chapter these terms will be used in order to approach the concept of cities of the future.

12.2.2 Blockchain Technology

Blockchain is a nomenclature attributed for a new distributed and information technologies that have the objective to save data on immutable and secure way [18, 19]. According to Galen et al. [15]:

> A blockchain is a digital, secure, public record book of transactions (a ledger). "Block" describes the way this ledger organizes transactions into blocks of data, which are then organized in a "chain" that links to other blocks of data. The links make it easy to see if anyone has changed any part of the chain, which helps the system protect against illegal transactions. [15, p. 06]

When the information reaches the blockchain, it becomes impossible to change its historical path, in this sense, the data became immutable [18]. Due to this, blockchain can be applied in different areas, offering a solution where reliable registration is required and becomes trusted [19]. Furthermore, there is also the decentralized concepts, which has the potential of bringing low cost solutions. In the context of digital cities, blockchain can be used for social impact and to record government decisions, monetary transactions, and state of the documents, projects, and plans. The blockchain involves three pillars and attributes: trust, transparency, and immutability. Figure 12.1 was drawn for representing them.

Figure 12.1 shows those aforementioned blockchain attributes, which can be applied in different areas in the city, as in the governance, health care, voting system, among others. In order to complement it, the concept of smart contract will be addressed in Sect. 12.2.3.

12.2.3 Smart Contracts

Along with the blockchain comes the idea of smart contracts, which involve electronic protocols that digitally check and automatically executed pre-defined contracts (negotiations, agreements, among others). The contracts evolve in an automatically fashion, executed by rules throughout a decentralized system that is

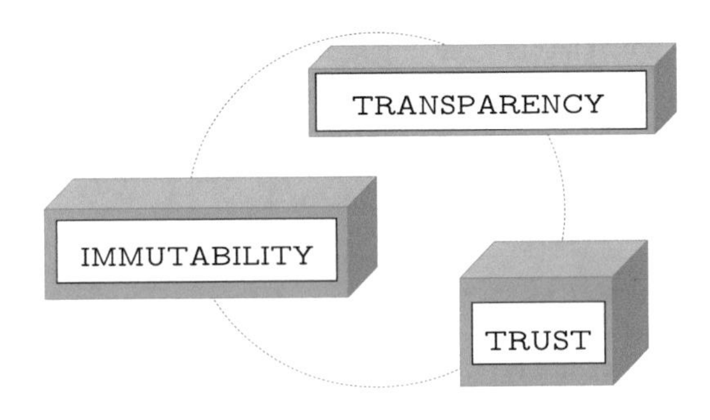

Fig. 12.1 Blockchain pillars

extremely hard to be stopped [15]. In summary, a smart contract works as a code container that encodes and reflects real-world contract in the cyber world.

Each contract represents an agreement made by two or more parts, where each part must fulfill its obligation under that pre-defined contract [19]. Figure 12.2 illustrates a smart contract and its interaction with different parties. First, someone needs to create the contract with some programming rules (written in the well-known programming languages), then, that contract is deployed into the network and becomes accessible for everyone connected. The contract is automatically executed by a network of computers, in this sense, public contracts become available for those who wants to use them. For example, a given contract C_y can be made available for a defined set of addresses $[A_1, A_2, \ldots, A_n]$ with pre-defined rules. In the case of private interests, private blockchain solutions can be used, creating a system which is accessible only for authorized agents. In Fig. 12.2, the entity can be a company, a person, among others.

The Smart contract advantages are:

- Automatic and autonomous execution;
- Fast and direct;
- Safe and trustless execution;
- Avoid manual error;
- "Cheap," according to the application;

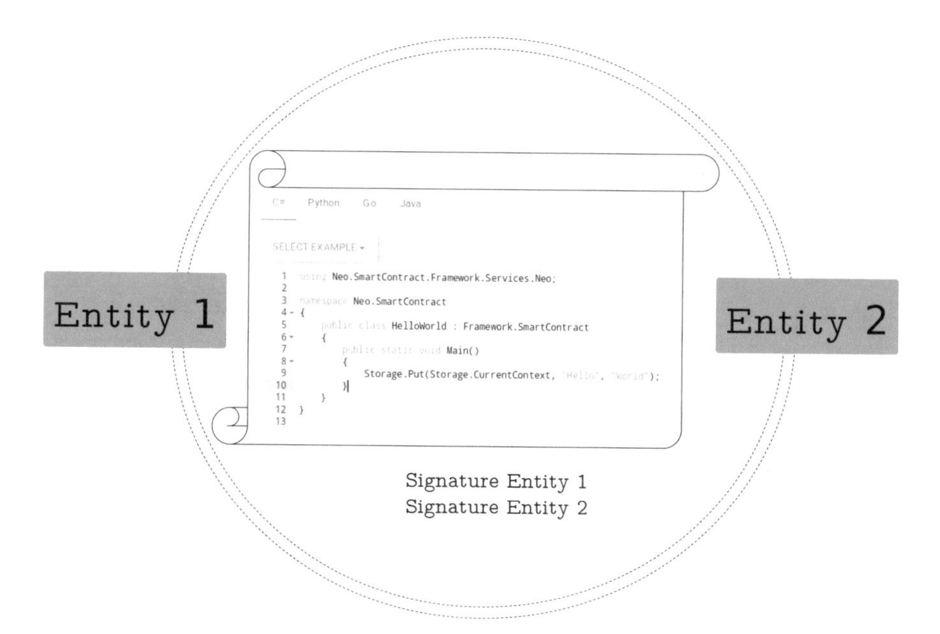

Fig. 12.2 Smart contract example. Figure drawn by the authors

- Transparent and backup by default;
- Code is law.

Macrinici et al. [19] complement:

> A key premise for contracts is that they represent a binding agreement between two or more parties, where every entity must fulfill their obligations according to the agreement. Another important element is that the agreement is enforceable by law, usually through a legal centralized entity (organization). However, smart contracts replace the trusted third parties; that is, the intermediaries between contract members. [19, p. 02]

12.2.4 Digital Democracy

Applications that promote democracy have a great potential for investments. In particular, blockchain based solutions present good aspects related to citizens transparent interaction with public sectors [15]. In this sense, it becomes an alternative for democratic governance, since the government could share political information, save and verify citizen identity, as well as other interesting applications. In addition, it has the potential of generating novel alternatives for a better citizens' participation in government decisions at the national, state, and municipality levels. Also, improved voting systems can be designed based on blockchain [16].

Figure 12.3 illustrates an example of digital democracy, where citizens can be more participative in government decisions, for increasing citizens' rights and promoting inclusive participation. One can say that the democracy exists when citizens have equal rights.

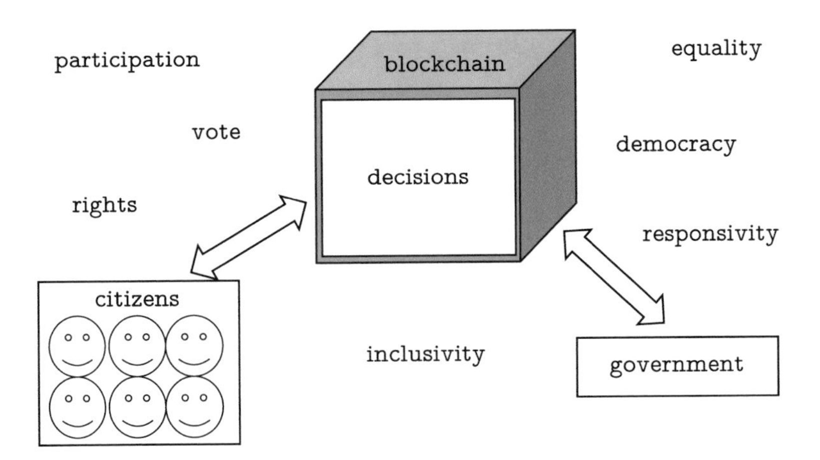

Fig. 12.3 Digital democracy

12.3 Bibliometric Search

By looking at the goal of understanding the relation between the terms explained in Sects. 12.2.1–12.2.4, a bibliometric search of keywords, in refereed world databases, is conducted. By using multiple databases it becomes possible to cover most part of the researches on the discussed topics [20].

In this sense, the following databases were considered (all of them accessed online on 28th of September, 2018): Springer Link, SCOPUS, IEEExplore, Science Direct, Web of Science, and Taylor & Francis Online. The bibliographic search shows the numbers of researchers done mentioning the selected keywords. The following keywords were combined and searched (Table 12.1 shows the obtained results):

1. Digital city (DC);
2. DC (digital city) + Blockchain;
3. DC (digital city) + Digital democracy;
4. DC (digital city) + Smart contract;

The concept of DC is showing a high number of publications with pioneer studies dating back to the 1990s [2, 5, 28], when the nomenclature and definition about a DC started to be uncovered and drawn. Both nomenclatures look for a city that attend citizens' necessities, promoting better life quality, urban development, mobility, and the search for an ideal/dreamed city [11, 26].

The E-government can be supported by the DC, in order that the Digital Government provides a virtual environment [3, 17]. According to Anthopoulos and Tsoukalas [3], "the digital city's definition is extended to the global Information Environment, focusing on the needs of a city area." [3, p. 91].

Bolívar [9] complements that the DC concept incentives the government to use information and communication technologies for promoting political participation. In this sense, the technology came to accomplish changes in different areas and in the government, to proportionate a transparent governance and look for political strategies, arising as a "smart governance" [9].

Table 12.1 Number of publications found in the well-known databases

	Keywords			
	DC +	DC + blockchain	DC + digital democracy	DC smart contract
Database search—2018				
1. Springer Link	124,691	375	8602	4204
2. SCOPUS	130,482	260	10,584	895
3. IEEEXplore	9730	17	7	4
4. Science Direct	110,570	174	2116	2377
5. Web of Science	5209	4	0	0
5. Taylor & Francis Online	68,484	57	12,903	1698

Finally, a voting system with blockchain can be inserted in the context of "smart governance," which breaks down some barriers and limitations of the current electronic voting system [16]. In this sense, the blockchain can be applied in different areas, as well as the smart contract tools together with smart governance. In the bibliometric search, the term presented the lowest number of publications, which is reasonable since it is an emerging technology along with the smart contract.

12.3.1 Terms Popularity

In the sense of complementing the bibliometric search, Fig. 12.4 was drawn from searchers done in the Google Trends (GT) platform. The GT is a Google's database that analyzes the popularity of some terms based on their private repository. In this system, it is possible to define the time period, local (country, region) and compare different terms.

The search was made between using the same labels chosen for the bibliographic search presented in the last section. Results of a worldwide search from the last 3 years were considered (from 22 October 2015 to 22 October 2018).

Table 12.2 analyzes the biggest and lowest peaks from each keywords search from the GT. By analyzing these numbers, it is possible to conclude that each label had a specific time of higher and lower demands. DC and smart contracts had constant peaks over the last 3 years, but the second one had more lowest peaks (only considering the year of 2016, the months of January, May, and July can be highlighted as lowest peaks). On the other hand, DC remained without lowest peaks and presented a higher peak on May 2017. Considering all searched labels, it is possible to affirm that the most stable period was between November 2017 and June 2018.

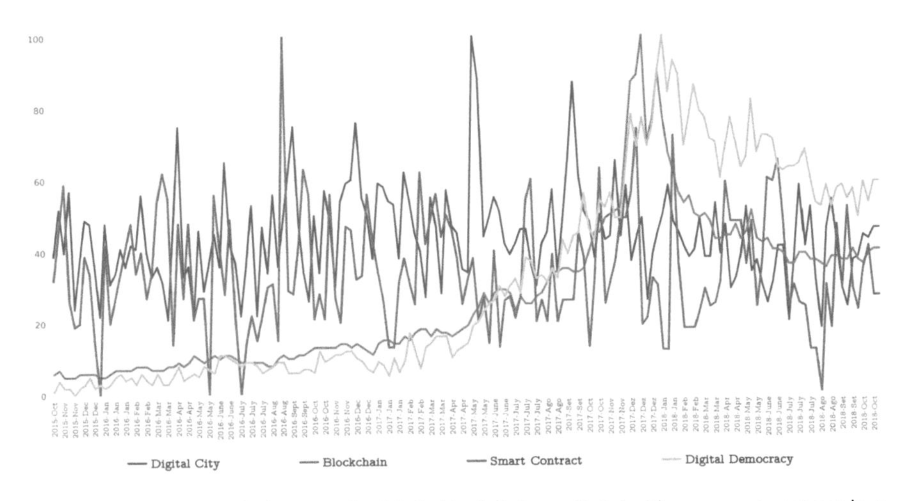

Fig. 12.4 Google Trends between the labels blockchain × digital cities × smart contract in a 3-year interval

Table 12.2 Lowest and highest peaks of searching according to Google Trend from 22 October 2015 to 22 October 2018

	Keywords			
	DC	Blockchain	Digital democracy	Smart contract
Google Trend search				
1. Higher peak	2018-Jan	2017-Dec	2018-Jan	2016-Sept
2. Lower peaks	2015-Nov	2015-Nov	2015-Nov	2015-Dec 2016-May/July

12.4 Conclusion and Future Works

Through the analysis of this study it was possible to conclude that the technologies approached here (blockchain, smart contract, and digital democracy) complement each other in order to provide better services and communication.

Beyond these technologies, it is noteworthy that IoT has an important contribution to DC, integrating the devices in the cities, as well as helping urban mobility. While these novel technologies have been approaching the citizens, some of them might not be aware about it. In this sense, this chapter had also this whole of increasing awareness. It is expected that society will evolve into a system where citizens are more participative and have a better relationship with the government, in particular, with more efficient and transparent voting systems. As a continuation of this study, a survey with citizens, with the purpose of comprehending concepts about SC, is proposed, which has the potential of pointing out what they agree about these new digital advances.

References

1. Ahvenniemi, H., Huovila, A., Pinto-Seppä, I., Airaksinen, M.: What are the differences between sustainable and smart cities? Cities **60**, 234–245 (2017)
2. Anthopoulos, L.G.: Understanding the smart city domain: a literature review. In: Transforming City Governments for Successful Smart Cities, pp. 9–21. Springer, Cham (2015)
3. Anthopoulos, L.G., Tsoukalas, I.A.: The implementation model of a Digital City. The case study of the Digital City of Trikala, Greece: e-Trikala. J. E-Gov. **2**(2), 91–109 (2006)
4. Anthopoulos, L.G., Vakali, A.: Urban planning and smart cities: interrelations and reciprocities. In: The Future Internet Assembly, pp. 178–189. Springer, New York (2012)
5. Anthopoulos, L.G., Gerogiannis, V.C., Fitsilis, P.: Supporting the solution selection for a digital city with a fuzzy-based approach. In: KMIS, pp. 355–358 (2011)
6. Barbosa, A.C., Oliveira, T.A., Coelho, V.N.: Cryptocurrencies for smart territories: an exploratory study. In: Proceedings of the IEE World Congress on Computational Intelligence (2018)
7. Batty, M.: The computable city. Int. Plan. Stud. **2**(2), 155–173 (1997)

8. Batty, M., Marshall, S.: Centenary paper: the evolution of cities: Geddes, Abercrombie and the new physicalism. Town Plan. Rev. **80**(6), 551–574 (2009)
9. Bolívar, M.P.R.: Smart cities: big cities, complex governance? In: Transforming City Governments for Successful Smart Cities, pp. 1–7. Springer, Cham (2015)
10. Cicero, M.T.: De Natura Deorum Libri Tres, vol. 3. University Press, Cambridge (1885)
11. Cocchia, A.: Smart and digital city: a systematic literature review. In: Smart City, pp. 13–43. Springer, Cham (2014)
12. Coelho, V.N., Oliveira, T.A., Coelho, I.M., Coelho, B.N., Fleming, P.J., Guimarães, F.G., Ramalhinho, H., Souza, M.J., Talbi, E.G., Lust, T.: Generic Pareto local search metaheuristic for optimization of targeted offers in a bi-objective direct marketing campaign. Comput. Oper. Res. **78**, 578–587 (2017)
13. Coelho, V.N., Veloso, I.F.O., Oliveira, T.A., Coelho, V.N., Veloso, I., Veloso, V.M., Souza, M.J.F., Filho, A.F.S.: A multi-criteria view about judicial and legislative decision making in digital cities and societies (in Portuguese). In: Proceedings of the XLIX Annual Brazilian Symposium on Operational Research, pp. 3979–3989 (2017)
14. Darwin, C.: Evolution by Natural Selection. Nelson, London (1963)
15. Galen, D., Brand, N., Boucherle, L., Davis, R., Do, N., El Ba, B., Kimura, I., Wharton, K., Lee, J.: Blockchain for social impact, moving beyond the hype. Graduate School of Stanford Business – Center for Social Innovation (2018)
16. Hjalmarsson, F.P., Hreioarsson, G.K., Hamdaqa, M., Hjalmtysson, G.: Blockchain-based e-voting system. In: 2018 IEEE 11th International Conference on Cloud Computing (CLOUD), pp. 983–986. IEEE, Washington (2018)
17. Hsieh, P.H., Chen, W.S., Lo, C.J.: An investigation of leadership styles during adoption of E-government for an innovative city: perspectives of Taiwanese public servants. In: Transforming City Governments for Successful Smart Cities, pp. 163–180. Springer, Cham (2015)
18. LaPointe, C., Fishbane, L.: The Blockchain Ethical Design Framework (2018). http://www.sciencedirect.com/science/article/pii/S0736585318308013
19. Macrinici, D., Cartofeanu, C., Gao, S.: Smart contract applications within blockchain technology: a systematic mapping study. Telematics Inform. **35**(8), 2337–2354 (2018). https://doi.org/10.1016/j.tele.2018.10.004. http://www.sciencedirect.com/science/article/pii/S0736585318308013
20. Mora, L., Bolici, R., Deakin, M.: The first two decades of smart-city research: a bibliometric analysis. J. Urban Technol. **24**(1), 3–27 (2017)
21. Nam, T., Pardo, T.A.: Conceptualizing smart city with dimensions of technology, people, and institutions. In: Proceedings of the 12th Annual International Digital Government Research Conference: Digital Government Innovation in Challenging Times, pp. 282–291. ACM, New York (2011)
22. Oliveira, T.A., Coelho, V.N., Ramalhinho, H., Souza, M.J., Coelho, B.N., Rezende, D.C., Coelho, I.M.: A VNS approach for book marketing campaigns generated with quasi-bicliques probabilities. Electron Notes Discrete Math. **58**, 15–22 (2017)
23. Oliveira, T.A., Coelho, V.N., Tavares, W., Ramalhinho, H., Oliver, M.: Operational and digital challenges to connect citizens in smart cities (in Portuguese). In: Proceedings of the XLIX Annual Brazilian Symposium on Operational Research, pp. 3819–3830 (2017)
24. Oliveira, T.A., Barbosa, A.C., Ramalhinho, H., Tavares, W., Oliver, M.: Citizens and information and communication technologies. In: Proceedings of the IEE World Congress on Computational Intelligence (2018)
25. Organization, I.S.: TSmart Cities Preliminary Report 2014. (2014). http://www.iso.org/iso/smart_cities_report-jtc1.pdf
26. Trindade, E.P., Hinnig, M.P.F., Moreira da Costa, E., Marques, J., Bastos, R., Yigitcanlar, T.: Sustainable development of smart cities: a systematic review of the literature. J. Open Innov.: Technol. Market Complexity **3**(3), 11 (2017)

27. Ubiquitous cities in Asia: a new concept of urban living. https://www.casaasia.eu/noticia/detalle/202459-ubiquitous-cities-in-asia-a-new-concept-of-urban-living. Accessed 15 July 2017
28. Wang, L., Wu, H., Song, H.: A Framework of Integrating Digital City and Eco-City. School of Business, Hubei University, Wuhan (2002)
29. Wright, A., De Filippi, P.: Decentralized Blockchain Technology and the Rise of Lex Cryptographia (2015)

Chapter 13
A Multicriteria View About Judicial and Legislative Decision Making in Digital Cities and Societies

Vitor N. Coelho, Thays A. Oliveira, Iara V. O. Figueiredo, Marcone J. F. Souza, and Iuri Veloso

Abstract The constant evolution of cities has driven the development of new tools for society. Applications inspired by operational research techniques, which aid decision making, can make viable dreams already dreamed up by philosophers. Among these, we highlight more participatory, legitimate, and reliable judicial and legislative systems. In this context, a multicriteria analysis seems necessary, balancing the different versions, beliefs, cultures, and consequent weights and measures desired by each citizen. In this paper, we present a new model for judicial/legislative processes in digital cities. The system proposes the use of sets of solutions, obtained from different weights adopted according to personal characteristics of those involved in the voting process. From a simple case of study, we highlight the possibilities, flexibility, and potential of the proposed system. The proposed framework shows up as promising tool for assisting decision making in other similar voting scenarios.

V. N. Coelho (✉)
Institute of Computer Science, Universidade Federal Fluminense, Niterói, RJ, Brazil
e-mail: satoru@ic.uff.br; http://vncoelho.github.io/

T. A. Oliveira
Department of Engineering and Information and Communication Technologies, Universitat Pompeu Fabra, Barcelona, Spain
e-mail: thaysaparecida.deoliveira01@estudiant.upf.edu

I. V. O. Figueiredo
Escola Nacional de Saúde Pública, Fiocruz, Rio de Janeiro, Brazil

M. J. F. Souza
Department of Computer Science, Universidade Federal de Ouro Preto, Ouro Preto, Brazil

I. Veloso
Faculdade Mineira de Direito, PUC Minas, Belo Horizonte, Brazil

V. N. Coelho et al. (eds.), *Smart and Digital Cities*, Urban Computing,
https://doi.org/10.1007/978-3-030-12255-3_13

13.1 Introduction

The evolution of cities to a paradigm of decentralized governance has been widely discussed [9]. It is expected that smart cities (SC) [1, 20] are going to bring the government closer to citizens [30]. Linked by new information and communication technologies, smart cities drive a strong trend towards decentralized democracies, as mentioned by Roberts [24]. Decentralizing decision-making process implies the sharing of decisions, a strategy that has been adopted in technologies inspired by multi-agent systems [8]. In this sense, studies in the field of SC [3, 6] can ensure better and more effective mobility aligned with access to opportunities, especially for urban populations.

In the same way, such systems are emerging as important allies to reach a more independent judiciary [26]. In this context, computational tools that define adequate limits and weights for different levels of governance are important resources for more sensible decision making. Initiatives to decentralize governance and promote local participation have been occurring even in rural areas, such as in the village of panchayats, India [13, 18].

Bi-directional and distributed communication, based on secure, permanent protocols, will undoubtedly be the wisest choice for the cities of the future. In particular, the use of blockchain based technologies [19, 28] can proportionate trust for those that want to follow this line of reasoning. The Brazilian institute ITS-RIO [15] is currently designing a tool for public petitions by discussing concepts of digital identity focused on the Brazilian scenario. More recently, a notorious project that focuses on digital identities is the one, namely The Key [32], developed inside the Ecosystem of the Neo Blockchain [14], formerly Antshares.

Smart devices [25], capable of obtaining information and interacting with the environment, will be an intrinsic part of SC. In this context, applications, models, and new paradigms aided by IoT equipment will be a great pillar of modern society [34]. However, reaching conclusions and processing the vast amount of information, extracted and shared in a SC is a task that requires operational research (OR), computational intelligence (CI), and optimization knowledge. As mentioned by Gibbard-Satterthwaite [12], every voting rule is subject to manipulation where there are more than two possibilities. The task of computing votes and reaching robust and concrete conclusions has been analyzed as a combinatorial optimization problem [2], of difficult resolution and, in some cases, belonging to the class of NP-hard problems [16]. There are different possibilities for reaching social agreements through voting, such as plurality with or without elimination, cumulative, approval, peer elimination, among others [27].

In the Brazilian context, reforms in the judiciary have been discussed by the population and by distinct works in the literature [4, 7, 17, 21, 22, 29, 31]. On the other hand, considering the slowness of transformations that require effective changes in laws and old paradigms, we highlight the potential that distributed and encrypted systems have [11]. In this context, we emphasize the role of OR and CI in motivating applications for the emergence of systems and approaches with these capacities and abilities. As can be expected, more impartial, participatory systems, in which citizens can optionally be anonymous, can become a reality.

In this brief study, we present a computational system, inspired by OR techniques and multi-objective optimization concepts, which could guide the steps towards a more participatory judicial and legislative system. The system described here introduces a weighted voting system. At the end of the process, a set of non-dominated solutions is returned, considering socio-demographic characteristics of those involved in voting. From a simple case study, we exalt the possibilities, flexibility, and power of the proposed methodology, inspired by the use of multicriteria decision-making tools [35]. Such proposal, despite computerizing the system, does not remove the "human" side of the process, it just gives strength to the participation of the population in more a transparent and low cost manner. The following contributions of the present work stand out:

- Introduction of a new distributed system in the context of digital cities;
- Design of a voting protocol that is robust against specific sets of weights for socio-demographic information of each individual's profile. It is achieved by generating random weights and presenting results in terms of sets of non-dominated solutions;
- Presentation of a multicriteria view to reach relevant facts in a judicial and legislative process;
- Assistance to the decision-making process made by judges;
- Motivation of researchers and government officials to invest and develop new technologies for society;
- Alert the population of the current possibilities that are emerging from the insertion of intelligent devices in our daily lives, especially in the context of smart cities.

The remainder of this paper is organized as follows. Section 13.2 presents the proposed system, as well as the concepts for a framework based on blockchain concepts (Sect. 13.2.3). A case of study, described in Sect. 13.3.1, with its results presented in Sect. 13.3.2, is considered in order to illustrate a scenario for the application of the proposed tool. Finally, Sect. 13.4 concludes this chapter and points out some possible extensions and future work.

13.2 Decentralized Solutions for Smart and Digital Cities

This paper proposes a decentralized and distributed system for judicial processes. A simplified model is described in Sect. 13.2.1. The decision-making process, which uses voting data, is described in Sect. 13.2.2. It is considered that the data collection and the entire framework are obtained by IoT devices within SCs and the data stored in a blockchain (as detailed in Sect. 13.2.3).

13.2.1 Proposed System

The interaction of citizens (who can be grouped or self-named in distinct classes) and other social agents can be made by intelligent devices, available in a decentralized way in the context of intelligent cities. Interactive and real-time platforms embedded within these devices, and available to citizens in the region under analysis, will be platforms for data entry/acquisition and dissemination of results. A blockchain inspired system will be able to store each information of the process in a transparent and permanent way.

A central coordinator, also chosen by similar protocols, could name a set of judges for the case, which would have weights pre-defined by the system. In the current Brazilian system, these jurors can be summoned in cases of intentional crimes against life [5].

After a certain period of survey of the facts, the prosecutor may present his replies as well as new facts. Among these, the system will be able to receive videos, photos, documents, and any other digital element. After the period of facts, citizens will have the possibility to vote and give weights for each information/fact of that particular process. In addition, a group of persons may participate jointly, when legally registered with a digital identity. In this same stage, it is emphasized that if there is more than one accused, the process can be divided into several strands with different replicas. On the other hand, a joint model could also be considered, opening the opportunity to access all versions in the same environment.

At any time in the process it will be possible to attach complementary information, even anonymously. The major advantage of integrating these platforms is to ensure a joint information access environment (including data such as those from the "Ranking of Politicians" portal [23]). Furthermore, it would be open a path for even trying to analyze how news and information that come during the process, as well as external influences, change the line of reasoning of those that are following or contributing to a given process.

Finally, a decision-making process returns possible facts relevant to the process in question, taking into account different aspects, such as:

- Socio-demographic information of each agent/citizen who participated in the voting;
- Personal data, optionally declared, such as: relative of the victim, witness, among others.

13.2.2 Multicriteria Decision Making

This section defines a strategy that can be used to compute each agent's votes (vide [33] for more information on agents and multi-agent systems) involved in the voting process. We describe a decision-making strategy based on weights for each citizen's profile variable, or jury (for cases where necessary); in particular, pre-defined before the process starts. For example, the following set of weights can be defined: 1, 40, and 2, respectively, for citizen, victim, and jury. In addition, each variable of the profile of individuals has a specific weight, and it is possible to analyze the impact of certain citizens on the final decision of the voting, for example:

- Three possible income levels with weights 4, 2, and 7, respectively, for low, medium, and high incomes. Therefore, the weighted final weight is: 0.31, 0.15, and 0.54.
- Two different education backgrounds, such as with weights 7 and 5;
- Two different gender categories, such as with weights 5 and 5.

Decision making would be done based on the weight for each fact ($n \in [-10, 10]$), multiplied by the weight of each involved agent. In this sense, a set of facts and replicas could be $FR = \{(f_1, r_1), (f_2, r_2), \ldots, fr_i, \ldots, (f_z, r_z)\}$ with each individual being able to provide its grade for each set of information fr_i. In this way, the final score of each set would be given by $n^{fr_i} = \sum_{c \in C}(w_c \times n_c^{fr_i})$, being C the effective set of citizens who gave a valid note to the question fr_i. Weights w_c are defined by the weighted average of the weight of each variable considered according to the profile of the individual, ($w_c^v = [0, 10]$), where v is one of the analyzed characteristics. In the case of a social group (such as a union or association), the entity could have a specific weight, adopted in view of the number of registered members or a bonus added to the weight of each sitting member who participated in the voting process.

For each grade in an information packet ($n^{fr_i} \mid fr_i \in FR$), positive values would point out facts in favor of the defendant; while, on the other hand, negative values would imply a weighty fact against the accused. In this way, the system would be able to classify the facts that are for and against.

In order to provide a multicriteria analysis, it is proposed to consider different combinations of weights for each socio-demographic variable. Thus, from a set of solutions, it becomes possible to verify the relevant facts for each set of weights, presenting possibilities that vary according to the profile of the people involved in the voting.

13.2.3 Trust and Transparency by Using the Blockchain

The information gathering, during all stages of the process, can happen through a reliable, transparent, and permanent platform, in which the insertion of new information would be published, permanently, in a blockchain. Blocks of news, authors, addresses, files, and timestamps tags would be blocked in a transaction and included in a digital registry of the process in question. This tool will enable future analyses that will relate news and the tendency of the votes on the facts and replicas leveraged in the initial stages (or even during the process).

In addition, by ensuring transparency and security in storing process data (in particular by promoting distributed approaches), future studies may reanalyze the same cases, but with other perspectives.

13.3 Case of Study

The case of study and software developed in the scope of this study can be found at https://github.com/vncoelho/judgmentssmartcities.

13.3.1 Description of the Scenario, Facts, Cases, and Motivation

For exemplifying the proposed system, a simple case study was designed, composed of distinct problems generated with random data. There are fictitious scenarios composed from 10 to 110 citizens, 1 to 100 facts, and 1 to 30 characteristics analyzed (each with two or three classes). Each possible characteristic of the profile of each involved individual in the voting is also defined in a random fashion.

In order to obtain a set of solutions relevant to the process, composed of different combinations of relevant facts, a large number of combinations of weights were randomly generated. For each characteristic of the profile of those involved in voting, a weight between 1 to 10 is generated for all possible classes (as detailed in Sect. 13.2.2). In the experiments described in the next section, 1000–10,000 random combinations of these weights were analyzed. The total number of experiments performed was 12,000. Among these, 2000 experiments were performed to verify the percentage of dominance between solutions with different sets of relevant facts, without considering the order given by the voting weights.

13.3.2 Obtained Results

As can be seen in Figs. 13.1, 13.2, and 13.3, the proposed model was able to obtain a large number of non-dominated solutions, defined as relevant. In particular, the method was able to find plenty of non-dominated solutions when the number of facts grows. For these cases, the percentage of non-dominated solutions increases because of the high number of possibilities for generating combinations of relevant facts (as well as different orders of relevance). In this sense, different combinations of weights easily result in different orders of the relevant facts.

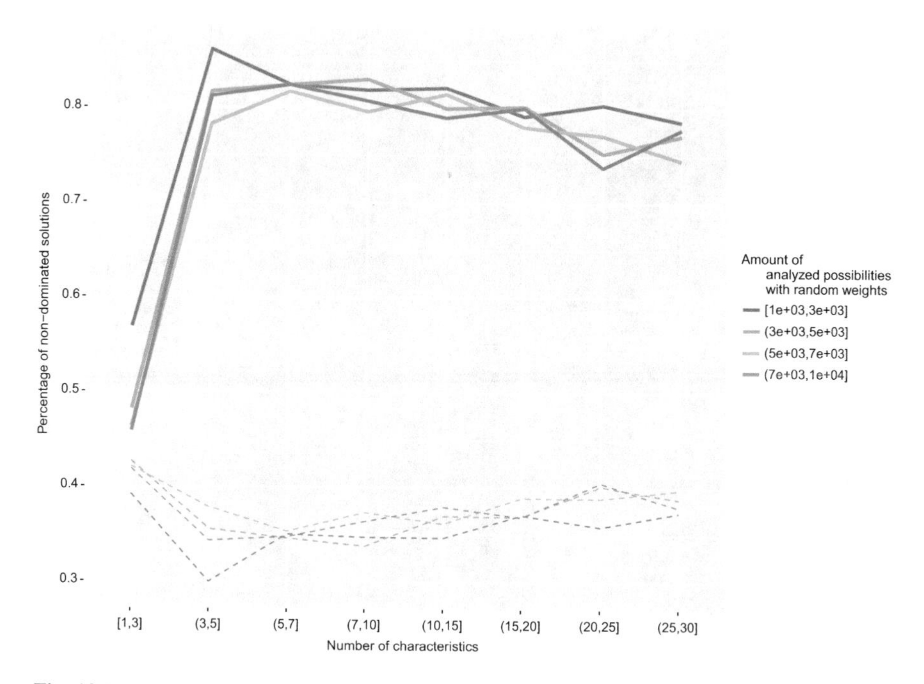

Fig. 13.1 Interaction graphs between the percentage of solutions obtained, the number of socio-demographic variables, and the combinations of weights

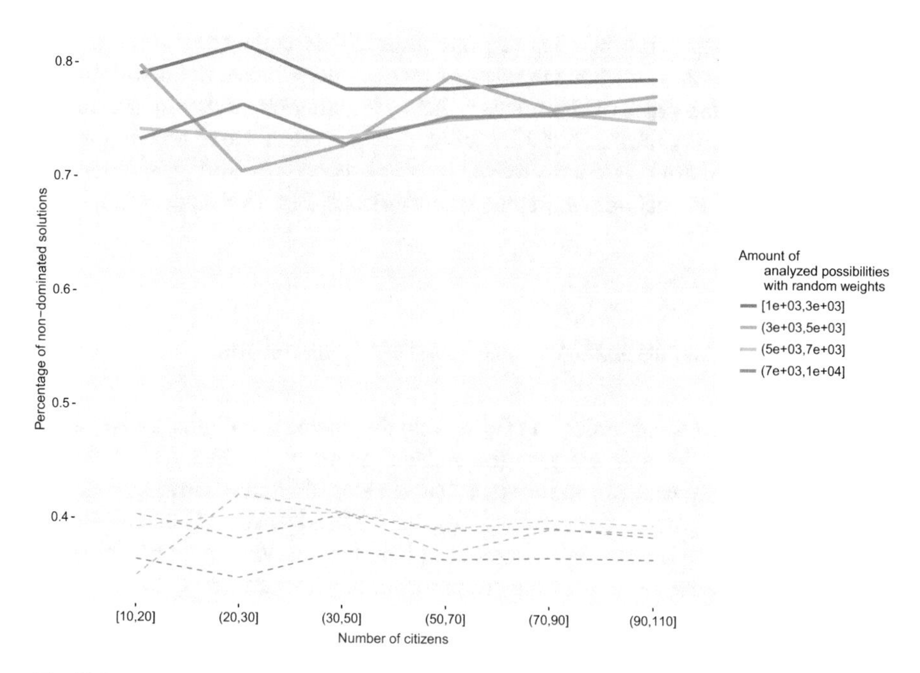

Fig. 13.2 Relationship between the percentage of non-dominated solutions and the number of citizens involved in the voting process

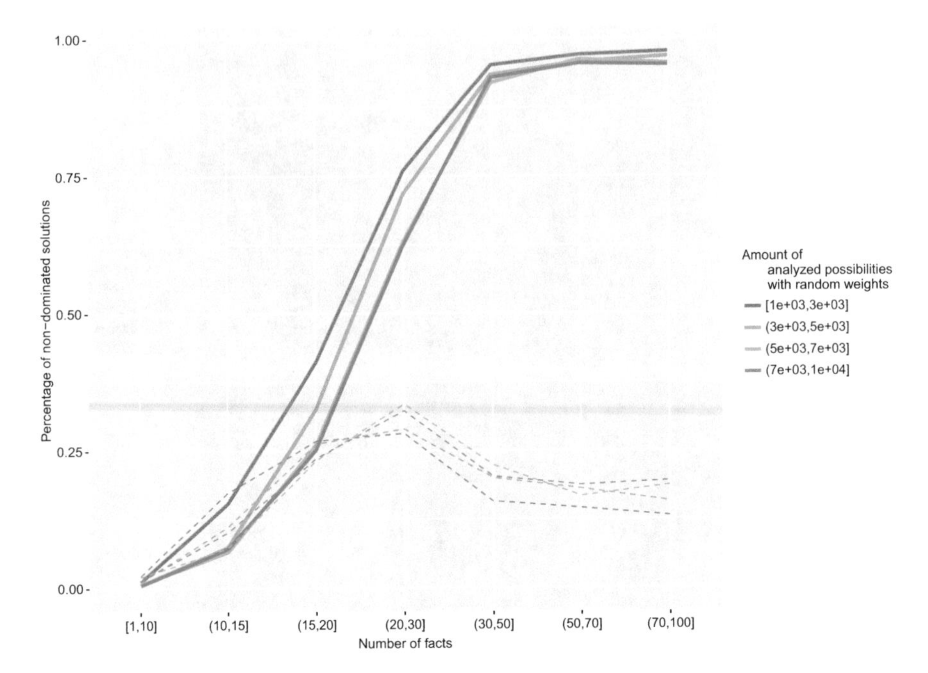

Fig. 13.3 Growth on the rate of non-dominated solutions with the increase in the number of considered facts

Considering this large number of possibilities, it is noteworthy the need for innovative approaches to filter the relevant facts. Among these, the possibility of not considering the order of the relevant facts as a criterion of dominance stands out. Thus, the graph shown in Fig. 13.4 shows that the percentage of non-dominated solutions obtained drops dramatically when order is not considered. In addition, the effects of increased socio-demographic characteristics, Fig. 13.5, appear in a more evident manner.

13.4 Final Consideration and Possible Extensions

Considering the constant evolution of cities into distributed paradigms, a new voting system that promotes and analyzes the citizens' participation in a digital system was proposed. In particular, the system was developed from a growing need for more balanced judgments and a multicriteria view. The advance and implementation of tools inspired by the strategies introduced here may provide more participatory and "fairer" systems. In view of the current technological advances, it is expected that decentralized and transparent systems will be the core of decision making in modern/intelligent societies.

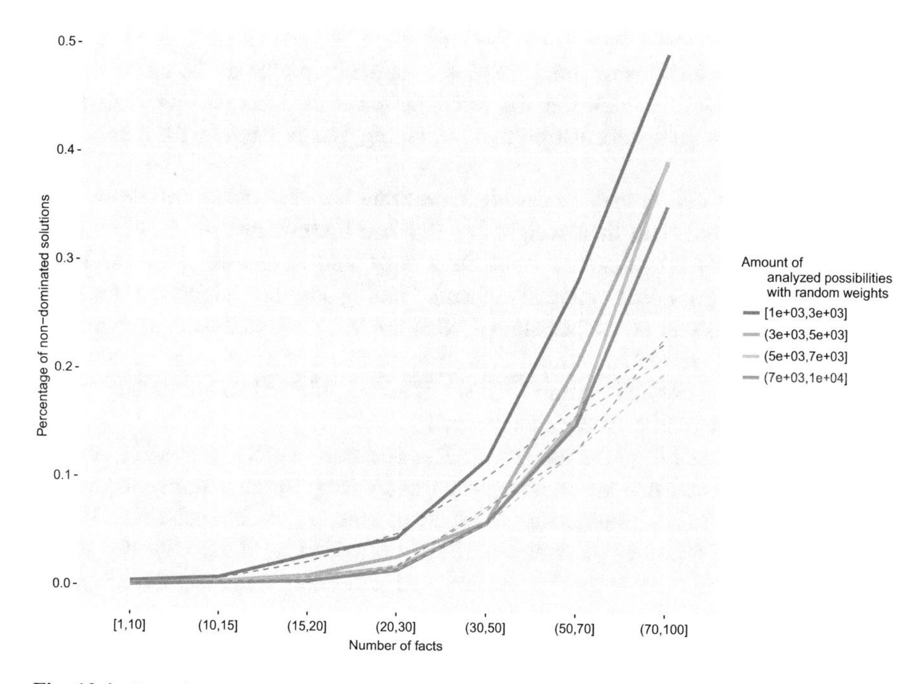

Fig. 13.4 Growth on the number of non-dominated solutions with the increase in the number of considered facts, considering only dominant solutions with different relevant facts

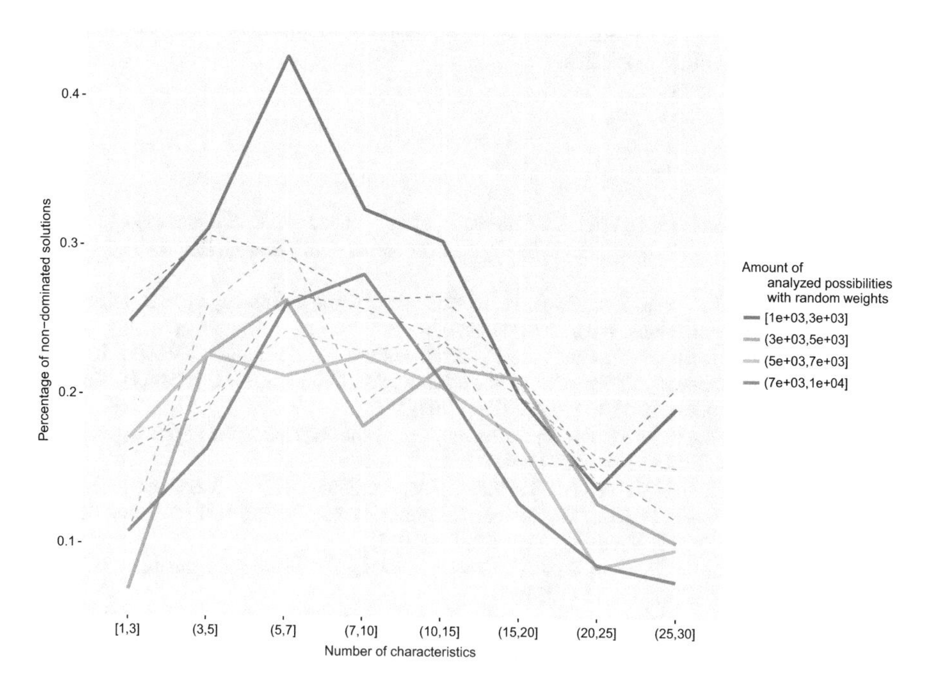

Fig. 13.5 Increase in the number of variables per individual and impact on the percentage of non-dominated solutions

The model proposed here could be embedded on several devices already in possession of the citizens or even installed in strategic points of the cities. Future work could consider the influence of external news and data, connecting publication and disclosure of this information against voting trends (during the period the process is open).

An additional step in the process could consider the exclusion of all those facts that were pondered with final weights ≥ 0, since these facts are in favor of the defendant or have no relevance (such facts may also have received these notes because they are possibly manipulated facts). Finding specific weights that give rise to certain conclusions is another strategy that could be adopted to understand the wishes of those involved in voting. Therefore, metaheuristic algorithms could seek solutions that increase and optimize the set of non-dominated solutions, expanding the range and power of possible conclusions.

As a future possibility, the use of classical decision-making techniques such as PROMETHEE II and analytic hierarchy process (AHP) [10] is recommended, which are usually used to find satisfactory alternatives among possible solutions. In this sense, one could consider the weights already provided by those involved in the voting and forward them to the AHP technique for defining the most relevant points of the process.

The proposal of this work motivates the advancement of novel voting protocols, not only for the challenging problems of our cities, but also for the scope of novel negotiation protocols for multi-agent systems. The idea of bringing the evolution of computational techniques closer to the devices used by citizens, with a focus on impartiality and transparency, could possibly impact on significant advances in the quality of life of future societies.

References

1. Almirall, E., Wareham, J., Ratti, C., Conesa, P., Bria, F., Gaviria, A., Edmondson, A.: Smart cities at the crossroads: new tensions in city transformation. Calif. Manag. Rev. **59**(1), 141–152 (2016)
2. Bartholdi III, J.J., Tovey, C.A., Trick, M.A.: The computational difficulty of manipulating an election. Soc. Choice Welf. **6**(3), 227–241 (1989)
3. Batty, M., Axhausen, K.W., Giannotti, F., Pozdnoukhov, A., Bazzani, A., Wachowicz, M., Ouzounis, G., Portugali, Y.: Smart cities of the future. Eur. Phys. J. Spec. Top. **214**(1), 481–518 (2012). https://doi.org/10.1140/epjst/e2012-01703-3
4. Brinks, D.: Judicial reform and independence in Brazil and Argentina: the beginning of a new millennium. Tex. Int'l L. J. **40**, 595 (2004)
5. Cady, M.C., Yin, C.W., de Araújo Filho, J.P., Vasconcelos, A., Nascimento, J.M.B., de Jesus Cerqueira, A., Olívia, M., Setúbal, S., Junior, O.P.S., Pessoa, R.B.: Tribunal do júri: uma breve reflexão. Jus Navigandi. Teresina. a **9** (2014)
6. Cano, J., Hernandez, R., Ros, S.: Distributed framework for electronic democracy in smart cities. Computer **47**(10), 65–71 (2014)
7. Carvalho, E., Leito, N.: The power of judges: the supreme court and the Institutional Design of the National Council of Justice [o poder dos juzes: supremo tribunal federal e o desenho institucional do conselho nacional de justia]. Revista de Sociologia e Politica **21**(45), 13–27 (2013). https://doi.org/10.1590/S0104-44782013000100003

8. Chalkiadakis, G., Elkind, E., Polukarov, M., Jennings, N.R.: The price of democracy in coalition formation. In: Proceedings of The 8th International Conference on Autonomous Agents and Multiagent Systems - Volume 1, AAMAS '09, pp. 401–408. International Foundation for Autonomous Agents and Multiagent Systems, Richland (2009). http://dl.acm.org/citation.cfm?id=1558013.1558068
9. Coe, A., Paquet, G., Roy, J.: E-governance and smart communities: a social learning challenge. Soc. Sci. Comput. Rev. **19**(1), 80–93 (2001)
10. Dağdeviren, M.: Decision making in equipment selection: an integrated approach with AHP and PROMETHEE. J. Intell. Manuf. **19**(4), 397–406 (2008)
11. Dai, W.: B-money (1998). http://www.weidai.com/bmoney.txt
12. Endriss, U.: Vote manipulation in the presence of multiple sincere ballots. In: Proceedings of the 11th Conference on Theoretical Aspects of Rationality and Knowledge, pp. 125–134. ACM, New York (2007)
13. Gokhale, S., Kapshe, C.: Review of decentralised planning initiatives and urban local government functions in India. In: Dynamics of Local Governments: A Comparative Study of India, UK and the USA. Local Government Quarterly, p. 85. Local Government Institute, Mumbai (2016)
14. Hongfei, D., Zhang, E.: NEO: A Distributed Network for the Smart Economy (2018). https://github.com/neo-project/docs/blob/master/en-us/whitepaper.md
15. Lemos, R.: Using the Blockchain for the Public Interest (2016). https://medium.com/positive-returns/using-the-blockchain-for-the-public-interest-2ed1f5114036
16. Mattei, N., Narodytska, N., Walsh, T.: How hard is it to control an election by breaking ties? (2013, preprint). arXiv:1304.6174
17. Melo Filho, H.C.: A reforma do poder judiciário brasileiro. In: Seminário sobre a Reforma do Judiciário (2003)
18. Mission, S.C.: Ministry of Urban Development, Government of India (2015). http://smartcities.gov.in/content/innerpage/strategy.php
19. Nakamoto, S.: Bitcoin: A Peer-to-Peer Electronic Cash System (2008)
20. Odendaal, N.: Towards the digital city in South Africa: issues and constraints. J. Urban Technol. **13**(3), 29–48 (2006)
21. Pinheiro, A.C.: A reforma do judiciário: uma análise econômica. Trabalho apresentado no Seminário Internacional "Sociedade e a Reforma do Estado", realizado em São Paulo de **26** (1998)
22. Pinheiro, A.C.: Judiciário, reforma e economia: a visão dos magistrados. Instituto de Pesquisa Econômica Aplicada (Ipea) (2003). http://hdl.handle.net/11058/2900
23. Políticos, R.: Ranking políticos (2014). http://www.politicos.org.br/
24. Roberts, N.: Public deliberation in an age of direct citizen participation. Am. Rev. Public Adm. **34**(4), 315–353 (2004)
25. Schaffers, H., Komninos, N., Pallot, M., Trousse, B., Nilsson, M., Oliveira, A.: Smart cities and the future internet: towards cooperation frameworks for open innovation. In: The Future Internet Assembly, pp. 431–446. Springer, Heidelberg (2011)
26. Shah, A.: Balance, accountability, and responsiveness: lessons about decentralization. World Bank Policy Research Working (2021) (1999). https://ssrn.com/abstract=623937
27. Shoham, Y., Leyton-Brown, K.: Multiagent systems: algorithmic, game-theoretic, and logical foundations. In: Aggregating Preferences: Social Choice. Cambridge University, Cambridge (2008)
28. Silva, L.: Smartcities on the Ethereum blockchain (2016). https://www.ethnews.com/the-future-is-now-smart-cities-on-the-ethereum-blockchain
29. Sinhoretto, J.: Law reform (case study) [reforma da justia (estudo de caso)]. Tempo Social **19**(2), 157–177 (2007)
30. Stepan, A.: Brazil's decentralized federalism: bringing government closer to the citizens? Daedalus **129**(2), 145–169 (2000)

31. Terra, L., Ventura, C., Medeiros, M., Passador, J.: Strategies for the distribution of power in Brazil: a proposal from the perspective of the viable system model (VSM). Syst. Res. Behav. Sci. **33**(2), 224–234 (2016). https://doi.org/10.1002/sres.2378
32. THEKEY: A Decentralized Ecosystem of an Identity Verification Tool Using National Big-Data and Blockchain (2017). White Paper, https://www.thekey.vip/sites/default/files/pdf/THEKEY_Whitepaper_171112.pdf
33. Weyns, D., Omicini, A., Odell, J.: Environment as a first class abstraction in multiagent systems. Auton. Agent. Multi-Agent Syst. **14**(1), 5–30 (2007). https://doi.org/10.1007/s10458-006-0012-0
34. Zanella, A., Bui, N., Castellani, A., Vangelista, L., Zorzi, M.: Internet of things for smart cities. IEEE Internet Things J. **1**(1), 22–32 (2014)
35. Zeleny, M., Cochrane, J.L.: Multiple Criteria Decision Making. University of South Carolina, South Carolina (1973)

Chapter 14
Smart Planning: Tools, Concepts, and Approaches for a Sustainable Digital Transformation

Alexandre C. Barbosa, Taciano M. Moraes, Danielle T. Tesima, Ricardo C. Pontes, Alex de Sá Motta Lima, and Barbara Z. Azevedo

Abstract The Smart Planning has been presented in the discussions in both urban computing, operational research, and digital cities and in the urban and regional planning field. The first can be referred as process of the integration, analysis, and processing of data in order to optimize city services, while the second group has a long trajectory in defining and understanding the urban dynamics. As a matter of fact, conflicts in identifying key similarities in approaches for a more sustainable use of those tools are constant and must be carefully handled. Through an exploratory perspective, the goal is to highlight the vastness of the planning field and the rise of several complexities when introducing modern technologies in the urban sphere.

A. C. Barbosa (✉)
Università Degli Studi di Padova, Padova, Italy

Katholieke Universiteit Leuven, Leuven, Belgium

Universitè Paris 1 Panthèon-Sorbonne, Paris, France

T. M. Moraes
UFG Media Lab, Goiania, Brazil

D. T. Tesima
Architecture and Urbanism UFOP, Sao Paulo, Brazil

R. C. Pontes
Graduate Program, UFSC, Florianópolis, Brazil

A. de Sá Motta Lima
Graduate Program, Université Paris 1 Panthéon-Sorbonne, Paris, France

B. Z. Azevedo
Geological Engineering UFOP, Belo Horizonte, Brazil

V. N. Coelho et al. (eds.), *Smart and Digital Cities*, Urban Computing,
https://doi.org/10.1007/978-3-030-12255-3_14

14.1 Introduction

The process of transforming a city or a territory into a smart one is complex and takes time. It means that not only by introducing novel technologies in territorial systems will transform the current urban problems and challenges into opportunities. Territorial systems here refers to the combine of actors from diverse segments of society sharing a common space with values, principles, and power relations [34]. Actually, the process of digital transformation must be embraced as an interdisciplinary field from developers, computer scientists, data scientists, urban and regional planners, and policy-makers. Only by exploring and aggregating different approaches, tools, and concepts into the same debate, we will be able to promote a sustainable digital transformation on smart cities or territories. Otherwise, we may even increase the enormous societal challenges we face by increasing complexity only aiming efficiency at the most. Shortening the digital divide must be at the core of urban computing development.

The point is that technology itself is not able to solve every single problem in a simplistic way. Nonetheless, it must be considered as a fundamental tool to guarantee an inclusive planning process as well as by assuring rational decision-making. Technology is a means. Nevertheless, it is relevant to point out here that political understanding of it is fundamental to promote smart territories [14]. It is not a technological issue, but it is a political and economical one.

In this paper, we intended to explore the relations between urban planning and smart cities in an exploratory approach. Based on that, we could better frame what really involves the smart planning scope. That is, framing the introduction of emerging technologies in a territory through the perspective of urban experts. In sequence, we explored four different fields of the smart planning in order to make clear the vastness of such field. The first is about the Geographic Information System, and its use and applications to urban computing. The second addresses the topic of underground planning as an emergent field of applied geosciences supported by information and communication technologies. The third is about the big data analytics dashboards and their roles in allowing efficient and real-time urban monitoring. Finally, the topic of interoperability is provided taken into account its different layers with regard to the ethics when designing digital infrastructures.

The main objective of this broad study could be described mainly in two ways. First, we aimed to support software and hardware development that effectively tackles societal challenges. Second, we intend to make clear that urban and territorial systems are complex and the introduction of modern technologies must bear this in mind. Efficiency above all is not necessarily sustainable. Rather, it may entail irreversible legacies due to path dependencies and network effects. Therefore, the idea of smart planning supported in this article is comprehensive. Does one that have a great picture of and how some technologies are being used in the city scope at the moment and the future potential supported by an in-depth critical thinking. Merging the debate by bridging narratives is a key component for urban computing stakeholders.

14.2 Urban (Smart) Planning

14.2.1 Urban Planning Overview

Think city as a work, a history of actors, or well-defined people and groups who idealized it and materialize it. As presented in one of the most important books of urban planning ever [28], the result of a city in the present is not mere chance, but a set of materialization of different urban planning doctrines, and their socio-spatial practices, understood by everyday actions. The older it is, the more incredible morphological diversity it will imply, with continuities and discontinuities of these doctrines in practice. Accordingly, is complex, but not impossible, to promote a sustainable digital transformation in a historic city with an intricate system of territories, result of its past activities, among actors from different generations, with unique values, principles, and power relations [19].

It is essential to obtain an overview of how different ideals of urban planning shape the city, to analyze how the ideal of smart city inserts in this panorama. In the previous centuries, the ideal of a modern city rose up, the core of urban planning consists of a modern city planning, an idealized city model committed to the "needs" of the modern inhabitant, in which technicians are responsible for describing these needs, consolidating a *heteronomy* relationship of the inhabitants with the technicians [28].

The ideals of urban planning have transformed during its history, taking on numerous forms, from the first modernist experience held in the city of Paris in France by Baron de Haussmann—in which the city is rethought and the first incursion is made of a practice based on centralizing urban planning supported on a rational logic; to the following philosophical reworkings [28]. The case is if this new paradigm of smart planning will reinforce this relation of *heteronomy*, or will, actually, bring an opportunity for a relation of autonomy towards a multistakeholderism governance model [19], with more citizen participation.

It is relevant to mention the Albrechts's work and the statement that they vary mainly inside four rationalities of planning [3]. It facilitates the comprehension of how the smart planning thought is incorporated in the social reform, the policy analysis, the social learning, and the social mobilization [2]:

- Value rationality: normative way for an alternative future. Focus on the long term.
- Communicative rationality: multistakeholder process. Collective vision and trust as a process.
- Instrumental rationality: emphasis on the most efficient way to solve problems.
- Strategic rationality: power relations are handled as a strategy.

There are several paradigms of planning that can provide an interesting historical overview of the topic in both research and practice [1]. The most known and traditional one is the rational comprehensive model. This one emerged in post second world war scenario (even though highly influenced by the nineteenth century ones) and it is based on the professional-scientific knowledge to solve urban and social problems. It is basically oriented to decision-making process that entails costs minimization and benefits maximization. However, it does not take into account who take those decisions, neither assess its consequences in different scales and scenarios [11].

As a response to the previous approach, the Bounded political rationality one emerged. Herbert Simon highlighted that rationality is limited by uncertainty [38]. The other main actor, Charles Lindblom, expressed that what brings the rationality to a decision-making process is the plurality of actors. This is what he called "the intelligence of democracy" [26].

The role of politics was supported with the rise of the Advocacy Planning Model. With this regards, the planning process is still an expert-led one. Nonetheless, the point here is the capacity in controlling initiatives rather than on technical expertise [16]. In the seventies, in parallel with the new order dynamics, the Political Economy Model was raised. In the sense that the planning process served mainly as a neoliberal strategy [27].

As a matter of reaction, Metzger supported a planning model that redistributes power. This one was called the Equity Planning Model [31] and its first step is to understand spatial inequalities. The planner here is more a territorial settings analyzer and communicator. In the nineties, the Social Learning and Communicative Action Model were strengthened. In this perspective, the urban planner was the responsible to bridge the experts and the community. The main objective in this paradigm is to systemically empower citizens. This model also fed the Radical Planning Model, in which the planner is much more an ally of the community, rather than outside [8]. That is, a social mobilizer.

Another important paradigm is the Strategic Spatial Planning (SSP). This one is supported in this paper and, therefore, the following subsection provides a great picture of its mechanisms and approaches. The idea here is that the SSP is meant to be the optimal between the bottom-up and the top-down approaches. One that is both scientific-oriented and community-based.

14.2.1.1 Strategic Spatial Planning

There are several stages in the planning process that public participation can be conducted. Such as the problem identification, the socio-spatial context, the goals and objectives, desired outcomes, and the legitimization of the proposed policies through cohesion and consensus building. However, this paradigm of planning has been criticized with regard to political and professional rationality as well as in driving long-term development. Furthermore, the success of decision relies on their effects in transforming conflicts by guaranteeing rationality [26]. In addition, a big variety of actors playing a direct role on the decision-making process is crucial, but also dangerous for such achievement [25], governance is a must.

Strategic spatial planning is an important alternative to counterbalance the limits of collaborative planning. Albrechts [2] argues that strategic spatial planning is a public process that aggregates different perspectives based on scientific knowledge, collaboration and socio-spatial interactions that builds a comprehensive frame for the development of a territory. In addition, this planning approach is based on a limited number of issues, taking into account the social, cultural, and political aspects of the context. This approach fosters multilevel governance and diversity of actors in the planning processes. It is also about creating new forms of understanding the socio-spatial dynamics. Moreover, the transparency and accountability of institutions are key issues that may enable an effective inclusiveness, eliminating uneven power structures. More recently, it has been associated with the "performance school" [5], that affirms that strategic spatial planning is about a normative (what should prevail); and epistemic (based on the effect of a project, program, or policy) consensus.

Healey's definition for strategic spatial planning [23] illustrates the issue assessed in this paper. She states that it is:

> A social process through which a range of people in diverse institutional relations and positions come together to design plan-making processes and develop contents and strategies for the management of spatial change. This process generates not merely formal outputs in terms of policy and project proposals, but a decision framework that may influence relevant parties in their future investment and regulation activities. It may also generate ways of understanding of building agreement, of organizing, and mobilizing to influence in political arenas.

14.2.2 Urban Computing, Planning Support Systems, and Smart Cities

Planning Support Systems, or Spatial Decision Support Systems, are a consolidated thematic in research and practice, while Smart City spatial applications is relatively new, but it has been manifested in territories more and more. According to Geertman [21], its use has been commonly agreed in the scientific environment as mainly due to increased efficiency and socio-technical complexity handling.

Geoinformation technology-based tools that facilitates regional and urban planners in identifying major issues, in understanding urban dynamics and in developing strategical solutions [40].

In this age of exponential growth of information, there are so many technologies and emerging at such a frenetic pace, that frequently the feelings of public administration regarding to adopting new tools are more related to overload, disinterest, infeasibility, frustration, and waste instead of improvement, progress, benefit, optimization, and enthusiasm. In spite of this, the number of initiatives of applied computing in the cities has grown year by year and generating more and more cases of success, both by means of public–private partnerships, as well as executed by municipalities' own teams and public agencies.

Numerous applications are possible with the maturation of Applied Computing areas that are promising for the context of Smart Cities, among which it is worth mentioning four:

- Internet of Things—Sensors that collect humidity, temperature, pressure, light, movement, velocity, gases, electrical resistivity, acoustics, etc., enabling, among other things, intelligent traffic lights and signals that adapt dynamically according to the meteorological conditions or even in the case of other major accidents, opening the way for firefighters, police, or ambulances.
- Cloud Computing—Storage and processing infinitely superior to what was possible with local datacenters, allowing for example facial recognition of real-time security camera images combined with other police information about outlaw criminals.
- Big Data Analytics—Interpretation of standards and data cross-referencing with other federal, state, and municipal data, ensuring better planning, decision-making, and use of public resources in cases such as the construction of infrastructure in more critical or cost-effective places.
- Machine Learning—Combining data collected from public, private, internal sources to the public administration, sensors, and other devices spread throughout the city, these can be used by algorithms such as Deep Learning in the preventive identification of problems, such as in electricity, telephony, sanitation, internet, and transportation, among other possibilities.

Through the use of the mentioned technologies and others more, government officials and other public agents will have more and more relevant information to their work. However, in order for them not to grow in a disorderly fashion and bring more confusion than understanding, they need to be organized and made available in a way that is simple and quick to visualize, understand, and act in a context where the use of dashboards is crucial.

14.3 Smart Planning: Tools, Concepts, and Approaches

For a smart planning process to truly happen, all the stakeholders carry responsibilities, which includes: governments, local authorities, private sector, universities, and community. In the following sections, we provide several components of the digital strategy of a city, from the use of georeferenced data to its display in monitoring dashboards. Again, the goal here is not to drive a one-size-fits-all guideline. Rather, the point here is to make explicit the comprehensive complexity that urban computing entails. Co-responsibility in the design and implementation is what this chapter aim to promote.

Michael Bloomberg's administration in New York in the year 2001 is an outstanding example of a successful technology initiative in this area. Among several ideas implemented, it is worth mentioning an introduction of an accountability portal within the city hall website, which served both for population and for public agents to accompany city services, through a simple interface and with indicators easily comparable to every other region of the city. Bloomberg eventually managed the city as a completely data-driven organization, setting performance targets for each department, prioritizing a colossal amount of existing indicators, and comparing them with nationally recognized standards of performance in other large cities [10].

Almost a decade later, what allowed even more advanced and interconnected technologies, the city of Rio de Janeiro built in 2010 an Operations Center in partnership with IBM—that developed the system. Its datacenter is able to integrate data from 30 different agencies into a single software, allowing the crossing of information from sensors, cameras, and public transport GPS devices, with other provided by the technicians who operate the system, thus allowing quicker and more efficient decisions about floods, accidents, crimes, service failures, traffic jams, and other types of problems, in real time or even with future predictions. The system was designed primarily to provide information about the city during the World Cup and the Olympics that would happen in the following years, but continued to prove very useful and with enormous return on investment even after the end of these two events [39].

14.3.1 Geographical Information System (GIS) and Its Use

Geographical Information Systems (GIS) is a set of tools with a particular value, because takes into account the need to answer or explore spatial questions. This system allows the relation of spatial information, such as location, patterns, trends, and conditions. Enhancing the possibilities of technology in understanding the geographical dynamics [37]. There are many examples of GIS applications, related to activities of the government, defense agencies, scientific research, commerce and business, agriculture, and environmental management.

Considering the use of geotechnologies in the last 30 years, the boom of information systems and the popularization of the internet is possible to say that the relation between humans and spatial information has changed [22, 35]. The technology development allows great performance in acquisition, storage, sharing, and management of data.

Geography has always been rich of information, and it is broad known that spatial data roles a strategical and historical importance for human kind. Since the first settlements, until nowadays: Information as a key to know the locations of resources, the safest places to be or to move faster [35]. The principles of spatial information importance in the past are the same in our days. The urban territory is full of spatial information, and it is knowledge could help us to better understand the social, political, or economical dynamics, so the local environmental problems and climate changes.

The future of GIS in the urban computing is related by the possibilities of spatial data modelling, analysis, and management. With the development of computer interfaces for handling these information, such as Webgis, the use and access becomes more popular [22]. The creation, maintenance, and share of spatial information are central topics into GIS use, because they facilitate the technical production of high value materials for urban planning, and so, to urban sustainable development, smart cities, and urban computing.

14.3.2 Underground Smart Planning

14.3.2.1 Importance of Underground Urbanization

According to an UN report [42], the current world population of 7.6 billion is expected to reach 8.6 billion in 2030 and 9.8 billion in 2050. In addition, the percentage of the world's population living in urban areas is growing steadily [44]. The figures rose quickly from 1960, when only 34% of the total population were living in urban areas to over 54% in 2016. To this extent, locating this increasing population in urban areas is becoming more and more challenging. Cities' territorial expansion is limited by the natural environment constraint and farmland protection. Although, advances in tunneling and excavating technology and increases in land prices have resulted in underground urbanization becoming a more feasible and economical option [30].

14.3.2.2 Underground Planning: An International Overview

In the context of underground planning, Helsinki is the pioneer. It was the first city worldwide to adopt an Underground Master Plan in 2009. Since 1960s, the city has been largely using the facilities of underground construction. From the beginning of the twenty-first century on the demand for underground space in the central area started growing rapidly, as did the need to control construction work [43]. Underground resources are the core of development of Helsinki's structure and nearby areas, aiding to build a more unified and eco-efficient structure. Underground planning improves the overall economy efficiency of facilities situated underground and enhances the safety of these facilities and their use [43]. In 2007, the Deep City research program was launched, a partnership between Switzerland and China. It divided underground resources into four types: space, geomaterials, groundwater, and geothermal energy. In this manner, developing the underground means to create new alternatives for transportation, utility infrastructures, new sources of mineral resources, and renewable thermal energy [30]. In China, Suzhou City was chosen as a pilot project based on its higher score of "Deep City applicability" over other Chinese cities (Beijing, Nanjing, Suzhou, and Shanghai) [29]. The Suzhou provincial and municipal governments were assisted by gathering resource database, with administrative consultation and the development of an information system to help decision-making. The authors [30] gathered and bench-marked the most desirable practices in management and administration of underground resources, looking at seven cities (Helsinki, Singapore, Hong Kong, Montreal, Minneapolis, Tokyo, and Shanghai). After analyzing many different factors for the supply side such as geo-risks, soil thickness, and aquifer outflow, among others, and for demand side like civil defence, commercial land prices, land use type, etc. The authors highlight that an integrated underground use and management can only be possible if all the city departments work together into decision-making [30]. Meanwhile in the United Kingdom, the British Geological Survey is researching about Urban Geoscience and how geoscientists can help to reshape future cities. The "Future of Cities" Project, launched in 2013 by the Government Office for Science, aimed to provide policy makers with the evidence, tools, and capabilities needed to support policy decisions in the short term to lead to positive outcomes for the UK cities in the long term [13, 41]. The subsurface is taken for granted when thinking about future cities, as its demand is growing continuously and it can provide many resources, as mentioned before. Sustainable utilization of urban underground requires a greater understanding of subsurface processes and a careful, well-considered, integrated planning approach [7].

14.3.2.3 Underground Planning: A Brazilian Overview

After researching in different databases (both Brazilian and international) about the theme, little information was found. However, two PhD theses stand out in this context. The first one is called "Urban Geology of São Paulo's Metropolitan Region," of 1998. This thesis presents a geological description of the area followed by the history of urbanization of the biggest Brazilian city. The author dedicates one chapter only for "Urban Geology," where he highlights the importance of geological, geomorphological, geotechnical, and hydrogeological knowledge in the process of urbanization. This work mapped the potential areas for landslides and floods in São Paulo's metropolitan area [36]. Although this work was presented on the relation between geology and urban planning, it was classified in the field of "sedimentary geology." The second thesis was written by an urban planner in 2009 and it is called "Underground Urbanism." The author discusses the ethical arguments for land use and occupation. She endorses the densification of cities by expanding downwards, enabling different activities to overlap. By doing this, the use of the ground is maximized without the requirement of skyscrapers. Her work found no urbanistic nor economic or technological restrictions for the use and occupation of the subsurface upon an indoor city [15].

14.3.2.4 Tools for Urban Planning

One of the simplest ways of communicating geoscience for urban planners and broad community is through maps. They are great at synthesizing information about geology, geomorphology, engineering geology, geotechnics, subsurface investigations, hydrology, hydrogeology, coastal zones management, planning, and land use. It is relevant to highlight the importance of an accurate ground field survey and inventory at different scales, Geographic Information Systems (GIS) mapping and databases, and seamless multidisciplinary urban studies as useful instruments for supporting a sustainable land use planning [7]. A key aspect of the smart urban geoscience concept must include Geographic Information Systems (GIS) as a means for digital mapping and communication to a larger audience. That framework includes merging numerous data about all elements of urban areas, such as transport, environment, economy, housing, culture, science, population, health, history, architecture, heritage, etc. Moreover, the format of a balanced multipurpose engineering geological map strongly relies on its main objective and the requirements of the end users, as well as the need to communicate information to all agents involved (practitioners, researchers, stakeholders, decision-makers, and the public) [15].

14.3.2.5 Future Prospects

Subsurface planning is a multidisciplinary task, so geoscientists will be required to work together with other professionals (e.g., architects, urban planners, and engineers), as well as local and national governments. If underground urbanization was used more in big cities worldwide, it could help to address sustainable development in terms of optimizing the land use, conserving supplies such as groundwater, geothermal energy, space, and mineral resources [30]. In this scenario, it is possible to state that mapping plays a central role in urban geoscience, specially for in situ geotechnical investigations, ground modelling, geological resources, heritage and geohazard assessments, and planning purposes. Once again, the importance of GIS is highlighted as a useful tool for urban planning management in order to assure a sustainable design with nature, environment, heritage, and society. At municipal level, urban geoscience studies assume major significance contributing to land/cover use management and planning. Therefore, innovative methodological approaches are required in the collection, analysis, design, and modelling of urban data [15].

14.3.3 Big Data Analytics Dashboards

With the massive amount of sensors and devices currently positioned in cities, it is inevitable to gather an exorbitant amount of information that ends up generating two great problems for the public administration: the first is infrastructure—increasing the amount and storage of servers or even your data is in the cloud, costs increase, and management is difficult; and the second in terms of viewing a colossal number of different information about the same subject (i.e., the city), which makes it difficult for people who need to analyze all this data to plan and take decisions.

The first problem turns out to be intrinsic and unlikely to be solved, perhaps only mitigated if the information produced grows linearly and technological evolution continues to grow exponentially, following Moore's famous Law [32]. However, a dashboard (or a set of them with different perspectives) can undoubtedly help solving the second problem or at least a significant part of it. However, in order for this to happen, this dashboard needs to follow some basic elements to ensure its effectiveness [24]:

1. Be simple and communicate in an uncomplicated way,
2. Possess minimal effects and distractions to avoid confusion,
3. Provide useful and meaningful information to stakeholders,
4. Adequate presentation of data to human visual perception.

There are numerous ways to implement a dashboard, depending on the context and the objectives it needs to fulfill. It can be implemented in a physical way through large displays in places accessible only to a restricted internal public, with more strategic and operational information. An example of this use is the dashboard of the aforementioned Operations Center of the City of Rio de Janeiro [18], which is considered one of the most modern in the world and played a key role during the Olympics and World.

Furthermore, it can be available online to an external public—so that it is accessible to a larger number of people—in order to ensure the accountability of municipal management. As an instance, the Inter-American Development Bank provides an integrated platform called Urban Dashboard [6] in which any person can visualize all indicators of the 23 themes that make up the three dimensions (urban, environmental and fiscal) of their methodology, and can also compare the indicators among all 77 cities participating in the program. Similarly but with a different focus, it can be a state or country initiative in the form of a benchmark between various cities of a region or even done by an NGO to compare cities from several countries, such as the Cities In Motion Index [9].

With such improvement in accountability provided by these dashboards, citizens can now understand the evolution of the city's main areas and also check whether their taxes are being well used, feeling not only better served by municipal management, but also more responsible, motivated, and empowered to proactively help solving city problems [4].

In addition to improving the city's accountability to the population, there is another reason why the use of dashboards is so interesting to smart cities and urban computing. Once two cities use the same assessment methodology, ensuring that information is standardized, even that they are on continents with different geographic conditions, have completely opposite cultures, or have different political-economic systems, they are still comparable through that set of parameters. Through dashboards you can establish a benchmarking and draw a parallel plan between them, allowing you to check which one is performing best in a particular area, through common indicators.

Such contrast is also interesting to stimulate among cities and administrations a certain amount of competition and ambition, characteristics usually absent (or present only on an individual perspective and in a counterproductive manner) in this context. Since cities well positioned in these rankings can attract even more investment and tourists, the stimulus becomes even more real for city managers.

Furthermore, other authors [12] argue that the use of dashboards also brings many advantages for city managers and public agents:

- Visual presentation of performance measures,
- Ability to identify and correct negative trends,
- Saves time compared to generating multiple reports,
- Measures efficiencies/inefficiencies,
- Ability to make more informed decisions based on Business Intelligence,
- Aligns organizational strategies and objectives,
- Gains full visibility of all systems instantly,
- Ability to generate detailed reports showing new trends,
- Quick identification of out-of-curve points and data correlations.

It is important to emphasize that these benefits on city management are also reflected to the population in terms of improved quality of life and with public services being offered more quickly and effectively. Ultimately, city dashboards benefit the population by giving better information access and visibility to public managers and investors.

14.3.4 Human and Computer Interoperability

When talking about software development, territorial planning, and governance models in a socio-technical environment, the issue of interoperability is fundamental. This topic is often mentioned in the media, nonetheless with few in-depth analysis about what is meant to be an efficient and resilient interoperable system [17].

As expected in the technological scenario, interoperability could be briefly defined as the capacity of devices, software, and data to easily communicate among them. However, according to Gasser and Palfrey [33] in the book *Interop: the promise and perils of highly interconnected systems*, interoperability could be subdivided in different ways. One of those is by dividing into layers: the technological, the data, the human, and the institutional ones. In this research, the idea is to provide a broad perspective of this theory in terms of conceptualization, current scenario, and normative future.

Another way to classify interoperability is as being vertical or horizontal. The first refers to the previous capacities mentioned inside the same organization, the case of *Apple®*. The latter is, rather, the possibility of exchange among other entities [20], the case of *Android®*.

The point is that interoperability to genuinely happen it should be able to optimally enable exchange and flow while preserving diversity. It is not about a one-size-fits-all standard approach to technology. Actually, it is about the ability of different entities to cooperate in order to promote network effects with utility for society as a whole [33]. Moreover, it should foster competition and innovation; but it must be based on open standards and transparency.

14.4 Final Considerations and Future Research

With the expansion and popularization of technology, now applied to all other areas of knowledge, more and more municipal managers, and citizens have been interested in the benefits of the so-called Smart Cities. However, numerous cities around the world are already titled in this way, these projects have still been very challenging, failing in planning (due to the complexity of variables), in their application (because they do not meet the real needs of the population) and in their budgets (implementing expensive and inefficient technologies with low return on investment).

This chapter enjoyed the opportunity to provide a comprehensive picture of the urban planning ecosystem. The reason for this was both in the sense of raising awareness of the diverse complexities that involves urban dynamics as well to support technology-based solutions for societal challenges.

First, we intended to provide an overview of the urban planning theory through a synthesized historical overview. The goal here was to highlight the *fluid* aspects of cities, one of unpredictable outcomes. Technology must support those systems, and technologists should take this into consideration. Solutions to indeed tackle the challenges we face today can be and should be designed in collaborative way. Technicalities and humanities must walk side by side.

This is the idea of Smart Planning supported in this paper, an approach that takes diverse urban intrinsic components into account supported by emerging technologies. Additionally, several studies and success stories have shown how much of Applied Computing concepts like Internet of Things, Cloud Computing, Big Data Analytics, and Machine Learning can revolutionize public administration and urban planning. The correct adoption of connectivity infrastructure; sensors and devices; integrated operating centers; and communication interfaces with the population can certainly bring countless gains to cities.

The paper also provided a broad picture of urban geosciences potentials. One by introducing the concepts of geographic information systems, since this source of data serves as the basis for urban computing software development. Moreover, the study also explores the emerging field of underground planning, through which the technology will increasingly play a fundamental role.

As an interface of those systems and the urban management and planning, the dashboards are also highlighted here. It functions above all as the translation of data and code into friendly visual screens. This is crucial for the technology complexity handling, hence more efficient decision-making.

In order to make it possible to visualize the infinity of data generated by both managers and the population, dashboards should be planned and implemented with a focus on the priority areas and information relevant to the context in question. These can be developed both with a focus on the external public (population and potential investors) as well as internal to the city hall and public agencies.

Through dashboards, both managers and municipal agents, as well as citizens and outside investors, have access to truly objective, factual, neutral, comparable, comprehensive, and reliable information that cannot normally be found in the information about the cities served by the media. In future, with its popularization, society will experience a profound change not only in the way cities are governed and managed, but also in the way we understand and experience urban life.

References

1. Albrechts, L.: Strategic (spatial) planning reexamined. Environ. Plann. B. Plann. Des. **31**(5), 743–758 (2004)
2. Albrechts, L.: Bridge the gap: from spatial planning to strategic projects. Eur. Plan. Stud. **14**(10), 1487–1500 (2006). https://doi.org/10.1080/09654310600852464
3. Albrechts, L.: Ingredients for a more radical strategic spatial planning. Environ. Plann. B. Plann. Des. **42**(3), 510–525 (2015). https://doi.org/10.1068/b130104p
4. Arnstein, S.R.: A ladder of citizen participation. J. Am. Inst. Plann. **35**(4), 216–224 (1969)
5. Bafarasat, A.Z.: In pursuit of productive conflict in strategic planning: project identification. Eur. Plan. Stud. **24**(11), 2057–2075 (2016). https://doi.org/10.1080/09654313.2016.1231800
6. Bank, I.D.: Urban dashboard. www.urbandashboard.org. Accessed 25 Jan 2018
7. Barkwith, A.: Optimising subsurface use for future cities. UK Government Office for Science (2015). https://assets.publishing.service.gov.uk/government/uploads/system/uploads/attachment_data/file/461763/future-cities-optimising-subsurface-use.pdf
8. Bencardino, M., Nesticó, A.: Urban Sprawl, Labor Incomes and Real Estate Values. In: Gervasi, O., et al. (eds.) Computational Science and Its Applications – ICCSA 2017. Lecture Notes in Computer Science, vol. 10405, pp. 17—30. Springer, Cham (2017). https://doi.org/10.1007/978-3-319-62395-5_2
9. Berrone, P., Ricart, J.E., Carraso, C., Ricart, R.: IESE Cities in Motion Index 2017. IESE Business School, University of Navarra, Navarra (2017)
10. Brash, J.: Bloomberg's New York: Class and Governance in the Luxury City, vol. 6. University of Georgia Press, Athens (2011)
11. Breheny, M.J.: A practical view of planning theory. Environ. Plann. B. Plann. Des. **10**(1), 101–115 (1983). https://doi.org/10.1068/b100101
12. Briggs, J.: Management reports & dashboard best practice. Target Dashboard. Retrieved 18 (2013). https://www.targetdashboard.com/site/Dashboard-Best-Practice/Management-Report-and-Dashboard-best-practice-index.aspx
13. British Geological Survey urban geoscience: future cities. http://www.bgs.ac.uk/research/engineeringGeology/urbanGeoscience/futureCities.html. Accessed 28 Feb 2018
14. Buck, N.T., While, A.: Competitive urbanism and the limits to smart city innovation: the UK future cities initiative. Urban Stud. **54**(2), 501–519 (2017). https://doi.org/10.1177/0042098015597162

15. Chaminé, H.I., Teixeira, J., Freitas, L., Pires, A., Silva, R.S., Pinho, T., Monteiro, R., Costa, A.L., Abreu, T., Trigo, J.F., et al.: From engineering geosciences mapping towards sustainable urban planning. Eur. Geologist J. **41**, 16–25 (2016)
16. Checkoway, B.: Paul Davidoff and advocacy planning in retrospect. J. Am. Plann. Assoc. **60**(2), 139–143 (1994). https://doi.org/10.1080/01944369408975562
17. Diallo, S.Y., Herencia-Zapana, H., Padilla, J.J., Tolk, A.: Understanding interoperability. In: Proceedings of the 2011 Emerging M&S Applications in Industry and Academia Symposium, EAIA'11, pp. 84–91. Society for Computer Simulation International, San Diego (2011). http://dl.acm.org/citation.cfm?id=2048513.2048530
18. do Rio de Janeiro, P.: Centro de operacoes do rio. http://cor.rio/. Accessed 17 Mar 2018
19. Friendly, A.: The right to the city: theory and practice in Brazil. Plann. Theory Pract. **14**(2), 158–179 (2013). https://doi.org/10.1080/14649357.2013.783098
20. Gasser, U.: Interoperability in the digital ecosystem (2015). http://dx.doi.or/10.213/ssrn.2639210
21. Geertman, S., Stillwell, J.: Planning Support Systems: Content, Issues and Trends. Springer, Dordrecht (2009)
22. Giacaglia, G.E.O.: A industria aeroespacial: questoes economicas, tecnologicas e sociais. Estudos Avancados **8**, 42–49 (1994)
23. Healey, P.: The treatment of space and place in the new strategic spatial planning in Europe. Int. J. Urban Reg. Res. **28**(1), 45–67 (2004). http://dx.doi.org/10.1111/j.0309-1317.2004.00502.x
24. Hetherington, V.: Dashboard demystified: what is a dashboard. Olszak & Ziemba (2004). Business intelligence systems as a new generation of decision support (2009)
25. Iassinovski, S., Artiba, A., Bachelet, V., Riane, F.: Integration of simulation and optimization for solving complex decision making problems. Int. J. Prod. Econ. **85**(1), 3–10 (2003)
26. Jean, L.: Dahl (robert a.) lindblom (charles e.) - politics, economics and welfare. Planning and politico-economic systems resolved into basic social processes. Revue économique **6**(2), 328–329 (1955)
27. Kantor, P., Savitch, H.V., Haddock, S.V.: The political economy of urban regimes: a comparative perspective. Urban Aff. Rev. **32**(3), 348–377 (1997). https://doi.org/10.1177/107808749703200303
28. Lefebvre, H.: Le Droit a la ville. Anthropos, Paris (1968)
29. Li, H.Q., Parriaux, A., Thalmann, P., Li, X.Z.: An integrated planning concept for the emerging underground urbanism: deep city method part 1 concept, process and application. Tunn. Undergr. Space Technol. **38**, 559–568 (2013)
30. Li, H., Li, X., Soh, C.K.: An integrated strategy for sustainable development of the urban underground: from strategic, economic and societal aspects. Tunn. Undergr. Space Technol. **55**, 67–82 (2016)
31. Metzger, J.T.: The theory and practice of equity planning: an annotated bibliography. J. Plan. Lit. **11**(1), 112–126 (1996). https://doi.org/10.1177/088541229601100106
32. Moore, G.: Gordon Moore: the man whose name means progress, the visionary engineer reflects on 50 years of Moore's law. IEEE Spectrum: Special Report, 50 (2017)
33. Palfrey, J.G., Gasser, U.: Interop: The Promise and Perils of Highly Interconnected Systems. Basic Books (2012)
34. Pike, A., Rodríguez-Pose, A., Tomaney, J.: What kind of local and regional development and for whom? Reg. Stud. **41**(9), 1253–1269 (2007). https://doi.org/10.1080/00343400701543355
35. Plewe, B.: GIS Online: Information Retrieval, Mapping, and the Internet, 1st edn. OnWord Press, Santa Fe (1997)
36. Rodriguez, S.K.: Geologia urbana da região metropolitana de São Paulo. Ph.D. thesis, Universidade de São Paulo (1998)
37. Saleh, B., Sadoun, B.: Design and implementation of a gis system for planning. Int. J. Digit. Libr. **6**(2), 210–218 (2006). https://doi.org/10.1007/s00799-005-0117-0
38. Scott, A.J., Roweis, S.T.: Urban planning in theory and practice: a reappraisal. Environ. Plan. A Econ. Space **9**(10), 1097–1119 (1977). https://doi.org/10.1068/a091097

39. Singer, N.: Mission control, built for cities: IBM takes 'smarter cities' concept to Rio de Janeiro. New York Times, March 3 (2012)
40. Sugumaran, R., Degroote, J.: Spatial Decision Support Systems: Principles and Practices, 1st edn. CRC Press, Boca Raton (2010)
41. UK Government Office for Science future of cities collection. https://www.gov.uk/government/collections/future-of-cities. Accessed 28 Feb 2018
42. UN Department of Economic and Social Affairs world population projected to reach 9.8 billion in 2050, and 11.2 billion in 2100. https://data.worldbank.org/indicator/SP.URB.TOTL.IN.ZS. Accessed 28 Feb 2018
43. Vahaaho, I.: 0-land use: underground resources and master plan in Helsinki. In: Zhou, Y., Cai, J., Sterling, R. (eds.) Advances in Underground Space Development, Copyright, p. 56-14 (2013)
44. World Bank urban population. https://data.worldbank.org/indicator/SP.URB.TOTL.IN.ZS. Accessed 28 Feb 2018

Part IV
Emerging Cities' Services and Systems

Chapter 15
When CI and Decentralized Systems Effectively Meet Smart Cities and Grids

Vitor N. Coelho, Yuri B. Gabrich, Thays A. Oliveira, Luiz S. Ochi, Alexandre C. Barbosa, and Igor M. Coelho

Abstract A global trend has been motivating programmers, investors, and the academia to go towards decentralized systems. Besides providing efficient solutions for complex problems faced in our daily life, these peer-to-peer communication protocols have been promoting greater freedom and transparency. In this paper, we point out open fields for researching, thinking, and developing in the context of Smart Cities and Smart Grids. Future cities will surely rely on efficient, autonomous, transparent, collaborative, and decentralized environment. In particular, we consider how renewable energy resources could be integrated with mini-/microgrids, which will be hearth of the future cities. Furthermore, we discuss possibilities for promoting territorial development, through the assistance of Blockchain-based platforms embedded with Computational Intelligence tools.

V. N. Coelho (✉) · L. S. Ochi
Institute of Computer Science, Universidade Federal Fluminense, Niterói, RJ, Brazil
e-mail: satoru@ic.uff.br; http://causahumana.org/vitor; https://creating.city

Y. B. Gabrich
Department of Computer Science, Universidade do Estado do Rio de Janeiro, Rio de Janeiro, RJ, Brazil
e-mail: yuri.gabrich.br@ieee.org

T. A. Oliveira
Department of Engineering and Information and Communication Technologies, Universitat Pompeu Fabra, Barcelona, Spain
e-mail: thaysaparecida.deoliveira01@estudiant.upf.edu

A. C. Barbosa
Università Degli Studi di Padova, Padova, Italy

Katholieke Universiteit Leuven, Leuven, Belgium

Universitè Paris 1 Panthèon-Sorbonne, Paris, France

I. M. Coelho
Instituto de Matemática e Estatística, Universidade do Estado do Rio de Janeiro, Rio de Janeiro, RJ, Brazil
e-mail: igor.machado@ime.uerj.br

V. N. Coelho et al. (eds.), *Smart and Digital Cities*, Urban Computing,
https://doi.org/10.1007/978-3-030-12255-3_15

15.1 Introduction

Academia and the industry are moving towards decentralization and scalable approaches for handling our modern systems [46]. Several decentralization levels exist in order to achieve specific goals, such as providing greater service stability and allowing the emergence of more scalable solutions to deal with computationally expensive problems. Public databases and lots of papers are now being stored into public decentralized storages, including ledgers [31]. Recent groundbreaking technological advances managed to solve the double-spending problem, creating the first public distributed ledger, called the Blockchain, supporting Bitcoin cryptocurrency [35]. Many other blockchain technologies have been developed since then, inspired by the success improvements on the open-source code of projects like: Ethereum [8], Neo, Ripple, and Litecoin, among others. The integration of blockchain technology with smart contract languages, such as Solidity [53], allowed the creation of trustless computing networks that are fully autonomous and capable of replacing centralized infrastructures. The case of Brooklyn Microgrid (BMG) [33] is an interesting recent example of connection between Smart Grids and Blockchain-based technologies [21]. Figure 15.1, inspired from [33], exemplifies the virtual and physical layers of BMG, which will intrinsically be the heart of distributed power trading. This combination of fully distributed components with high-performance computing is also going to be part of cities' environment. As can be seen, deterministic and reliable engines should take care of market operations, while private blockchains can play the role of the transparency and reliability.

Discussions surrounding what are the Smart Cities (SC) [41] have been recently reaching several strategical areas. In particular, works from the literature have been highlighting the importance of the energy grid in these idealized cities [29]. As pointed out by Hofman et al. [25], better energy quality may even lead to greater availability of high-quality fresh water. Renewable energy resources may even promote remigration to rural areas, as well as preventing urban exodus [26], due to the possibility of offering "urban services and facilities" in isolated spots and promoting smart territories. In addition, with the rise of these aforementioned distributed databases, in connection with peer-to-peer cash systems [35], a novel structure for payments between autonomous entities will be reality, changing the way that citizens will interact with cities and services [42]. The idea of establish and transform communities for using these systems can turn some points emphasized by Wei Dai in 1998 into reality, reproduced below [16]:

> I am fascinated by Tim May's crypto-anarchy. Unlike the communities traditionally associated with the word "anarchy," in a crypto-anarchy the government is not temporarily destroyed but permanently forbidden and permanently unnecessary. It is a community where the threat of violence is impotent because violence is impossible, and violence is impossible because its participants cannot be linked to their true names or physical locations.
>
> Until now it is not clear, even theoretically, how such a community could operate. A community is defined by the cooperation of its participants, and efficient cooperation requires a medium of exchange (money) and a way to enforce contracts. Traditionally, these

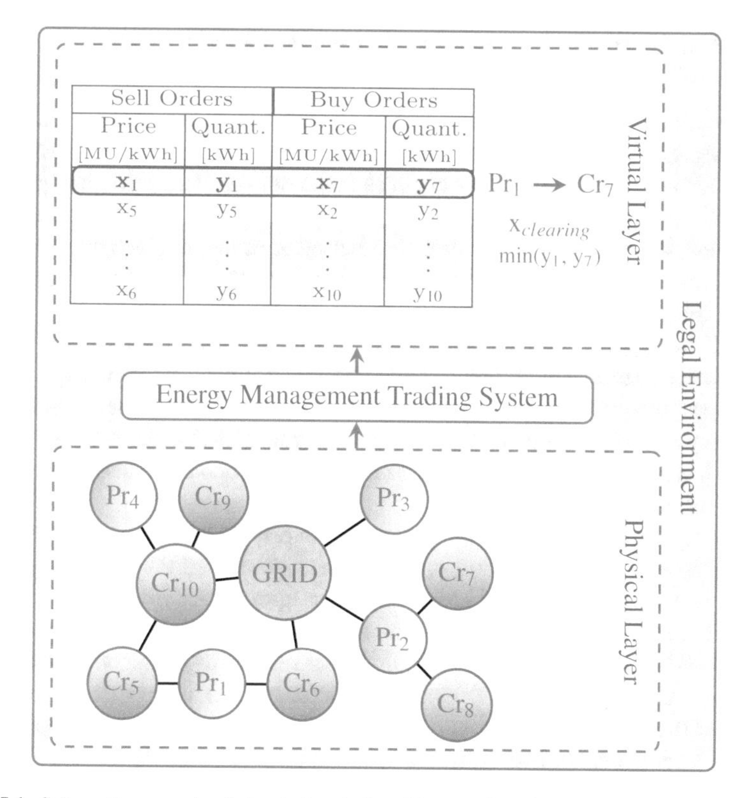

Fig. 15.1 Schematic example of physical and virtual layers of a microgrid

services have been provided by the government or government sponsored institutions and only to legal entities. (…)

In this paper, we contribute to the literature by discussing the importance of evolving the energy grids and how decentralized technologies can promote a stable and fair system, throughout the use of Computational Intelligence (CI) tools, in connection with the future cities. Through an interdisciplinary perspective, we highlight citizens' possibilities and their important role in this transformation in order to promote the balance between the efficiency and resilience of sociotechnical systems.

This paper is organized in five sections, including this Introduction. Section 15.2 describes the state of the art for energy systems in fully distributed environments and smart cities, also pointing out current challenges. Section 15.3 discusses the impact of decentralization for urban and rural areas, and how smart grids and decentralized systems are reshaping society organization. Section 15.4 presents open challenges and how they can be possibly solved by applying distributed technologies combined with computational intelligence. Finally, Sect. 15.5 concludes the work, with a brief discussion and future perspectives on the topic of smart cities, computational intelligence, and distributed systems.

15.2 Current Challenges and State-of-the-Art Systems

The concept of decentralization is commonly referred to any activity that used to be centralized, in terms of both private businesses and government policies, for instance. An example with supply chain case can help us clarify the change that has happened. Firstly, the distribution center used to be centralized for better control and supervision of the whole process, but it used to have a big lack on velocity and high delivery cost. So, some improvements were made to adjust those two variables such as inserting other distribution centers on strategic places to deal with all related demands. The constant evolution of distributed distribution centers has been improved by Artificial Intelligence (AI) and CI data analysis, being the measure of its impact the success by customers satisfaction, a real market thermometer. Self-adaptive systems have gained much improved learning capabilities even for complex problems, such as AlphaGo, that performed the near impossible task of beating the world champion of Go [43].

Similarly, expansion on power grids was shaped by consumers' needs, distinguishing by the way studies and planning were conducted though. On electricity market, stakeholders used to be divided into groups accordingly with similar technical characteristics and this is considered to determine future power demand and generation, based simply on linear growth planning [18]. On supply chain case, stakeholders have a more important role since their inputs are considered to determine the best solution for each occasion, given to its future plan operation a dynamism and complex behavior.

In our society, it is broadly accepted the importance of electricity in any activity. Not just for country economic growth/stability but as well for a rising on humanity quality of life [27]. In this way, decentralization of power generation and redesign of energy market sounds trivial, if not for difficulty in integration of small power customers into metering data sharing and technical issues [28].

Solar cells are becoming more affordable and, consequently, deployment of residential photovoltaic microgrids is increasing [24, 45]. Furthermore, optimization and novel simulation tools are assisting the achievement of efficient and resilient solar cells [23]. Coelho et al. [13] pointed out possibilities for using electric vehicles in connection with citizens wishes, designing a novel optimization model for planning energy power dispatching in a multicriteria view. On the other hand, a centralized agent cannot handle this scenario in an effective manner. Not only because it would be subjected to optimize its own profits, but also because high costly computational activities are requested. In this sense, a fully distributed system sounds reasonable. On the other hand, these systems require smart devices to interact and reach agreements, as highlighted by Coelho et al. [15]. MAS paradigms are an interesting and groundbreaking solutions for mini-/microgrid and smart cities' services that rely on autonomous agents.

The mini-/microgrid concept is part of this change, in which Distributed Energy Resources (DER) have been gradually implemented, promoting traditional power grids in regard to local electricity needs and desire [2]. Different cases around the world have been facing several technical challenges and resolutions [19]. Either for differences in power access on urban and rural areas or in levels of power capacity installed.

Despite technical subjects, however, some approaches are arising to connect people directly and reshape the way energy market behaves by optimizing territorial supply and demand. Although mini-/microgrid communities with DER act off-grid as an outcome to the lack of power lines in rural areas [27], the communication between its adopters used to be *in locu* and defined as a one-way-direction problem solving. In addition, these systems have been used by citizens to generate green energy locally on the existing power network. Moreover, blockchain technology is allowing a broad opportunity for those groups to exchange electricity as a custom business model instead.

Decentralization is a well-known term on computer environment. Computer technologies and architectures have been a "trend of fluctuation between the centralization and subsequent decentralization of computing power, storage, infrastructure, protocols, and code" [20]. Blockchain, in its turn, is impacting directly on how computing, storage, and processing are used relying on "a peer-to-peer distributed ledger forged by consensus, combined with a system for 'smart contracts' and other assistive technologies" [20]. In which, it "give(s) explicit control of digital assets to end users and remove the need to trust any third-party servers and infrastructure" [5].

On this scope, individuals have been experiencing a diverse return beyond the usual (and expected) profit. For instance, it is already possible to send power credits via cryptocurrency payment to a remote intercontinental smart meter for a limited energy-consuming period, working as a prepaid power bill plan, and a way to charity safely [48]. Additionally, it is relevant to point out that sustainable development is territorially embedded and in an anthropological way it is about enhancing individual competences and autonomy [37]. Therefore, DER can promote social innovations by empowering local rural users to become solar entrepreneurs, to deal with solar power as an investment program, to connect households and businesses in a diverse approach [44]. Through the same approach, although under a limited group in a big city, the raised flag is both sustainability and independence. Either in order to choose a preferred power source or to stay powered on during a blackout [7]. In a more broad spectrum of initiatives, by partnering with local power utility, individuals take part to be co-creators and co-responsible for their energy needs and supply, no matter where you are from, because blockchain tokens can be acquired by anyone in the world, i.e., individuals can be a nature activist, a local grid user, or a remote investor. Within this environment, all views commented before are possible and new ones keep welcome too [38]. In this regard, it is important to highlight the issue of governance of distributed ledger technologies [12, 51]. Crossing out borders, the same project has aided to create the term Microgrid-as-a-Service (MaaS), which is the generation and management of each individual electricity open space to trade it in case of excess energy [47].

The development of these new approaches has been improving solutions and creating new local challenges, specifically on cities, where social, economical, and political structures are tighter. Nevertheless, it is only on the embryonic stage, what may explain a certain enthusiasm.

15.3 Connecting Individuals and Services in Urban and Rural Areas

Smart cities should be designed and receive a special attention in handling decentralized energy sources. Policy-makers should be aware about the possibilities and the role of decentralizing infrastructure. Therefore, think tanks and territorial technology centers implementation are of extreme importance. These establishments are the geographical representation of the bridge for interdisciplinary applied research. Several countries around the globe are embarking on this journey [1, 4, 9, 17, 30, 36, 49, 52]. A question that has been bordering the energy planner is: "Will an adapted energy grid become better than a new designed one, coming from the sand?" Several novel startups are being developed in countries in the "Global South," motivating studies and innovation on these cities under development.

Energy commercialization is in the heart of the future cities, allowing more efficient, reliable, and cheaper energy. It is relevant to point out the great potential for these distributed energy exchange scenarios. It may enact the sharing economy development and territorial cohesion being interpreted by society as an exchange agreement. By localizing this supply and demand challenges, the process of resilience building is much more likely to happen. The idea of using a blockchain-based software for peer-to-peer energy trading has been quoted in Australia [50], tracking energy trade from point to point.

Citizens' houses and buildings will respond to price signals from the market and interact with the energy grid [34]. Demand Response has been advocated as a new Virtual Energy Storage System (VESS), able to intelligently manage energy consumption of loads [10]. Large amount of available data will surely assist these aforementioned tasks [6].

To guarantee the shared control of these DER, by who is being directly benefited from it, a set of rules and guidelines must be clearly defined in advance. Only by doing so, we can indeed promote a smart and decentralized governance system in terms of urban/territorial collaboration [32]. This structure can be supported, for example, by the Ethereum-based Decentralized Autonomous Organizations (DAO). This tool can enable the participation of the stakeholders, consolidate a transparent voting system, and solve conflicts without the need of high levels of human capital [11].

15.4 Open Challenges and Future Research Directions

Renewables penetration as mini-/microgrid will surely affect power system reliability, because power grid operation used to be stable with synchronous machines correlation between generation and consuming balances, and now and beyond this pattern will be more discrete [2]. All variables that imply on power quality will demand particular attention to keep on acceptable levels. Studies of the effect of distributed energy have already been discussed to surpass those challenges [28]. Specifically, those concerning load forecasting [14, 18] and economics behavior were compared to energy balance with DER and electricity storage [13, 54].

Even though DER have a positive return for its adopters and blockchain could expand these enthusiasm, "the success, failure, and reasons behind its adoption can stem from a multitude of different policies that are often strongly rooted in the individual context of each country" [39]. Ramalho et al. [39] complement that "decentralized and socialized provision of energy will clear the path for an aspired participatory form of democracy," but financial and economical crises have a huge influence on policy motivation and adoption to support it. As an example, for some utilities' business model, current rate structure is inadequate for dealing with such diffusion, creating cross subsidies and threatening the important economic equilibrium of the power services [22].

Moreover, the smartness expected for the city of the future is open to develop not only power grid innovation but social aspects as well. In this sense, architecture, social geography, regional, and urban planning are the main disciplines related to understand the incoherent performance of cities, crafted by its own citizens [40] in a run to urban areas. Therefore, experts from these fields can contribute to better comprehension of negative gentrification processes; that is, consequences that can lead to population migration and displacement forced by community restructuring or law imposing [3]. CI abilities should bare this issue in mind.

15.5 Final Considerations

We have discussed in this paper topics involving decentralized systems (mainly smart grids) and their impact on society. Based on dozens of references in literature, we strongly believe that support fully distributed paradigms is the key behind several recent groundbreaking technologies (including cryptocurrencies and blockchain), becoming a necessary tool for society transformation. These systems can provide secure infrastructure with redundant information and high service availability, also becoming a path for freedom and transparency. The transparency is a fundamental feature of Smart Cities, in order to empower citizens with real-time information and capability to take informed decisions that avoid corruption.

One industry that will suffer greater impact in the following years is the energy sector. Smart grids are already being implemented to connect small customers (even in Brazil after recent regulation changes). For instance, neighbors in Brooklyn already have the capability to exchange energy. Renewable energy (mainly solar) is now attracting the interest of small customers, which in turn are demanding a stronger participation in the free markets (although some third-party blockchain technologies are already allowing such exchanges), creating larger grids with local generators/prosumers. Although it is not possible to know in advance which technologies will become "standards" for Smart Cities in a near future, it is inevitable to realize that many of these changes are coming to stay. Therefore, the next steps to be taken will shape society considerably and the precautionary principle guides them. More now than never, everyone is responsible for technology use and development.

Acknowledgements Vitor N. Coelho would like to thank the support given by FAPERJ (grant E-26/202.868/2016). Alexandre C. Barbosa would like to thank the support given by the University of Padova.

References

1. Acharjee, P.: Strategy and implementation of smart grids in india. Energ. Strat. Rev. **1**(3), 193–204 (2013). http://dx.doi.org/10.1016/j.esr.2012.05.003. Future Energy Systems and Market Integration of Wind Power
2. Ackermann, T., Prevost, T., Vittal, V., Roscoe, A.J., Matevosyan, J., Miller, N.: Paving the way. IEEE Power Energy Mag. **15**(6), 61–69 (2017). https://doi.org/10.1109/MPE.2017.2729138.
3. Albright, C.: Gentrification is sweeping through america. Here are the people fighting back (2017). https://www.theguardian.com/us-news/2017/nov/10/atlanta-super-gentrification-eminent-domain
4. Alemán-Nava, G.S., Casiano-Flores, V.H., Cárdenas-Chávez, D.L., Díaz-Chavez, R., Scarlat, N., Mahlknecht, J., Dallemand, J.F., Parra, R.: Renewable energy research progress in mexico: a review. Renew. Sust. Energ. Rev. **32**, 140–153 (2014). http://dx.doi.org/10.1016/j.rser.2014.01.004
5. Ali, M.: The next wave of computing (2017). https://medium.com/@muneeb/the-next-wave-of-computing-743295b4bc73
6. Batty, M.: Big data, smart cities and city planning. Dialogues Hum. Geogr. **3**(3), 274–279 (2013)
7. Brooklyn microgrid project - lo3 energy. https://www.brooklyn.energy/
8. Buterin, V.: Ethereum white paper: a next generation smart contract & decentralized application platform (2013). http://www.the-blockchain.com/docs/Ethereum_white_paper-a_next_generation_smart_contract_and_decentralized_application_platform-vitalik-buterin.pdf
9. Chen, Y.H., Lu, S.Y., Chang, Y.R., Lee, T.T., Hu, M.C.: Economic analysis and optimal energy management models for microgrid systems: a case study in Taiwan. Appl. Energy **103**, 145–154 (2013). http://dx.doi.org/10.1016/j.apenergy.2012.09.023
10. Cheng, M., Sami, S.S., Wu, J.: Benefits of using virtual energy storage system for power system frequency response. Appl. Energy **194**, 376–385 (2016). http://dx.doi.org/10.1016/j.apenergy.2016.06.113
11. Chohan, U.: The decentralized autonomous organsations and governance issues. SSRN Electron. J. (2017). http://dx.doi.org/10.2139/ssrn.3082055
12. Christidis, K., Devetsikiotis, M.: Blockchains and smart contracts for the internet of things. IEEE Access **4**, 2292–2303 (2016)

13. Coelho, V.N., Coelho, I.M., Coelho, B.N., de Oliveira, G.C., Barbosa, A.C., Pereira, L., de Freitas, A., Santos, H.G., Ochi, L.S., Guimarães, F.G.: A communitarian microgrid storage planning system inside the scope of a smart city. Appl. Energy **201**, 371–381 (2016). http://dx.doi.org/10.1016/j.apenergy.2016.12.043. Final publication expected to 2017
14. Coelho, V.N., Coelho, I.M., Rios, E., Filho, A.S.T., Reis, A.J.R., Coelho, B.N., Alves, A., Netto, G.G., Souza, M.J., Guimaraes, F.G.: A hybrid deep learning forecasting model using gpu disaggregated function evaluations applied for household electricity demand forecasting. Energy Procedia **103**, 280–285 (2016)
15. Coelho, V.N., Cohen, M.W., Coelho, I.M., Liu, N., Guimarães, F.G.: Multi-agent systems applied for energy systems integration: state-of-the-art applications and trends in microgrids. Appl. Energy **187**, 820–832 (2017). http://dx.doi.org/10.1016/j.apenergy.2016.10.056
16. Dai, W.: b-money (1998). http://www.weidai.com/bmoney.txt
17. Díaz, P., Masó, J.: Evolution of production and the efficient location of renewable energies. The case of China. Energy Procedia **40**, 15–24 (2013). http://dx.doi.org/10.1016/j.egypro.2013.08.003. European Geosciences Union General Assembly
18. Dobschinski, J., Bessa, R., Du, P., Geisler, K., Haupt, S.E., Lange, M., Möhrlen, C., Nakafuji, D., de la Torre Rodriguez, M.: Uncertainty forecasting in a nutshell. IEEE Power Energy Mag. **15**(6), 40–49 (2017). https://doi.org/10.1109/MPE.2017.2729100
19. Ela, E., Wang, C., Moorty, S., Ragsdale, K., O'Sullivan, J., Rothleder, M., Hobbs, B.: Electricity markets and renewables. IEEE Power Energy Mag. **15**(6), 70–82 (2017). https://doi.org/10.1109/MPE.2017.2730827
20. Foundation, T.L.: Blockchain for business - an introduction to hyperledger technologies (2017). https://www.edx.org/course/blockchain-business-introduction-linuxfoundationx-lfs171x. EdX MOOC
21. Gabrich, Y.B., Coelho, I.M., Coelho, V.N.: Tendências para sistemas microgrids em cidades inteligentes: Uma visão sobre a blockchain. In: XLIX Simpsio Brasileiro de Pesquisa Operacional, Blumenau, pp. 1–12 (2017)
22. Gianelloni, F., de Azevedo Dantas, G., Alves, J.F., de Castro, N.: The distributed electricity generation diffusion impact on the brazilian distribution utilities. In: 3rd International Conference on Energy and Environment: Bringing Together Engineering and Economics, Porto (2017). https://goo.gl/5bmqU3
23. Green, M.A., Hishikawa, Y., Warta, W., Dunlop, E.D., Levi, D.H., Hohl-Ebinger, J., Ho-Baillie, A.W.: Solar cell efficiency tables (version 50). Prog. Photovolt. Res. Appl. **25**(7), 668–676 (2017). https://doi.org/10.1002/pip.2909. PIP-17-089
24. Hameiri, Z.: Photovoltaics literature survey (no. 138). Prog. Photovolt. Res. Appl. **25**(12), 1077–1083 (2017). https://doi.org/10.1002/pip.2967
25. Hofman, J., Hofman-Caris, R., Nederlof, M., Frijns, J., Van Loosdrecht, M.: Water and energy as inseparable twins for sustainable solutions. Water Sci. Technol. **63**(1), 88–92 (2011)
26. Horio, M., Shigeto, S., Ii, R., Shimatani, Y., Hidaka, M.: Potential of the 'renewable energy exodus' (a mass rural remigration) for massive {GHG} reduction in Japan. Appl. Energy **160**, 623–632 (2015). http://dx.doi.org/10.1016/j.apenergy.2015.03.087
27. IEA, I.E.A.: Energy access outlook 2017, from poverty to prosperity (2017). https://www.iea.org/publications/freepublications/publication/weo-2017-special-report-energy-access-outlook.html
28. Lew, D., Asano, M., Boemer, J., Ching, C., Focken, U., Hydzik, R., Lange, M., Motley, A.: The power of small. IEEE Power Energy Mag. **15**(6), 50–60 (2017). https://doi.org/10.1109/MPE.2017.2729104
29. Li, M., Xiao, H., Gao, W., Li, L.: Smart grid supports the future intelligent city development. In: 2016 Chinese Control and Decision Conference (CCDC), pp. 6128–6131 (2016). https://doi.org/10.1109/CCDC.2016.7532097
30. Lin, C.C., Yang, C.H., Shyua, J.Z.: A comparison of innovation policy in the smart grid industry across the pacific: China and the {USA}. Energy Policy **57**, 119–132 (2013). http://dx.doi.org/10.1016/j.enpol.2012.12.028

31. Mainelli, M., Smith, M., et al.: Sharing ledgers for sharing economies: an exploration of mutual distributed ledgers (aka blockchain technology). J. Financ. Perspect. **3**(3), 38–69 (2015)
32. Meijer, A.J., Gil-Garcia, J.R., Bolívar, M.P.R.: Smart city research: contextual conditions, governance models, and public value assessment. Soc. Sci. Comput. Rev. **34**(6), 647–656 (2016). https://doi.org/10.1177/0894439315618890
33. Mengelkamp, E., Gärttner, J., Rock, K., Kessler, S., Orsini, L., Weinhardt, C.: Designing microgrid energy markets: a case study: the brooklyn microgrid. Appl. Energy **210**, 870–880 (2018). https://doi.org/10.1016/j.apenergy.2017.06.054
34. Morvaj, B., Lugaric, L., Krajcar, S.: Demonstrating smart buildings and smart grid features in a smart energy city. In: Proceedings of the 2011 3rd International Youth Conference on Energetics (IYCE), pp. 1–8. IEEE, Piscataway (2011)
35. Nakamoto, S.: Bitcoin: A Peer-to-Peer Electronic Cash System (2008)
36. Pao, H.T., Fu, H.C.: Renewable energy, non-renewable energy and economic growth in Brazil. Renew. Sust. Energ. Rev. **25**, 381–392 (2013). http://dx.doi.org/10.1016/j.rser.2013.05.004
37. Pike, A., Rodríguez-Pose, A., Tomaney, J.: What kind of local and regional development and for whom? Reg. Stud. **41**(9), 1253–1269 (2007). http://dx.doi.org/10.1080/00343400701543355
38. Power Ledger Whitepaper (2017). https://powerledger.io/media/Power-Ledger-Whitepaper-v3.pdf
39. Ramalho, M.S., Câmara, L., Sílva, P.P., Pereira, G., Dantas, G.: Assessing energy policies drivers of the deployment of distribution generation: a review of influencing factors. In: 3rd International Conference on Energy and Environment: Bringing Together Engineering and Economics, Porto, Portugal (2017). https://goo.gl/EaLzYs
40. Sennett, R.: The open city. https://goo.gl/m2u85a
41. Shetty, V.: A tale of smart cities. Commun. Int. **24**(8), 16 (1997)
42. Silva, L.: Smartcities on the ethereum blockchain (2016). https://www.ethnews.com/the-future-is-now-smart-cities-on-the-ethereum-blockchain
43. Silver, D., Huang, A., Maddison, C.J., Guez, A., Sifre, L., van den Driessche, G., Schrittwieser, J., Antonoglou, I., Panneershelvam, V., Lanctot, M., Dieleman, S., Grewe, D., Nham, J., Kalchbrenner, N., Sutskever, I., Lillicrap, T., Leach, M., Kavukcuoglu, K., Graepel, T., Hassabis, D.: Mastering the game of go with deep neural networks and tree search. Nature **529**, 484–489 (2016). https://doi.org/10.1038/nature16961
44. Solshare. https://www.me-solshare.com/
45. Swaleh, M., Green, M.: Effect of shunt resistance and bypass diodes on the shadow tolerance of solar cell modules. Solar Cells **5**(2), 183–198 (1982). https://doi.org/10.1016/0379-6787(82)90026-6
46. Swan, M.: Blockchain: Blueprint for a New Economy. O'Reilly Media, Sebastopol (2015)
47. Tech Mahindra and Power Ledger unite to unleash the power of MaaS (2017). https://www.techmahindra.com/media/press_releases/TechMahindra-PowerLedger-unite-to-unleash-the-powerofMaaS.aspx
48. The Blockchain: Enabling a Distributed & Connected Energy Future (2016). http://www.mitforumcambridge.org/event/the-blockchain-enabling-a-distributed-and-connected-energy-future/
49. Tugcu, C.T., Ozturk, I., Aslan, A.: Renewable and non-renewable energy consumption and economic growth relationship revisited: evidence from G7 countries. Energy Econ. **34**(6), 1942–1950 (2012). http://dx.doi.org/10.1016/j.eneco.2012.08.021
50. Vorrath, S.: Bitcoin-inspired peer-to-peer solar trading trial kicks off in perth: renew economy. http://reneweconomy.com.au/bitcoin-inspired-peer-to-peer-solar-trading-trial-kicks-off-in-perth-29362/. Accessed Feb 17, 2017
51. Walport, M.: Distributed ledger technology: beyond blockchain. UK Government Office for Science (2016)

52. Welsch, M., Bazilian, M., Howells, M., Divan, D., Elzinga, D., Strbac, G., Jones, L., Keane, A., Gielen, D., Balijepalli, V.M., Brew-Hammond, A., Yumkella, K.: Smart and just grids for sub-Saharan Africa: exploring options. Renew. Sustain. Energy Rev. **20**, 336–352 (2013). http://dx.doi.org/10.1016/j.rser.2012.11.004
53. Wood, G.: Ethereum: a secure decentralised generalised transaction ledger (2013). http://gavwood.com/paper.pdf
54. Zheng, M., Wang, X., Meinrenken, C.J., Ding, Y.: Economic and environmental benefits of coordinating dispatch among distributed electricity storage. Appl. Energy **210**, 842–855 (2018). https://doi.org/10.1016/j.apenergy.2017.07.095

Chapter 16
Optimal Energy Trading Policy for Solar-Powered Microgrids: A Modeling Approach Based on Plug-in Hybrid Electric Vehicles

Hendrigo Batista da Silva and Leonardo P. Santiago

Abstract Battery management is key to enact the widespread use of microgrid-connected electric vehicles. We thoroughly review the literature and tackle the role of battery in the process of energy commercialization between the microgrids and utilities. In particular, by considering the battery management as a stochastic inventory control problem, we develop a dynamic programming model and we obtain an optimal policy for it. Then, we further explore the baseline model by investigating a scenario in which the microgrid is constrained by a budget defined a priori. Such a budget constraint captures situations when the microgrid profile is risk averse. We end by discussing the main issues that stem from such a budget constraint scenario.

16.1 Introduction and Related Literature

Plug-in hybrid electric vehicles (PHEVs) and the electric vehicles (EV) have a large potential for reducing fossil fuels consumption and CO_2 emissions [41]. In line with that, PHEVs and EV enable a higher penetration of renewable sources in the energy matrix of a country.

Current technology allows PHEV or EV owners to generate their own energy from solar photovoltaic panels or other distributed generation sources, installed on their own homes. The energy generated from distributed sources can be commercialized with the grid or used by the homeowner. Such flexibility allows homeowners to act like energy producers and consumers, creating the so-called

H. B. da Silva (✉)
Federal University of Minas Gerais, Belo Horizonte, MG, Brazil
e-mail: hendrigobatista@ufmg.br

L. P. Santiago
Copenhagen Business School, Frederiksberg, Denmark
e-mail: ls.om@cbs.dk

V. N. Coelho et al. (eds.), *Smart and Digital Cities*, Urban Computing,
https://doi.org/10.1007/978-3-030-12255-3_16

prosumer (producer + consumer). By commercializing the energy with the grid, the value of energy over time in a microgrid becomes object of concern to prosumers. Therefore, an integrated system that manages multiple energy resources will play a key role in the years to come, specially with the increasing penetration of PHEVs and EVs in the automotive market. Moreover, it should also be highlighted that vehicles are usually parked for around 90% of time and the batteries are a significant capital investment [25]. Security and reliability issues also increase the need for a microgrid to operate in a centralized management system.

The integration of PHEVs and EVs to the grid, known as "vehicle-to-grid" or V2G in the literature, presents a large potential for improving the management of the entire electric systems [17]. V2G technology can boost the income for owners, fostering its adoption, in addition to improving the macrogrid stability [39]. In that regard, the period of connection and disconnection from the grid gains special importance. The charging of the PHEV and EV fleet can overlap with the peak load time and, if there is no adequate planning of such connections, additional investments will be required [15]. Such management becomes essential with the diffusion of the microgrid and its inherent features that are associated with intermittent solar generation. Therefore, a formal approach to deal with such a complex scenario gains importance for managing the integration of PHEVs and EVs to the grid in a dynamic manner. In that regard, the dynamic programming approach is deemed to manage such an integration over time, since it defines optimal controls over a certain time horizon. Furthermore, this approach presents the flexibility of offering optimal policies in closed loop systems. Thus, with such an approach, the connection of vehicle batteries to the grid can be optimized throughout time.

The integration of battery systems with microgrids, such as PHEVs and EVs, has been discussed in the academic literature over the past years. It should be highlighted the study of [21], which presents the conceptual structure for a successful V2G integration dealing with both technical and market environments. Kramer et al. [20] reviews the main electric vehicles with possibility of V2G integration. Su et al. [38] makes a general review of charging facilities, PHEVs and EVs batteries, intelligent management systems, V2G, and communication requirements for such integration. Guille and Gross [13] presents a proposal of integration structure of EVs in the grid, in a way that vehicles could level demand in off-peak periods and provide loads when the grid presents a high demand, working as a generator. Madawala et al. [24] also presents a conceptual model to facilitate the V2G integration and [35] shows the potential benefits and impacts of unidirectional V2G. It should be highlighted the work of [23], which proposes a transference system that facilitates the charging and discharging of multiple EVs simultaneously.

Regarding the integration with eolic sources, [9] proposes a grid model in which EVs act as buffers of the differences between the prediction and actual generation of eolic turbines. Lund and Kempton [22] shows that the addition of EVs to the grid allows the integration of a large amount of eolic energy, reducing gas emissions in the atmosphere. Pillai and Bak-Jensen [29] studies the capacities of V2G regulation in the highly eolic electric system in Western Denmark, using a model of frequency control.

In the ancillary services provisioning context for PHEVs and EVs, some important studies should be highlighted. Kempton and Tomić [18] offers a formulation to estimate revenues and costs for a vehicle that provides energy in peak markets and ancillary services. It presents the V2G integration as an important factor to the stability and reliability of the grid and an important way of large scale storage of renewable energy. In that regard, [1] also presents studies of vehicles in Sweden and Germany as ancillary services providers. White and Zhang [42] explores the financial returns potential of the use of PHEVs as a resource for frequency control. Dallinger et al. [4] studies the impacts that different mobility patterns of such vehicles could generate in the delivery of ancillary services, presenting insights about the optimal size of the fleet and possible adaptation in regulation. Galus et al. [11] also proposes ancillary service provisioning by aggregating PHEVs and controllable residential loads. Kisacikoglu et al. [19] studies the impact in the bidirectional charger of PHEVs with the provisioning of loads and ancillary services to the grid. Ota et al. [27] proposes a control scheme that provisions these services based on frequency deviation in the plug-in terminal, which could be easily integrated in the electronic circuits of vehicles or units of residential loads to facilitate plug and play operation. Quinn et al. [31] proposes new models for V2G availability, reliability of the energy company, and time series of ancillary services prices, presenting more feasible results for its implementation in the short term. Sortomme and El-Sharkawi [36] develops a V2G algorithm for optimizing the load and ancillary services scheduling, as reserves and load regulation. Wu et al. [43] proposes a game theory model to understand the interactions between EVs and aggregators in the V2G market, where EVs provide frequency regulation services to the grid.

Some papers also discuss technical aspects of placing charging stations in parking lots, as in [5], which proposes a model for municipal parking lots. For instance, [16] proposes a scheduling model to efficiently use available loads in order to maximize the profit of the PHEVs and EVs owner, when a fleet is in a parking lot and taking into account system constraints. Saber and Venayagamoorthy [33] proposes a scheduling model for charging of vehicles to obtain the maximum benefit in scenarios with limited parking space.

The impact of PHEVs and EVs on the grid has also been tackled in recent studies. For instance, [12] presents a review of the main consequences that PHEVs can cause in distribution networks, through an approach that considers the combination of driving and charging patterns, charging time and penetration of such vehicles in the market. Farmer et al. [7] proposes a model to estimate the impact of an increasing number of PHEVs or EVs in transformers and cables in systems with medium voltage distribution. Mitra and Venayagamoorthy [26] presents a model that assesses stability gains for the electric system connected with vehicles, through the use of transient simulations and Prony Analysis in a WAC controller, implemented in a real time digital simulator. Considering the impact in gas emissions, [34] studies the impact in the dispatches of traditional generators that PHEVs caused in Texas, estimating a reduction in emission rates with the increase of PHEVs fleet.

Regarding the papers related to the difficulties of such integration, [28] studies the effects of the combined use of batteries for driving and V2G in the lifetime of the battery. Galus et al. [10] proposes details for the integration of PHEVs to the grid in the current technology, with all the technical difficulties. Sovacool and Hirsh [37] makes a critical analysis of a large scale integration of vehicles to grid, presenting social-technical barriers resulting from this process.

Several papers approach the V2G integration through sequential decision models in a finite time horizon. It should be highlighted the work of [8] that offers an integrated approach in a way it could be properly managed by a load aggregator, which allows a proper management of the charging process and the definition of optimal bids for this energy. Rotering and Ilic [32] proposes a sequential decision model to optimize the time of charging and the energy flow based on future prices of electricity, reducing the costs in a significant way without increasing the battery degradation. Such a paper also considers the participation of prosumers in providing ancillary services, obtaining extra revenues. Clement-Nyns et al. [3] proposes, through stochastic dynamic programming, a charging model in a coordinated fashion, in such a way to minimize grid losses and optimize the load factor. Last but not least, [14] tackles charging issues, through dynamic programming, obtaining an optimal control for each vehicle.

However, in spite of noteworthy attempts to manage V2G integration, the management of the energy stored in the PHEVs and EVs connected to the microgrids still has a large potential to be explored. Since it deals with a costly asset, which will be used in a finite time horizon, the use of inventory management techniques presents a high potential for optimization of such systems. We remark that the efficient management of such system needs to take into account the costs and risks inherent to this process. Tulpule [40] is one of the few papers that tackles the amount of energy stored in EV's batteries as a microgrid inventory to be managed over time. In spite of its relevance, there are some points not tackled in that paper that could foster the engagement of consumers and avoid their exposition to price volatility, in scenarios with high energy costs—especially in real-time pricing environments.

The novelty of this book chapter is to offer a formulation that takes a holistic approach to the uncertainties involved in a microgrid environment and explicitly considers their inherent costs. As such, through our approach, we expect a smaller resistance from the final consumers. This fact can enact the potential use of microgrids in buying and selling markets, by coping with the minimum volatility principle (for more details on minimum volatility principle, see, e.g., [6]). We define six sources of uncertainties and their relation with different risks, which are associated with inherent costs in this trading modeling:

1. Risk of low energy level in the batteries. For instance, if PHEVs leave before expected without charging as expected, the vehicle might use gas instead incurring the cost of it (gas). In the case of EVs, the cost is associated with not being able to use the vehicle. The higher the cost of gas for PHEVs or the cost of not using the vehicle, the bigger the impact of such risk.

2. Risk of low energy generation on the spot. This risk is associated with the price of energy in the grid, responsible for charging the vehicles and for providing energy to local loads in the absence of local generation. The higher the price of electricity from the grid, the bigger the impact of this scenario.
3. Risk of substantial increase of electricity prices from the grid. This scenario is more common under real-time pricing, when the local generation does not match the local demand. The risk is also associated with the price of electricity from the grid. As before, the higher the price, the higher the impact of this scenario.
4. Risk of substantial decrease of electricity prices from the grid. Similarly, such scenario is more common under real-time pricing, when the local generation/-supply is higher than local demand. This risk is associated with the opportunity cost of storing energy in periods with high prices and have to sell it in lower price periods.
5. Risk of peaks of energy demanded from the microgrid. As in the case of low local generation, this is associated with the expectation of electricity price from the grid. The higher the price, the higher the impact.
6. Risk of demand response programs to poorly manage its budget due to volatility aversion, which can be more common in scenarios with real-time pricing. The cost associated can be considered as the cost of feasibility of such programs.

In a nutshell, our contribution through this book chapter is to offer a formulation that simultaneously takes into account the principle of minimum volatility, and the costs associated with the six risks previously discussed. The formulation draws on the inventory management problem and is adapted in order to capture specific features required to manage microgrids.

16.2 Optimal Energy Trading Policy

This section describes the formulation for an optimal energy trading policy between the microgrid and the utility grid. The model considers managerial aspects of an energy trading control systems throughout a time horizon. We consider energy, not the power, as the unity to be optimized in this period. For the conceptual presentation of the model, we consider that technical aspects concerning energy exchanges and voltage quality do not constrain the control and state spaces. We propose a stochastic dynamic programming model, which considers the utilization of the electricity stored in PHEVs as the system states, similar to the inventory control problem. The charging of the vehicles is managed in such a way to minimize the purchasing costs for the microgrid and the opportunity costs of storage. A numerical example illustrates the applicability of such model in a scenario in which vehicles are disconnected in the morning and connected towards the end of the day. In the numerical example, we consider a microgrid with solar generation characterized by generation peaks around noon. This section considers the modeling features discussed in the previous section, which will then be explored in the next section.

16.2.1 Model

To improve the readability of the variables, we list in Table 16.1 the model notations. The index k represents the decision stage in each variable.

In the sequence, we discuss the assumptions, the system dynamic, the cost function, and the optimal policy.

16.2.1.1 Assumptions

Since the prosumers can trade the energy with the grid, they have to deal with a trade-off regarding what to do with the energy generated and stored in the batteries of the vehicles. This trade-off is similar to the problem of the stochastic inventory control, in which the purchase decisions are taken based on the expected value of demand and the purchase cost, an assumption of this model. We also consider that prosumers are grouped into microgrids in order to facilitate the joint management of the local energy generated.

16.2.1.2 Stochastic Factors

There are four sources of uncertainty in this microgrid management system: intermittent generation of PV panels, local demand of the microgrid, electricity prices in a real-time pricing (RTP) scenario, and the arrivals and departures of vehicles from the microgrid.

Intermittent generation v_k is an important feature of alternative sources of electricity, specially solar energy. Since generation is directly dependent on radiation

Table 16.1 Notation of variables

Variables	Notation
x_k	Stored energy in the microgrid
u_k	Total energy traded
v_k	Local energy generation
d_k	Local energy consumption
w_k	Balance between generation and consumption
a_k	Arrival and departure factor of vehicles
$J_k(x_k)$	Cost function per state
c_k	Purchased energy cost in the beginning of each stage
p_k	Non-planned energy cost
h_k	Stored energy cost
y_k	Augmented state vector
π_k	Internal price of energy in demand response events
z_k	Available budget to be managed in the time horizon

and therefore, dependent on weather conditions to be realized in the future, it presents a stochastic value. Values concerning expected meteorological conditions should be considered in such a way to support decisions that concern these random variables.

The second source of uncertainty is the local demand or consumption in the microgrid d_k, which is not necessarily known. However, it can be estimated based on historical patterns that vary depending on temperature, weekday, or time of the day.

The third stochastic factor, price of electricity p_k, has an important role in the decision system, since it impacts directly the amount of energy that should be commercialized. In a classical RTP scenario, consumers are exposed to electricity price variations, reflecting better the market conditions of supply and demand. Information from the utility or from the system operators about critical network conditions can help in the treatment of this stochastic factor.

The last source of uncertainty is the arrivals and departures a_k of PHEVs or EVs in the microgrid. As the local demand, historic patterns or even a priori information given by the prosumer to the management system could be utilized to obtain a more efficient decision.

16.2.1.3 System Dynamics

The variation of the quantity stored between consecutive decision stages is influenced by some factors: total amount of vehicles that connected or disconnected from the microgrid, total amount of energy purchased or sold to the utility grid, local generation and consumption inside the microgrid. Hence, the equation that captures the system dynamics is as follows:

$$x_{k+1} = a_k x_k + u_k + v_k - d_k \tag{16.1}$$

In this equation, the state x_k represents the total amount of energy stored inside the microgrid in stage k.

The factor a_k is associated to the vehicles. Since the total energy stored is related to the connection and disconnection of vehicles, this factor considers this influence over x_k. If $a_k < 1$, it is expected that less vehicles will be connected to the grid in the next decision stage. This situation can occur, for example, in the beginning of mornings in residential microgrids, when it is often observed departures of vehicles due to the daily commute in weekdays. For the same reason, it is expected $a_k > 1$ in the end of the afternoon.

The decision or control variable is u_k, which represents the total amount of energy to be sold or purchased in the transactions with the energy grid. In this model, we consider that there is a purchase when $u_k > 0$ and there is a sale when $u_k < 0$.

The expected value of local generation of electricity is associated with the random variables v_k. As most of the distributed generation sources are intermittent,

v_k is associated with an estimative inferred a priori based on solar radiation forecasts and meteorological conditions expected for the microgrid locality.

The local energy consumption is represented by the random variable d_k, whose probability distribution inferred a priori can be based on demand forecast methodologies. These methodologies can utilize historic data or seasonal trends.

For simplification purpose, we consider that model stochasticity can be represented by a new random variable w_k, which represents the net balance between the energy generated and consumed locally in a period. This variable is described by a probability distribution resulted from the union of two other random variables in the model:

$$w_k = v_k - d_k \tag{16.2}$$

Therefore, the dynamic systems equation can be replaced by the equation:

$$x_{k+1} = a_k x_k + u_k + w_k \tag{16.3}$$

16.2.1.4 Cost Function

The energy generated supplies instantly the local load. Energy surplus could be stored in the batteries of PHEVs or EVs or even sold back to the grid. This decision depends essentially on energy price expectations. Therefore, there is a clear clash of uncertainties. If higher prices are expected in the near future, it is preferable to store energy in the batteries and take advantages of future periods to earn more revenues to the microgrid. However, the time of departure is also an uncertainty factor that impacts directly this decision, as well as the expectation of local generation in the microgrid in the near future.

We can therefore observe that the optimal decision that a microgrid control system should take depends on the magnitude of the costs associated with each risk. The energy price of the grid and the gas price are decisive features for such a system for PHEVs.

Considering the capacity of energy storage as the state space, we defined expected costs for each state and control. From meteorological data, expected values for the PV electricity generation are inferred for the next stage. We also defined expected values for the local demand, utility electricity price, and also arrivals and departure patterns in a typical weekday.

We considered a stochastic dynamic programming model to obtain optimal decisions for the system. A control policy is obtained in closed loop for each stage of decision. This optimal control is the amount of energy that should be bought or sold back to the grid in each stage.

For each decision stage k, it is associated a recursive cost function $J_k(x_k)$. The objective is to minimize this function in respect to the controls for all the stages and states, obtaining an optimal decision for each stage. In general, the cost function will increase with the purchase of electricity from the grid and will decrease with the sales.

The cost function penalizes the non-balancing in a same decision stage, that is, the lack of energy stored in the microgrid summed with the purchase or sale planned in the beginning of each stage. This penalization occurs through the electricity price of the grid p_k, and it is given in this inventory model in monetary units per non-planned energy units in the beginning of the stage. This cost can impact the definition of contracts between microgrids and utilities. The function also penalizes the energy surplus stored in the end of each period that could be sold to the grid. This penalty is given in monetary units per stored energy units and is represented by the parameter h_k.

Positive and negative imbalances are penalized asymmetrically, that is, are mutually exclusives. Therefore, we have for all k:

$$r(x_k) = p_k \max(0, -x_k) + h_k \max(0, x_k) \tag{16.4}$$

Considering Eq. (16.4), we defined the following penalty function $H_k(a_k x_k + u_k)$:

$$\begin{aligned} H_k(a_k x_k + u_k) = E_{w_k}[r(a_k x_k + u_k + w_k)] = E_{w_k}[p \max(0, -a_k x_k - u_k - w_k) \\ + h \max(0, a_k x_k + u_k + w_k)] \end{aligned} \tag{16.5}$$

In order to minimize the total costs, we derived a cost function considering the purchase costs or the "negative costs" of the sales given by the term $c_k u_k$, besides the penalty function previously presented. Therefore, we have for $k = 0, \ldots, N-1$:

$$J_k(x_k) = \min_{u_k} \left\{ E_{w_k}[c_k u_k + H(a_k x_k + u_k) + J_{k+1}(a_k x_k + u_k + w_k)] \right\} \tag{16.6}$$

We consider that the terminal cost is given by the cost of equivalent gas to the amount not charged in all the PHEVs in the end of the time horizon, or by the cost of non-utilization in all the EVs. Both cases are captured by the variable g. Therefore, we have

$$J_N(x_N) = (x^{sup} - x_N)g \tag{16.7}$$

In this function, x^{sup} represents the maximum state of charge of the system.

16.2.1.5 Optimal Control Policy

The optimal policy represents which are the optimal controls (purchase or sale) that should be taken between the microgrid and the grid. Its derivation is based on [2].

The optimal policy μ_k^* for all k is given by:

$$\mu_k^* = u_k^* = S_k - a_k x_k \tag{16.8}$$

in which:

$$S_k = arg \min_{y_k} G_k(a_k x_k + u_k) \tag{16.9}$$

$$G_k(a_k x_k + u_k) = c_k(a_k x_k + u_k) + H_k(a_k x_k + u_k) + E_{w_k}[J_{k+1}(a_k x_k + u_k + w_k)] \tag{16.10}$$

16.2.2 Numerical Example

In order to illustrate the applicability of this model, we present a numerical example with the optimal policy and simulation of the microgrid behavior under this optimal control considering PHEVs connections.

We specified a random walk for the distributed generation that matches the typical pattern of a sunny day and another for the local demand, as specified in Fig. 16.1. The distribution generation curve has a peak around noon, considering the peak in PV panels properly oriented. On the other hand, the local demand curve has a peak in the end of the afternoon and beginning of evening, a typical pattern for residential consumption in a weekday.

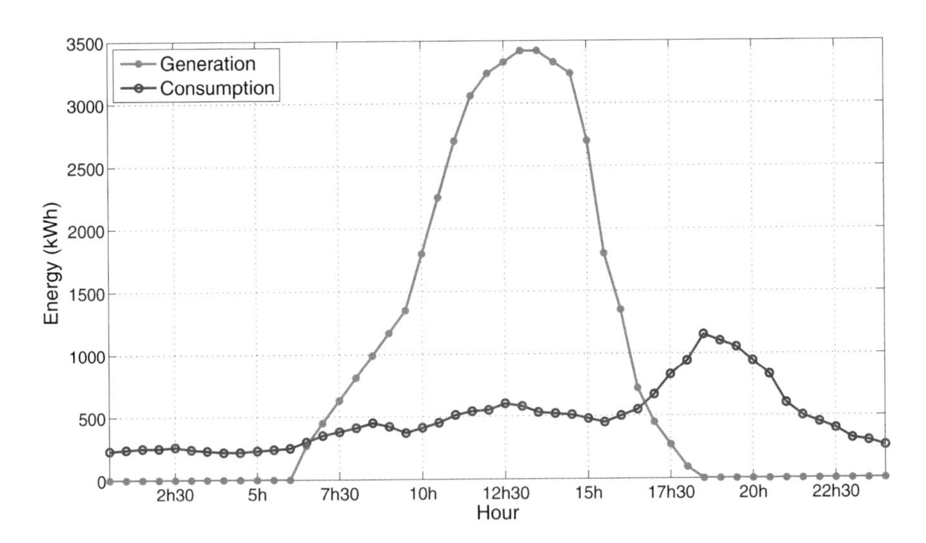

Fig. 16.1 Sample of generation and consumption curves in a hypothetic microgrid

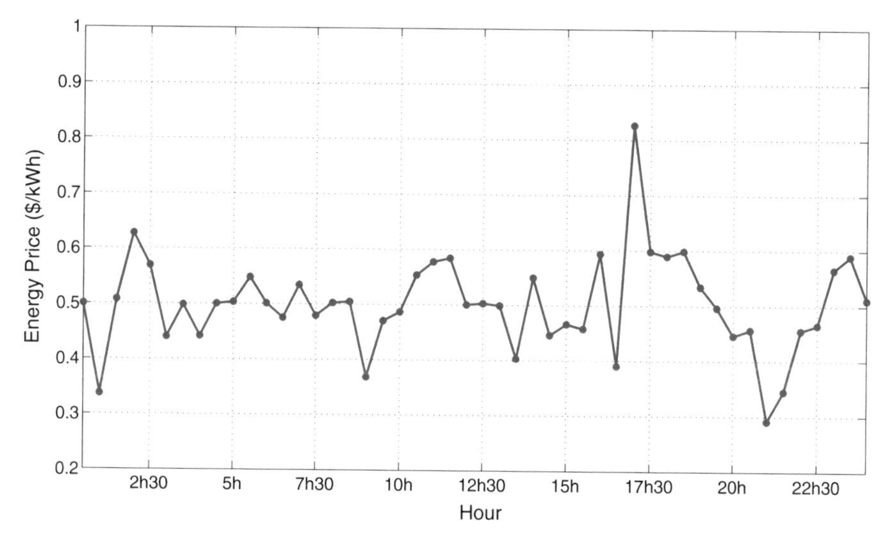

Fig. 16.2 Energy price over a day

We also consider the unitary prices of energy shown in Fig. 16.2. This price scenario is characterized by a grid in the modality RTP. We specified a random walk for these prices for all the planning horizon in order to illustrate the applicability of the model.

The penalty cost for the non-planned amount in the beginning of stage was considered $p = \$1000/\text{kWh}$ and the storage cost $h = \$500/\text{kWh}$. The cost p should be expressive in magnitude in order to not observe a lack of energy in the microgrid with the sale of all the energy generated between decision stages.

A one-day horizon was discretized in intervals of 30 min for the definition of the amount that should be purchased or sold by the microgrid. This discretization implies that the total amount of energy traded should be equal to the optimal control in this interval. We do not specify in each period this trade should happen or if it should be uniform along the 30-min interval. It could actually be processed in a small fraction of that time, as long as technical constraints allow this operation to occur. This issue is not approached in the scope of our study and an aggregated proposal could be explored in future studies.

The availability rate of PHEVs for the next stage of decision is defined in the sample path presented in Fig. 16.3. This curve captures a higher disconnection of vehicles in the morning and a higher connection in the end of the afternoon, as previously discussed.

The system state evolution represents the total energy stored along a day in the batteries of PHEVs. The energy stored, as well as the energy traded or optimal controls are specified in Fig. 16.4.

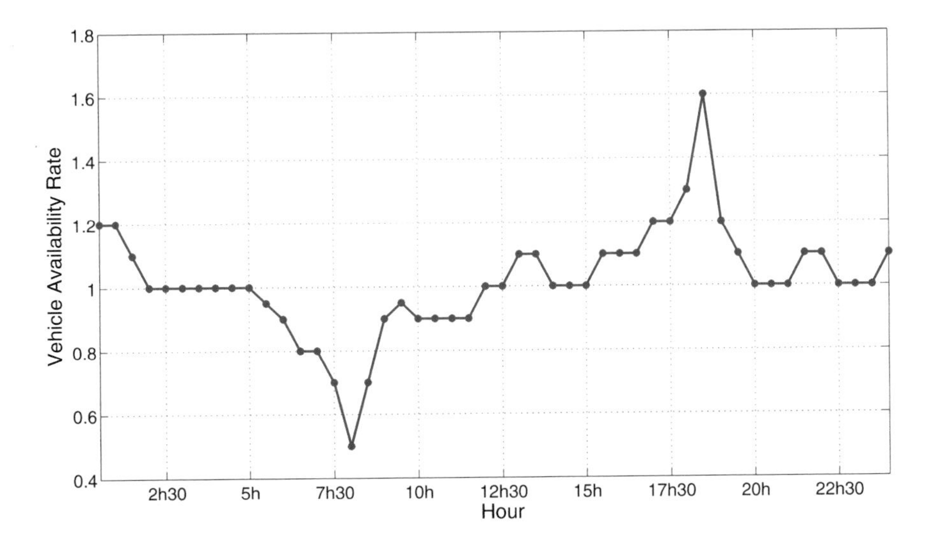

Fig. 16.3 Vehicle availability rate over a day

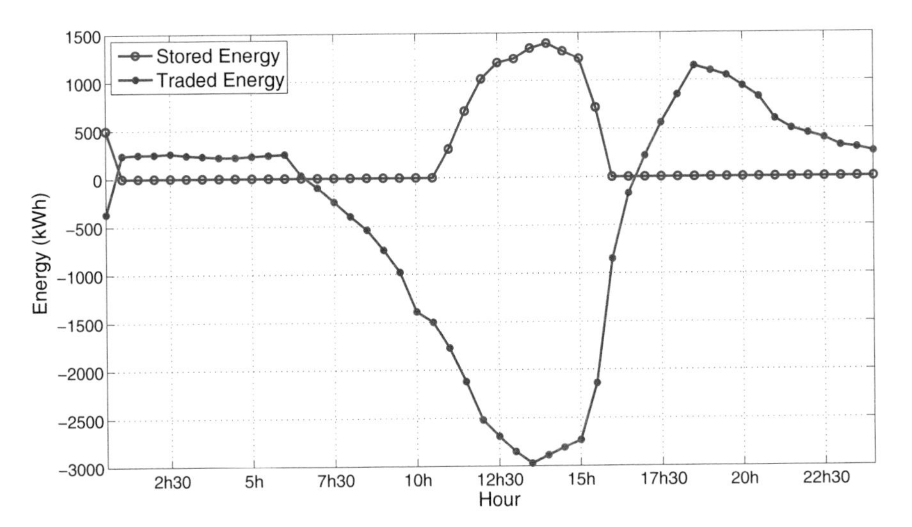

Fig. 16.4 Stored and traded energy in the vehicles

16.2.3 Discussion

The convexity of functions G_k previously presented can be observed in the transversal section in Fig. 16.5, obtained from the numerical example. For each decision stage, if we consider a section along this dimension, one can realize that function $G_k(y_k)$ has a minimum value S_k, increasing monotonically for values above or below. For a cross section at 8 am, we have the convex function presented

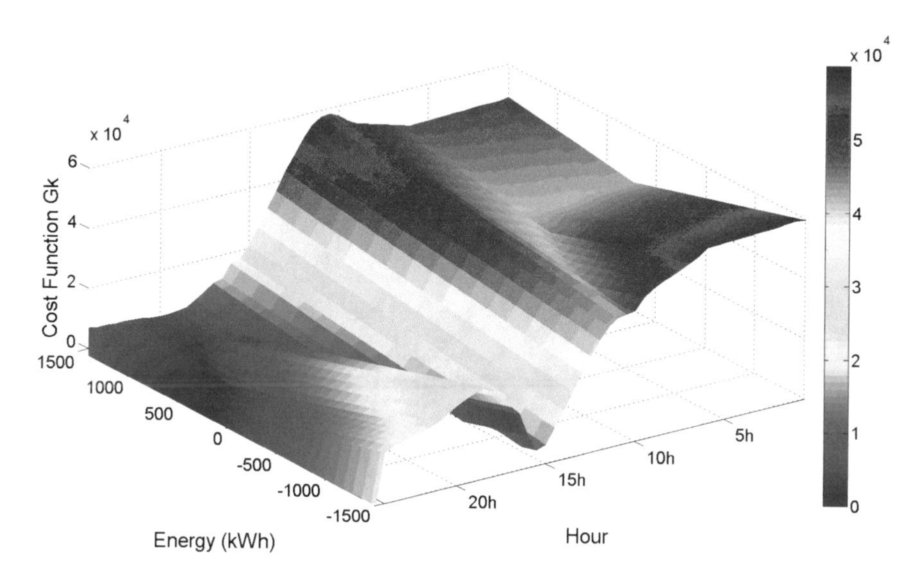

Fig. 16.5 Cost function G per state and stage

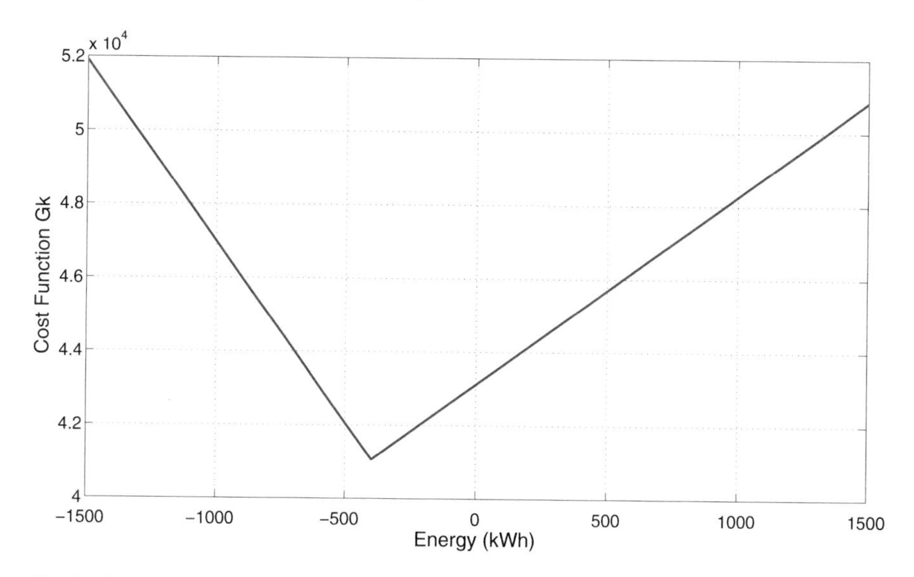

Fig. 16.6 Cross section of cost function G at 8 am along energy dimension

in Fig.16.6. Note that the function is convex by hour or decision stage, by not along the day.

With these parameters, the model presents a system state with null value in different periods of the day. Since the states represent the stored energy in the batteries of PHEVs, it is important to highlight the importance of a correct dimensioning of the cost parameters. In this example, the cost function leads the

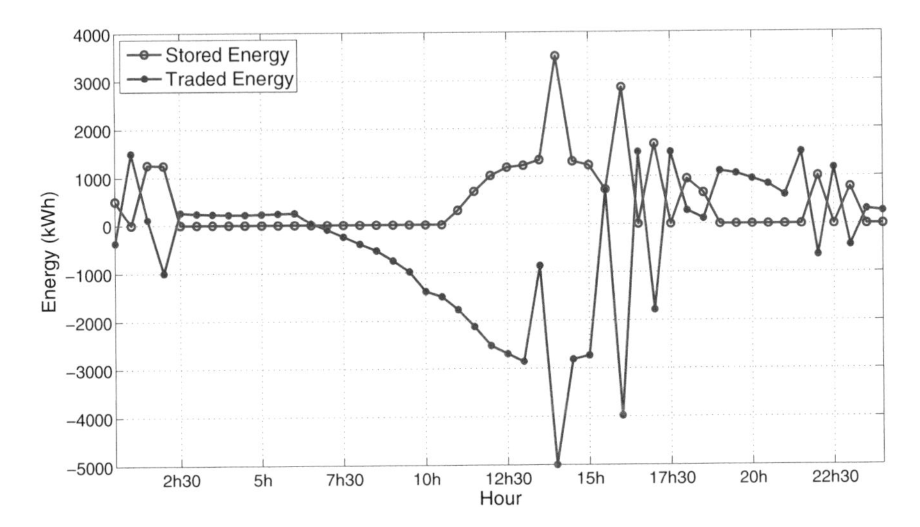

Fig. 16.7 Stored and traded energy considering a lower value for h

model to prefer to sell large part of the energy. With the storage parameter decreased to $h = \$0, 10$/kWh, the trade-off between to storage or sell energy presents a different behavior, as presented in Fig. 16.7. It shows more instability between purchase and sale, but it contains several stages with positive storage.

This formulation does not tackle the principle of minimum volatility presented in [6], letting the consumers exposed to price variations in real-time. In the next section, an integrated model is proposed as an addition to this formulation, defining a budget to be managed in the time horizon and discussing the impacts of such approach.

16.3 Model Integration with Demand Response

This section extends the previous approach by proposing a model for management of microgrid with distributed generation sources, V2G connection, and demand response events. The focus is to shed light on the budget management throughout the time. In particular, we consider the microgrid has a predetermined budget for buy energy from the energy company. In this case, optimal controls specifies the total amount of energy that should be bought from the energy company, and also reactions to demand response events, in order to better use the budget. In particular, the microgrid manages the energy stored in the vehicles, which are connected and disconnected from the grid, in order to obtain the minimum cost for its operation.

16.3.1 State Space

The energy stored x_k and budget z_k are represented by the state variables of the system in each stage k. It is noted by the 2-dimension variable y_k.

$$y_k = \begin{pmatrix} x_k \\ z_k \end{pmatrix} \tag{16.11}$$

16.3.1.1 Energy Stored Level

The variable x_k represents the amount of energy stored in the microgrid in the vehicles. Considering PHEVs, they are also moved by gas and they have more flexibility to provide energy to the microgrid, since there is not a vital commitment with a state of charge near 100% in the moment of disconnection. Therefore, PHEVs are more suitable for participating in a microgrid management system. The value of state at $k = 0$ is known a priori by the system.

16.3.1.2 Available Budget

The variable z_k represents the budget available up to the end of a period for purchase of external energy. It was proposed based on the principle of minimal volatility [6] for greater consumer engagement in the proposed program to ensure the long-term success of the approach. The total budget is set a priori to the system at $k = 0$.

16.3.2 Control Space

We consider the control represents the amount of energy that should be purchased or sold to the utility grid, as well as an internal price of electricity of demand response events at stage k. Therefore, it is represented by the 2-dimension column vector U_k. So, the control space is the combination of trade energy u_k with internal price π_k.

$$U_k = \begin{pmatrix} u_k \\ \pi_k \end{pmatrix} \tag{16.12}$$

16.3.3 System Dynamics

The equation, that represents the system dynamics, captures how the state y_k evolves along two subsequent decision stages. We evaluate the behavior of each component of such state vector.

16.3.3.1 Level of Energy Stored

Regarding the first component of the state vector, the dynamics is similar to the inventory control problem discussed in the previous section. The variation of the level of energy stored between two consecutive decision stages in the microgrid is influenced by the following factors: amount of vehicles that disconnected or connected to the microgrid, amount of energy purchased or sold to the utility grid, internal consumption inside this interval, and internal generation of electricity by distributed generation sources. Then, we define an equation to capture the system dynamics, regarding the stored energy level x_k, as follows:

$$x_{k+1} = a_k x_k + u_k - d_k + v_k \tag{16.13}$$

In this equation, the factor a_k is associated with the availability of vehicles in the microgrid. Since the storage is directly related to the connection of vehicles, x_k was weighted by the multiplication by a_k, as we did in the previous section. If $a_k < 1$, it indicates that less vehicles were connected to the grid in the next decision stage, as in the mornings on a typical weekday for residential microgrids. By the same reason, it is expected $a_k > 1$ in the end of afternoon.

The control variable u_k represents the amount of energy to be purchased or sold to the utility grid at stage k. This trade decision is represented by the variable signal. We consider that there is a purchase if $u_k > 0$ and a sale if $u_k < 0$.

The internal electricity consumption is represented by the variable d_k, and is associated with the consumption inside the interval between decision periods. In critical conditions of internal generation that results in excessive external energy purchases, like a sequence of cloudy days for example, the consumption d_k is influenced by the control variable π_k. This variable represents a higher internal price due to demand response events. This consumption variation is caused due to energy price variation for internal loads in relation to a fixed price π_0, charged in the absence of demand response programs. The variable d_k can be estimated based on the price-elasticity equation applied to the electricity market and to the mean values of demand d_k^m for a specific time at a price π_0. Therefore, we have:

$$\epsilon_k = \frac{(d_k - d_k^m)}{d_k^m} \frac{\pi_0}{(\pi_k - \pi_0)} \tag{16.14}$$

Isolating variable d_k, we have:

$$d_k = \frac{\epsilon_k d_k^m}{\pi_0} \pi_k - \epsilon_k d_k^m + d_k^m \tag{16.15}$$

It is important to highlight that final consumers are more inclined to respond to a price modification in some periods than in others. For example, the daytime period has a response capacity inferior to the evening period, due to the fact that during the day, in general, consumers are working and less available to actively respond to RTP. Therefore, it is interesting to consider elasticity as a variable that depends on the decision stage k.

The estimation of internal generation of electricity is associated with the random variable υ_k. Since the majority of distributed sources are intermittent, like solar and wind, υ_k is associated to an estimative obtained based on local meteorological conditions in the microgrid for the time horizon.

Replacing (16.15) in the dynamic systems equation (16.13), we have the representation of how the stored energy changes over time:

$$x_{k+1} = a_k x_k + u_k - \frac{\epsilon_k d_k^m}{\pi_0}\pi_k + \epsilon_k d_k^m - d_k^m + \upsilon_k \tag{16.16}$$

16.3.3.2 Available Budget

Regarding the second component of the vector of states, the dynamics is influenced by the cost or revenue of the traded energy. Therefore, we define a dynamic systems equation regarding the available budget z_k:

$$z_{k+1} = z_k - c_k u_k \tag{16.17}$$

The parameter c_k represents the unitary price of energy in the utility grid. The multiplication by u_k subtracts the value of the purchased energy from the available budget for that horizon or sums the sold energy, when u_k is negative.

16.3.3.3 Augmented State

Grouping the dynamics of both states in the augmented space y_k, we have the following joint dynamic systems equation:

$$y_{k+1} = \begin{pmatrix} x_{k+1} \\ z_{k+1} \end{pmatrix} = \begin{pmatrix} a_k & 0 \\ 0 & 1 \end{pmatrix}\begin{pmatrix} x_k \\ z_k \end{pmatrix} + \begin{pmatrix} 1 & -\dfrac{\epsilon_k d_k^m}{\pi_0} \\ -c_k & 0 \end{pmatrix}\begin{pmatrix} u_k \\ \pi_k \end{pmatrix}$$
$$+ \begin{pmatrix} (\epsilon_k d_k^m - d_k^m) \\ 0 \end{pmatrix} + \begin{pmatrix} \upsilon_k \\ 0 \end{pmatrix} \tag{16.18}$$

In order to simplify, we adopt the following matrices. Since part of some matrices will be used in the formulation of cost equations, we subdivided them by line. For example, A_k^1 refers to line 1 from matrix A_k and A_k^2 refers to line 2:

$$A_k = \begin{pmatrix} A_k^1 \\ A_k^2 \end{pmatrix} = \begin{pmatrix} a_k & 0 \\ 0 & 1 \end{pmatrix} \tag{16.19}$$

$$B_k = \begin{pmatrix} B_k^1 \\ B_k^2 \end{pmatrix} = \begin{pmatrix} 1 & -\frac{\epsilon_k d_k^m}{\pi_0} \\ -c_k & 0 \end{pmatrix} \tag{16.20}$$

$$D_k = \begin{pmatrix} D_k^1 \\ D_k^2 \end{pmatrix} = \begin{pmatrix} (\epsilon_k d_k^m - d_k^m) \\ 0 \end{pmatrix} \tag{16.21}$$

$$V_k = \begin{pmatrix} V_k^1 \\ V_k^2 \end{pmatrix} = \begin{pmatrix} v_k \\ 0 \end{pmatrix} \tag{16.22}$$

Therefore, the dynamic systems equation of this model is given by:

$$y_{k+1} = A_k y_k + B_k U_k + D_k + V_k \tag{16.23}$$

16.3.4 Cost Function

A cost function $J_k(y_k)$ is defined for each decision stage. It aims to minimize the purchase of energy from the utility grid and consequently, maximize the sales. Moreover, control variables U_k minimize the penalties related to: demand response events dispatches, non-planned consumption, storage surplus, and extrapolation of a priori defined budget. Then, we have a cost function per stage k defined as a sum of different functions:

$$G_k(y_k) = L_k^1(U_k) + L_k^2(y_k) + L_k^3(y_k) \tag{16.24}$$

Each term of the cost function will be discussed in detail.

16.3.4.1 Trading Energy with the Grid

One of the factors that compose the first term is the cost or revenue obtained with the energy trading activity between the microgrid and the utility grid. It is determined by the multiplication of the total energy traded by the electricity spot price c_k. Since a purchase is represented by $u_k > 0$, this term represents a cost that should be minimized.

16.3.4.2 Dispatch of Demand Response Events

The cost function also penalizes in the term $L_k^1(U_k)$ the dispatch of demand response events. In an extension of the formulated approach, this term aims to avoid events to be dispatched when the microgrid defines a higher budget for purchase of external energy from the grid, in line with the principle of minimum volatility explained in [6]. This penalization is done through the parameter $r_k > 0$, given in monetary units per price unit, penalizing higher internal prices. Therefore, the first function is given by:

$$L_k^1(U_k) = C_k U_k = \begin{pmatrix} c_k & r_k \end{pmatrix} \begin{pmatrix} u_k \\ \pi_k \end{pmatrix} \tag{16.25}$$

16.3.4.3 Non-planned Consumption of Energy

The function $L_k^2(y_k)$ penalizes the non-planned consumption of electricity through the parameter p_k, given in monetary units per non-planned energy units. This term can be compared with the cost of an emergency purchase of energy. Therefore, it should be necessarily higher than the purchase cost $p_k > c_k$. This function is given by:

$$L_k^2(y_k) = p_k \max\left(0, -A_k^1 y_k - B_k^1 U_k - D_k^1 - V_k^1\right) \tag{16.26}$$

16.3.4.4 Storage Surplus of Energy

The storage surplus stored in the end of each stage is penalized by the function $L_k^3(y_k)$. This penalization is given in monetary units per stored energy units and it is represented by h_k. The function is given by:

$$L_k^3(y_k) = h_k \max\left(0, A_k^1 y_k + B_k^1 U_k + D_k^1 + V_k^1\right) \tag{16.27}$$

16.3.4.5 Budget Overrun

In the final stage, the terminal cost $J_N(y_N)$ is given by the penalization of the budget overrun in the time horizon. Matrices P and Q determine how each final state is penalized. This term is given by:

$$J_N(y_N) = P y_N + Q = \begin{pmatrix} -g & -1 \end{pmatrix} \begin{pmatrix} x_N \\ z_N \end{pmatrix} + \begin{pmatrix} x^{sup} g \\ 0 \end{pmatrix} \tag{16.28}$$

The negative signal -1 aims to penalize this budget extrapolation, since z_N will also be negative in this situation. As in the previous section, we consider x^{sup} as the maximum state of the system.

16.3.4.6 Cost Function per Stage

The cost function per stage $J_k(y_k)$ is given by the sum of the functions L_k^1, L_k^2, L_k^3, and the expectation of cost for the next stage, given the recursive feature of the dynamic programming model. Then, we have:

$$J_k(y_k) = \min_{U_k} \{G_k(y_k) + E[J_{k+1}(y_{k+1})]\} \tag{16.29}$$

16.3.5 Discussion and Future Research

The next steps of this research will be to improve the modeling in the penalization of demand response and to consider technical and quality constraints for energy exchanges as battery voltage. Moreover, we will study which discretization time is better for this approach, as well as explore the convexity of cost functions considering the bi-dimensional aspect for two controls.

16.3.5.1 Mathematical Formulation

Part of the penalization of demand response will be explored in the next steps of this research. We will verify if the current linear penalization with the price should be modified. Moreover, we will explore in the modeling possible technical constraints like voltage control and considerations for reactive power and cost of acquisition of new batteries. Given some physical limitations of the microgrid infrastructure and market features, it will also be studied the inclusion of constraints in the control space, considering limits for energy exchanges, as well as limits in the tariff prices.

16.3.5.2 Optimal Policy

Regarding the optimal policy, we will explore how bi-dimensional controls can affect the convexity of the cost function, impacting in the optimal policy. Moreover, the proposed constraints in the control space can impact directly the optimal policy. Therefore, these considerations should be evaluated when deriving the optimal policy.

16.3.5.3 Computational Limitations

One of the major problems that arise with dynamic programming is the "curse of dimensionality." The increase of state and control spaces can directly impact the solution time of the algorithms and the amount of computational memory required. Powell [30] shows how the increase of the state and control spaces can be computationally problematic.

1. If the state variable $x_t = (x_{t1}, x_{t2}, x_{t3}, \ldots\ldots, x_{tI})$ has I dimensions, and if x_{ti} can have L possible values, then it can have a total of L^I possible states.
2. If the control variable $u_t = (u_{t1}, u_{t2}, u_{t3}, \ldots\ldots, u_{tJ})$ has J dimensions, and if u_{ti} can have M possible values, then it can have a total of M^J possible controls.

Exploring these questions will be extremely important in models with vectorial variables, as proposed in this chapter.

16.4 Conclusion

This book chapter presents an approach for the management of the energy stored in vehicles connected to the microgrid via inventory control problem, and it was structured in three main parts, which reflect our contributions. Firstly, to position our work, we present a thorough review of the literature of V2G connections. Then, by considering the battery management as a stochastic inventory control problem, we develop a dynamic programming model and we obtain an optimal policy that characterizes it. Thirdly, we further explore the baseline model by investigating a scenario in which the microgrid is constrained by a budget defined a priori. Such a budget constraint captures situations when the microgrid profile is risk averse and we discuss the implications of such a scenario. There are several possibilities for future research. For instance, to each of the uncertainties and risks pointed out in our literature review, the decisions in a microgrid can be optimized if uncertainties are reduced or if management can improve its ability to react to their impact (risks) in a timely manner.

Acknowledgements This research was partially supported by the Research Foundation of Minas Gerais State (FAPEMIG), Brazil (Grant Number: PPM-00149-12), by a scholarship supported by the Brazilian Federal Agency for Support and Evaluation of Graduate Education (CAPES) within the Ministry of Education of Brazil, and by the National Council for Scientific and Technological Development (CNPq), Brazil (Grant Number: 473966/2013-1 and 303906/2013-8).

References

1. Andersson, S.L., Elofsson, A., Galus, M., Göransson, L., Karlsson, S., Johnsson, F., Andersson, G.: Plug-in hybrid electric vehicles as regulating power providers: case studies of Sweden and Germany. Energy Policy **38**(6), 2751–2762 (2010)
2. Bertsekas, D.P.: Dynamic Programming and Optimal Control. Athena Scientific, Belmont(1995)
3. Clement-Nyns, K., Haesen, E., Driesen, J.: The impact of charging plug-in hybrid electric vehicles on a residential distribution grid. IEEE Trans. Power Syst. **25**(1), 371–380 (2010)
4. Dallinger, D., Krampe, D., Wietschel, M.: Vehicle-to-grid regulation reserves based on a dynamic simulation of mobility behavior. IEEE Trans. Smart Grid **2**(2), 302–313 (2011)
5. Du, Y., Zhou, X., Bai, S., Lukic, S., Huang, A.: Review of non-isolated bi-directional dc-dc converters for plug-in hybrid electric vehicle charge station application at municipal parking decks. In: 2010 Twenty-Fifth Annual IEEE Applied Power Electronics Conference and Exposition (APEC), pp. 1145–1151. IEEE, Piscataway (2010)
6. Dupont, B., De Jonghe, C., Olmos, L., Belmans, R.: Demand response with locational dynamic pricing to support the integration of renewables. Energy Policy **67**, 344–354 (2014)
7. Farmer, C., Hines, P., Dowds, J., Blumsack, S.: Modeling the impact of increasing PHEV loads on the distribution infrastructure. In: 2010 43rd Hawaii International Conference on System Sciences (HICSS), pp. 1–10. IEEE, Piscataway (2010)
8. Foster, J.M., Caramanis, M.C.: Optimal power market participation of plug-in electric vehicles pooled by distribution feeder. IEEE Trans. Power Syst. **28**, 2065–2076 (2013)
9. Galus, M.D., La Fauci, R., Andersson, G.: Investigating PHEV wind balancing capabilities using heuristics and model predictive control. In: 2010 IEEE Power and Energy Society General Meeting, pp. 1–8. IEEE, Piscataway (2010)
10. Galus, M.D., Zima, M., Andersson, G.: On integration of plug-in hybrid electric vehicles into existing power system structures. Energy Policy **38**(11), 6736–6745 (2010)
11. Galus, M.D., Koch, S., Andersson, G.: Provision of load frequency control by PHEVs, controllable loads, and a cogeneration unit. IEEE Trans. Ind. Electron. **58**(10), 4568–4582 (2011)
12. Green, R.C., Wang, L., Alam, M.: The impact of plug-in hybrid electric vehicles on distribution networks: a review and outlook. Renew. Sustain. Energy Rev. **15**(1), 544–553 (2011)
13. Guille, C., Gross, G.: A conceptual framework for the vehicle-to-grid (v2g) implementation. Energy Policy **37**(11), 4379–4390 (2009)
14. Han, S., Han, S., Sezaki, K.: Development of an optimal vehicle-to-grid aggregator for frequency regulation. IEEE Trans. Smart Grid **1**(1), 65–72 (2010). https://doi.org/10.1109/TSG.2010.2045163
15. Heydt, G.: The impact of electric vehicle deployment on load management strategies. IEEE Trans. Power Apparatus Syst. **PAS-102**(5), 1253–1259 (1983)
16. Hutson, C., Venayagamoorthy, G.K., Corzine, K.A.: Intelligent scheduling of hybrid and electric vehicle storage capacity in a parking lot for profit maximization in grid power transactions. In: 2008 IEEE Energy 2030 Conference, pp. 1–8. IEEE, Piscataway (2008)
17. Kempton, W., Tomić, J.: Vehicle-to-grid power fundamentals: calculating capacity and net revenue. J. Power Sources **144**(1), 268–279 (2005)
18. Kempton, W., Tomić, J.: Vehicle-to-grid power implementation: from stabilizing the grid to supporting large-scale renewable energy. J. Power Sources **144**(1), 280–294 (2005)
19. Kisacikoglu, M.C., Ozpineci, B., Tolbert, L.M.: Examination of a PHEV bidirectional charger system for v2g reactive power compensation. In: 2010 Twenty-Fifth Annual IEEE Applied Power Electronics Conference and Exposition (APEC), pp. 458–465. IEEE, Piscataway (2010)
20. Kramer, B., Chakraborty, S., Kroposki, B.: A review of plug-in vehicles and vehicle-to-grid capability. In: 34th Annual Conference of IEEE Industrial Electronics, IECON 2008, pp. 2278–2283. IEEE, Piscataway (2008)

21. Lopes, J.A.P., Soares, F.J., Almeida, P.M.R.: Integration of electric vehicles in the electric power system. Proc. IEEE **99**(1), 168–183 (2011)
22. Lund, H., Kempton, W.: Integration of renewable energy into the transport and electricity sectors through v2g. Energy Policy **36**(9), 3578–3587 (2008)
23. Madawala, U.K., Thrimawithana, D.J.: A bidirectional inductive power interface for electric vehicles in v2g systems. IEEE Trans. Ind. Electron. **58**(10), 4789–4796 (2011)
24. Madawala, U.K., Schweizer, P., Haerri, V.V.: "Living and mobility" - a novel multipurpose in-house grid interface with plug in hybrid blueangle. In: IEEE International Conference on Sustainable Energy Technologies, ICSET 2008, pp. 531–536. IEEE, Piscataway (2008)
25. Markel, T., Kuss, M., Denholm, P.: Communication and control of electric drive vehicles supporting renewables. In: IEEE Vehicle Power and Propulsion Conference, VPPC'09, pp. 27–34. IEEE, Piscataway (2009)
26. Mitra, P., Venayagamoorthy, G.: Wide area control for improving stability of a power system with plug-in electric vehicles. IET Gener. Transm. Distrib. **4**(10), 1151–1163 (2010)
27. Ota, Y., Taniguchi, H., Nakajima, T., Liyanage, K.M., Baba, J., Yokoyama, A.: Autonomous distributed v2g (vehicle-to-grid) satisfying scheduled charging. IEEE Trans. Smart Grid **3**(1), 559–564 (2012)
28. Peterson, S.B., Apt, J., Whitacre, J.: Lithium-ion battery cell degradation resulting from realistic vehicle and vehicle-to-grid utilization. J. Power Sources **195**(8), 2385–2392 (2010)
29. Pillai, J.R., Bak-Jensen, B.: Integration of vehicle-to-grid in the western Danish power system. IEEE Trans. Sustain. Energy **2**(1), 12–19 (2011)
30. Powell, W.B.: Approximate Dynamic Programming: Solving the Curses of Dimensionality, vol. 703. Wiley, Hoboken (2007)
31. Quinn, C., Zimmerle, D., Bradley, T.H.: The effect of communication architecture on the availability, reliability, and economics of plug-in hybrid electric vehicle-to-grid ancillary services. J. Power Sources **195**(5), 1500–1509 (2010)
32. Rotering, N., Ilic, M.: Optimal charge control of plug-in hybrid electric vehicles in deregulated electricity markets. IEEE Trans. Power Syst. **26**(3), 1021–1029 (2011)
33. Saber, A.Y., Venayagamoorthy, G.K.: Optimization of vehicle-to-grid scheduling in constrained parking lots. In: IEEE Power & Energy Society General Meeting, PES'09, pp. 1–8. IEEE, Piscataway (2009)
34. Sioshansi, R., Denholm, P.: Emissions impacts and benefits of plug-in hybrid electric vehicles and vehicle-to-grid services. Environ. Sci. Technol. **43**(4), 1199–1204 (2009)
35. Sortomme, E., El-Sharkawi, M.A.: Optimal charging strategies for unidirectional vehicle-to-grid. IEEE Trans. Smart Grid **2**(1), 131–138 (2011)
36. Sortomme, E., El-Sharkawi, M.A.: Optimal scheduling of vehicle-to-grid energy and ancillary services. IEEE Trans. Smart Grid **3**(1), 351–359 (2012)
37. Sovacool, B.K., Hirsh, R.F.: Beyond batteries: an examination of the benefits and barriers to plug-in hybrid electric vehicles (PHEVs) and a vehicle-to-grid (v2g) transition. Energy Policy **37**(3), 1095–1103 (2009)
38. Su, W., Eichi, H., Zeng, W., Chow, M.Y.: A survey on the electrification of transportation in a smart grid environment. IEEE Trans. Ind. Inf. **8**(1), 1–10 (2012)
39. Tomić, J., Kempton, W.: Using fleets of electric-drive vehicles for grid support. J. Power Sources **168**(2), 459–468 (2007)
40. Tulpule, P.: Control and optimization of energy flow in hybrid large scale systems-a microgrid for photovoltaic based PEV charging station. Ph.D. Thesis, The Ohio State University (2011)
41. Tulpule, P.J., Marano, V., Yurkovich, S., Rizzoni, G.: Economic and environmental impacts of a PV powered workplace parking garage charging station. Appl. Energy **108**, 323–332 (2013)
42. White, C.D., Zhang, K.M.: Using vehicle-to-grid technology for frequency regulation and peak-load reduction. J. Power Sources **196**(8), 3972–3980 (2011)
43. Wu, C., Mohsenian-Rad, H., Huang, J.: Vehicle-to-aggregator interaction game. IEEE Trans. Smart Grid **3**(1), 434–442 (2012)

Chapter 17
UAVs and Their Role in Future Cities and Industries

Bruno Nazário Coelho

Abstract Unmanned aerial vehicles (UAVs) have been used in several different fields. Each day, new applications for these devices arise. Also called RPA (remotely piloted aircraft) or popularly known by drones, they are becoming cheaper and more accessible to the whole population. Undoubtedly, the UAVs will play a fundamental role in the everyday life of the smart cities, urban areas, and industries, contributing to the improvement of the quality of life in its most different aspects. These devices can provide information acquisition about the industrial process and its equipment, data of raw materials yard, and also information about remote areas. Companies can benefit from UAVs applications either acting directly in the industrial process or simply with the acquisition of aerial images for specific purposes. In this chapter, some of the main applications of UAVs and their role in smart cities and industries as well as their characteristics and some of the most promising developments are presented. Several examples of applications along the production chain of heavy industries and for the primary sector of the economy are approached.

17.1 Introduction

A new era has dawned with the arrival of new technologies and new ways of seeing the world, which is now much more seen from above due to the popularization of portable and low-cost unmanned aerial vehicles (UAVs), which have been used in the most diverse applications, from transportation to the acquisition of important data and information. When remotely piloted, these equipment are also called RPA (remotely piloted aircraft), being a more commonly used term when facing the industrial and professional areas. These devices are popularly known as drones, with their characteristic buzzing sound, these small flying objects have brought joy for children, fun for young and old people, and astonishment for others. They are

B. N. Coelho (✉)
UFSJ - Federal University of São João del-Rei, Ouro Branco, Brazil
e-mail: brunocoelho@ufsj.edu.br

V. N. Coelho et al. (eds.), *Smart and Digital Cities*, Urban Computing,
https://doi.org/10.1007/978-3-030-12255-3_17

also the source of income for many people nowadays, the research object of many studies, and the solution of industrial problems in various sectors.

Its history is older than many may think [1]. The first UAVs arose in the first decades of the last century, aimed mainly at military applications. Since then, its development has been widely diffused, but only in the last decades have it gone beyond these applications.

The wide spreading of portable electronics in the last two decades has brought a new phase to the history of mankind. Together with internet access, it provided a great ease of access to information and knowledge, previously restricted to the most favored classes. Over time, several sensors have been added to these portable electronic devices—especially to smartphones—in order to facilitate their use and provide new and better experiences of navigation and interactivity for the consumers. Consequently, it encouraged the development and cheapening of the sensors (accelerometers, gyroscopes, magnetometer/digital compass, barometer, etc.), which later enabled the production of UAVs at a very affordable cost to a large part of the population.

These aerial vehicles became rapidly popular, initially by hobbyists of aeromodelling, but soon after, their use was extended by companies specialized in filming and photography. They have finally taken over the stores and can be easily purchased by anyone for the most diverse purposes. Today, these equipment are the source of sustenance for numerous independent professionals and companies that provide services that depend on the skills provided by these aerial vehicles.

In a more general aspect in the cities, the administrators and the population itself have sought to integrate the new technologies with daily activities, promoting a great diversification in their uses and applicabilities. For the UAVs it is not different, every day new studies on the use of these equipment in new situations emerge, which often put their capacities on test and push their limits to the extreme.

The industries follow the same pattern, constantly seeking new technologies in order to get ahead of their competitors, due to the high competitiveness and market demand. Although the use of drones in the industry is not directly related to the industrial internet of things (IIoT) and the Industry 4.0, its use in the industrial environment often appears to be the first step towards the use of innovations. This is probably for the reason that drones attract attention, generating a much stronger visual marketing than if compared with the implementation of cyber-physical systems that characterize IIoT and Industry 4.0.

In this chapter, some of the main applications of UAVs will be presented, as well as their role in smart cities and industries, their characteristics, and the most promising developments related to this technology. Several applications along the production chain will be approached to exemplify its use in heavy industries and in the primary sector of the economy.

17.2 Smart Cities

The term smart city has been increasingly used in the last two decades and the number of related works has also expanded exponentially. A large part of the researches emphasizes mainly the technological aspects of the smart city systems, but other studies highlight the social aspects in question. An interesting and detailed bibliometric analysis on smart cities can be seen in [2], where the authors address their growth in the most diverse aspects, their divergences, and subdivisions.

Numerous applications for the use of UAVs in urban centers have been evaluated, and large companies have invested huge amounts of money in research related to incorporating these aerial vehicles to perform routine tasks. In smart cities, UAVs can contribute to achieve greater efficiency in different sectors, working in a distributed way and with task exchange between multiple vehicles, adding technological value while also directly influencing social aspects.

Based on a published work which presents six characteristics of a smart city [3], it is possible to make a correlation between the application of UAVs in smart cities according to these characteristics:

- Smart economy,
- Smart people,
- Smart governance,
- Smart mobility,
- Smart environment, and
- Smart living.

These characteristics are, in fact, six sectors of a society, which can be adopted as umbrella terms that represent several other areas. Correlating the possibilities of UAVs applications with these characteristics of a smart city presented by the authors, it is observed that this technology can certainly influence and play an active role in all of them. The use of UAVs can contribute to the smart economy by promoting competitiveness in the market and providing agility and flexibility in collecting and delivering small products over longer distances and in a short term. Consequently, it stimulates the economy and enables small local businesses to reach a larger public and a larger area of coverage. Increasingly, large retail companies are investing huge amounts of money in the development of new transportation possibilities for small consumer goods, streamlining the purchases and deliveries of their products. Regarding the smart people, the UAVs will help in the acquisition of data and information that will contribute to solve social problems, influencing social and human capital. In relation to the smart mobility aspect, they will support traffic services and local accessibility, as well as be applied in mapping and analyzing the coverage area and signal quality of antennas, expanding the availability of information and communication technology (ICT) infrastructure. In the scope of smart governance, the UAVs will contribute with data and information of scenarios for decision making. Concerning the smart living feature, they may be used in several applications to improve health conditions, as well as facilitating humans daily life. For instance, acting in the transportation of equipment for emergencies, in accidents in remote places, in search operations, in disaster aid, etc. They will provide rapid delivery of medicines to hard-to-reach places. In addition, they will contribute to individual safety with air monitoring, and patrols, providing greater flexibility and coverage area for security officers. The smart living concept also involves houses and buildings. In this sense, it is noteworthy the applications of UAVs for autonomous cleaning of building glass facades and house painting. Finally, with regard to the smart environment, they will act directly in the monitoring and accomplishment of tasks in nature, in parks, in isolated and remote areas, and natural resources. They will provide assistance and rapid relief in case of environmental disasters and search and rescue operations. They will also contribute with the acquisition of information to control pollution, forest fires, and environmental protection.

These are just a few examples of how drones can be applied in the most diverse aspects of a smart city. In some cases, they will act more directly than in others, but in any case they will contribute to a wide range of factors. There is still much to be discussed in relation to norms and legislation for the use of UAVs within cities. Each country has been working on its own legislation, according to its interests, but the main aspect that can limit their use are safety issues.

Drones applications currently occur in an isolated way, with practically no integration and information sharing between the systems. The trend is increasingly for multi-agent systems, where multiple UAVs will work in an integrated and joint manner, being able to be reallocated to several applications in numerous systems in a dynamic way. The complete integration of the equipment and information of several areas will be obtained only over time and after the consolidation of the uses in each of them.

17.3 Smart Industries

The industries are increasingly opening their gates for the application of new technologies that can bring benefits to the productive processes, which currently includes the use of UAVs within the industrial area. Many of them have invested in research for the development of the most diverse applications that can bring more safety and reduce production costs. These applications required more robust and more reliable drones, with higher payloads that meet the requirements for safe flights within the industrial environment. These devices can provide the acquisition of information about the process equipment, data of raw materials yard, and information from remote areas and hard-to-reach locations.

Most companies can benefit from UAV applications, both directly, acting in their process, and indirectly, with simple acquisition of aerial images for specific purposes. In this article, the main focus is on applications for primary and heavy industries, more specifically for applications in the mining production chain and throughout the production process in the steel industry.

17.3.1 Mining Industry

Currently, there are several projects and ongoing research involving the use of UAVs in the mining industry. Most current applications are related to the acquisition of aerial imagery, but new ideas for different applications have been investigated.

Most of the applications are focused on open-pit mining, but researches on mapping underground areas and underground mining can also be found [4]. In underground situations, the major difficulty is the limitations of antennas and GPS signal range, and in simultaneous localization and mapping (SLAM) during autonomous flights.

17.3.1.1 Open-Pit Mining

In the open-pit mining, there is a wide range of possibilities for the use of UAVs. Several researches have been developed in search of improvement in information acquisition and safety aspects.

Currently, the main application of UAVs in open-pit mining is in the mapping of the mine and the mining fronts, through aerial images and the generation of a 3D model of the mine, where several information can be analyzed for blasting, routes for the vehicles, areas of risk, volume of material, among others. In safety matters, inspection routines may be conducted prior to dismantling operations to ensure that there are no people and vehicles in hazardous areas that may be affected by debris material projection during the blasting. Subsequently, after the blasting in the open-pit mine, UAVs can be used for rapid information acquisition with the mapping of the mining front.

The growth in the use of autonomous off-road trucks and vehicles within the mining areas requires reliability in data transmission systems. In this sense, adding signal receivers in UAVs allows the mapping of signal area of antennas and WiFi signal in the mining area, ensuring the quality of the communication between the equipment. Guarantee of good communication systems is essential for autonomous vehicles in the mine.

The autonomous off-road trucks and cars that run in the mines currently require a previous map with a high degree of detail and precision of the areas in which they can move. Before a new area of locomotion is enabled, in order to ensure that this area is safe and ready to be used, a driver usually goes to the location along with a support car and maps it with the vehicle sensors, passing through the place to be released. Although it ensures that unsafe areas are not released for autonomous vehicles, it depends on a human going to the site to do the mapping. In this aspect, UAVs can assist in this work of mapping and release of new areas, reducing the need for people to circulate in the mine areas.

17.3.1.2 Materials Transportation

Among the main applications of UAVs in industrial plants is the monitoring and inspection of material transport systems in all stages of the production process, both raw materials and processed products. One of the most used material transport equipment is conveyor belts, which can reach tens of kilometers in a single industrial installation. These systems require routine inspections, often at remote and high locations. These characteristics highlight the possibility of using UAVs to perform tasks in these equipment, and several researches have been done in this field [5], using many different sensors for this task of inspecting the conveyor belt structure and the rollers.

For long distances, some companies use the transportation carried out by pipelines. In this case, the UAVs can assist inspection operations in remote areas (similar to oil and gas pipelines [6]), being part of the set of equipment transported in support cars for inspection routines.

Monitoring routines in logistics systems for material transport, such as railway inspections [7, 8], can benefit from the use of UAVs carrying sensors for analysis and maintenance of the permanent way.

17.3.1.3 Processing Plant

At processing plants, UAVs are currently used for volume measurement in ore piles [9], inspections, and data acquisition at heights. They can contribute with routine inspections, corrosion, and cracks analysis in building structures, bridges, towers, antennas, large equipment, among others.

17.3.1.4 Dams and Reservoirs

In both water and tailing dams and reservoirs, the UAVs contribute to monitoring along the banks and points of interest, as well as periodic inspections of structures and sensors [10, 11].

Effluent monitoring systems can benefit from the implementation of autonomous sample collection in rivers and dams, using UAVs equipped with sample holders to acquire materials along a trajectory. This equipment promotes a flexible and agile collection, increasing the speed of analysis, from the water collection systems for the production process to the effluent analysis and monitoring of the water quality generated after the process in the industry.

17.3.2 *Steel Industry*

At steel industry, UAVs have been used in several areas throughout the production process, assisting in the acquisition of information from various industrial equipment during their inspection.

17.3.2.1 Raw Materials Yard

In the raw materials yard, one of the main applications of UAVs, besides aerial monitoring, is in the calculation of material volume in the stockpiles. With the acquisition of aerial images through autonomous flight and subsequent processing in a photogrammetry software, the amount of material in the cells is estimated with the required precision.

The same has been applied in scrap yards, scanning the area and estimating the volume of material in the piles, which are usually separated by piles of materials with similar chemical composition.

17.3.2.2 Material Transportation

The materials transport systems at steel industry are very similar to those of mining, mainly composed of rail transport, conveyor belts, and cranes, according to Sect. 17.3.1.2.

The main difference is that in the steel industry, monitoring the temperature of the transported material is a very important information for the production process. Therefore, the use of sensors and thermal cameras in the UAVs provides the acquisition of very relevant data to the production process.

17.3.2.3 Blast Furnace

In the blast furnace area, the UAVs have already been used for structural monitoring, and image acquisition of the pipes in height.

One of the major health risks for people who move around the structures near the blast furnace is the leakage of carbon monoxide (CO). The monitoring and mapping of possible leaks of toxic gases can be done using UAVs equipped with sensors to detect these gases, avoiding the risk of poisoning accidents.

Mechanisms that can benefit from the use of UAVs for routine inspection are the bleeders, which are pressure valves located at the top of the blast furnace, aimed to relieve the high pressure inside of it. As it is a high-risk area for people, the use of equipment for monitoring and inspection brings great benefits.

With the use of thermal cameras, it is possible to map potential structural problems in the blast furnace and in the pipes, both in the heated oxygen pipes from the regenerators for injection in the blast furnace by the tuyeres, as in the exhaust gases pipes at the top of the blast furnace for the cyclones and gas cleaning systems.

17.3.2.4 Steelmaking

Primary and secondary steelmaking, continuous casting, rolling, and other subsequent processes in the steel industry usually occur within large enclosed sheds. Although much more limited, the indoor use of UAVs can also enable the acquisition of information and inspect assets in height. In this complex environment, the most common is to use navigation based on vision systems.

As examples of UAVs, applications in indoor areas, inspection on cranes, winches, etc., as well as corrosion analysis in the structural part of equipment and buildings can be carried out. In addition, it is also possible to perform small tasks inside buildings, in the lighting systems, and sensors in high places.

17.3.3 Cement Industry

In the cement industry, much of what has been discussed in the previous sections on the use of UAVs, in various applications and areas, can also be applied in a quite similar way because of the similarity of much of the equipment throughout the production process. In this sense, application can be related to the entire production and supply chain, such as those of the mining industry: mining; storage of raw materials; production systems; and transportation.

Thermal analysis in height can be made in the heat exchangers in the preheating tower and in the rotary kiln for the production of clinker. It is possible to identify potential sources of issues by temperature variations, indicating areas of erosion, wear, corrosion, refractory collapse, among other common problems.

The monitoring to control the emission of polluting gases can also be done with sensors coupled to the UAVs, contributing to meeting environmental standards along the production chain.

17.3.4 Energy Industry

Numerous applications of UAVs are already in use in the power generation and distribution industry, from structural inspections on offshore oil platforms to the ethanol industry, with monitoring of sugarcane plantations. It also includes the use of UAVs for inspections in chimneys and gas pipelines, monitoring gas leak detection, corrosion, exposed pipes, erosion, among other possible issues.

In the electricity generation and distribution industry many applications can be found nowadays, several of them in the early stages of research, but still showing up as a very promising topic.

17.3.4.1 Electric Power Distribution

The electric power distribution system has a grid with thousands of kilometers of extension, which periodically goes through inspection routines. Some of the data from these inspections can be acquired by monitoring using vertical take-off and landing (VTOL) UAVs, starting from support points or support vehicles and having a good coverage area in flight.

Among the applications in the electric energy transmission sector, it is worth mentioning the image monitoring and infrared (thermographic) camera in the transmission lines (LTs), in their connections and substations of energy. Moreover, structural and corrosion analyses in the towers contribute to reduce the risk of accidents.

An application that interests the concessionaires of transmission and distribution of energy is in the monitoring of vegetation overgrowth, controlling the vegetation under transmission lines.

17.3.4.2 Hydroelectric Plants

In hydroelectric plants, one of the main applications is routine monitoring in dams, performing actions similar to those described in Sect. 17.3.1.4. UAVs can be used for structural inspections at dams and spillways, detecting cracks, corrosion, and other potential issues in structures.

17.3.4.3 Thermoelectric Plants

Sensor-equipped UAVs have now been used for tower and chimney inspections in many industries, including thermoelectric power plants. When the production is stopped or reduced, inspections of the chimney coating can be made, being a much simpler operation than those involving human beings.

Monitoring the emission of pollutant gases can also be performed by UAVs, including mapping the area of gas dispersion in the region.

17.3.4.4 Wind Power Plants

In wind power plants, UAVs can assist height inspections in a very practical and efficient way. With various sensors, they can detect erosion, cracks, and other defects in blade surfaces.

Unusual but useful applications of UAVs come up every day, solving problems in a simple and practical way, like using an industrial drone to de-ice a wind turbine [12].

The structure of the generation tower can also be analyzed with sensors embedded in UAVs, performing inspection operations quickly and safely.

17.3.4.5 Photovoltaic

The giant farms of solar panels require constant inspection and maintenance of these panels. A single section of a damaged panel can completely hamper the production of that panel.

From aerial images with thermal cameras it is possible to detect faults in the panels in a simple and agile manner, reducing the losses of energy generation by defects in the panels.

17.3.5 Agriculture

One of the first applications of UAVs in the industry occurred in the agriculture sector, initially with acquisition of simple aerial images, and later with multi-spectral cameras and other tools of remote sensing and monitoring of the plantations, allowing a fast and efficient control in all phases of the plantation cycle.

Several studies have been developed with applications of UAVs for precision agriculture [13, 14].

The identification of affected areas for pest control in large plantations can be aided by information acquired from aerial images. On a small scale, the control itself can be aided by the localized spray of pest control products by UAVs.

17.4 Conclusions

Nowadays, the influence of UAVs is evident, which becomes clear with the great diversity of applications and researches in progress in many different areas. The growing trend of UAVs usage is also evident due to the gains made with its applications, enabling the acquisition of information in a fast and flexible manner and performing tasks without the need of human involvement.

Undoubtedly, the UAVs will play a fundamental role in the everyday life of the smart cities, urban areas, and industry, contributing to the improvement of the quality of life in its most different aspects.

In the industries, the UAVs contribute to the acquisition of data about the productive processes and the detection of leaks, security, and potential sources of issues. For long-distance monitoring and inspection, the trend is for fixed-wing UAVs, or VTOLs, which can take off from suport cars in the regions of inspections.

Therefore, UAVs can bring numerous advantages, but all this will only be possible if adequate rules and legislation are developed, controlling their use in a way that can contribute to society.

Acknowledgements The author would like to thank the CAPES / ITV support—call 20/2016.

References

1. Valavanis, K.P., Kontitsis, M.: A historical perspective on unmanned aerial vehicles. In: Valavanis, K.P. (ed.) Advances in Unmanned Aerial Vehicles. Intelligent Systems, Control and Automation: Science and Engineering, vol. 33. Springer, Dordrecht (2007)
2. Mora, L., Bolici, R., Deakin, M.: The first two decades of smart-city research: a bibliometric analysis. J. Urban Technol. **24**(1), 3–27 (2017). https://doi.org/10.1080/10630732.2017.1285123
3. Giffinger, R., Ferter, C., Kramar, H., Kalasek, R., Pichler-Milanovic, N., Meijers, E.: Smart cities: ranking of European medium-sized cities, report from Centre of Regional Science, Vienna (2007)
4. Freire, G.R., Cota, R.F.: Capture of images in inaccessible areas in an underground mine using an unmanned aerial vehicle. In: Hudyma, M., Potvin, Y. (eds.) Proceedings of the First International Conference on Underground Mining Technology. Australian Centre for Geomechanics, Crawley (2017)
5. Nascimento, R., Carvalho, R., Delabrida, S., Bianchi, A., Oliveira, R., Garcia, L.: An integrated inspection system for belt conveyor rollers - advancing in an enterprise architecture. In: Proceedings of the 19th International Conference on Enterprise Information Systems V.2: ICEIS, pp. 190–200 (2017). ISBN 978-989-758-248-6. https://doi.org/10.5220/0006369101900200
6. Gomez, C., Green, D.R.: Small-scale airborne platforms for oil and gas pipeline monitoring and mapping. UCEMM - University of Aberdeen Report (2015)
7. Flammini, F., Pragliola, C., Smarra, G.: Railway infrastructure monitoring by drones. In: 2016 International Conference on Electrical Systems for Aircraft, Railway, Ship Propulsion and Road Vehicles & International Transportation Electrification Conference (ESARS-ITEC), France (2017). https://doi.org/10.1109/ESARS-ITEC.2016.7841398
8. Wu, Y., Qin, Y., Wang, Z., Jia, L.: A UAV-based visual inspection method for rail surface defects. Appl. Sci. **8**, 1028 (2018). https://doi.org/10.3390/app8071028
9. Raeva, P.L., Filipova, S.L., Filipov, D.G.: Volume computation of a stockpile - a study case comparing GPS and UAV measurements in AN open pit quarry. In: XXIII ISPRS Congress, Czech Republic (2016). https://doi.org/10.5194/isprsarchives-XLI-B1-999-2016
10. Ridolfi, E., Manciola, P.: Water level measurements from drones: a pilot case study at a dam site. Water **10**, 297 (2018). https://doi.org/10.3390/w10030297
11. Wang, K.L., Huang, Z.J.: Discover failure mechanism of a landslide dam using UAV. In: ICCCBE2016 16th International Conference on Computing in Civil and Building Engineering, Japan (2016)
12. Aerones: Wind turbine de-ice (2018). https://www.aerones.com/eng/wind_turbine_maintenance_drone. Cited 16 Oct 2018
13. Zhang, C., Kovacs, J.M.: The application of small unmanned aerial systems for precision agriculture: a review. Precis. Agric. **13**, 693. https://doi.org/10.1007/s11119-012-9274-5
14. Mogili, U.R., Deepak, B.B.V.L.: Review on application of drone systems in precision agriculture. Procedia Comput. Sci. **133** (2018). https://doi.org/10.1016/j.procs.2018.07.063

Chapter 18
Critical Systems for Smart Cities: Towards Certifying Software

Erick Grilo and Bruno Lopes

Abstract Critical systems require high reliability and are present in many applications. Standard techniques of software engineering are not enough to ensure the absence of unacceptable failures and/or that critical requirements are fulfilled. Verifying and certifying systems for Smart Cities is one of the challenges that still require some effort. Smart Cities models may be seen as Cyber-Physical Systems and they may be formalized as Finite State Machines. We discuss how to reason over these models as Finite State Machines formalized in a logical background from which it is possible to provide certified software for the Smart Cities domain.

18.1 Does Software Certification Matters?

Critical systems are systems in which failure may result in loss of life, significant destruction, high financial loss, or environmental damage [16]. In short, systems that need a high level of reliability. There are many examples of critical systems applied in a wide range of areas, from medical devices to nuclear systems.

Standard software engineering techniques are not designed to deal with fault non-tolerant systems. In many domains such systems need a way to ensure its safety in order to guarantee that the system indeed meet the required reliability. Formal systems compose a theoretical and implemented background able to model and reason about systems, ensuring (mathematically) that requirements are fulfilled and that systems behave as expected.

Formal systems were used to certify Paris Metro line 1 (Paris's subway system in France) [11], leading it to a fully automated subway system, eliminating the need of building another subway line, saving millions of dollars. Airbus uses formal methods in order to certify avionics control systems of its families of aircrafts A318 and A340-500/60 [5, 27].

E. Grilo (✉) · B. Lopes
Instituto de Computação, Universidade Federal Fluminense, Niterói, Brazil
e-mail: simas_grilo@id.uff.br; bruno@ic.uff.br

V. N. Coelho et al. (eds.), *Smart and Digital Cities*, Urban Computing,
https://doi.org/10.1007/978-3-030-12255-3_18

The absence of such certification in critical systems may lead to catastrophic scenarios: between 1985 and 1987, the Therac-25 medical electron accelerator was involved in (at least) six radiation overdoses occurrences [22]. This episode led several people to death and injured many others. This happened due to a combination of factors, in which we can cite overconfidence of the software engineers and the lack of certification [21]. In March 2018, a Uber's driverless car initiative was involved in an accident that killed a pedestrian in Tempe, Arizona.[1]

Many modern systems are becoming safety critical. Financial loss and even deaths can result from their failure [16]. Hence, since the end of the 1980s and early 1990s, researches have been directed on applying formal methods for critical systems [16, 24, 26].

A common approach for Smart Cities is to look at them as Cyber-Physical Systems [6, 12, 28]. Cyber-Physical Systems propose the integration of computational elements and sensors to interfere in the "cyber" and in the "physical" environments.

Ensuring safety and reliability of Cyber-Physical Systems is one of their main challenges [18]. Modelling them in logic-based systems makes possible to reason about them and take advantage of a wide theoretical and software framework.

The usage of logic systems to model and reason about Cyber-Physical Systems seems to be a promising approach [13]. Proof assistants [23], like Coq [4, 9] and Isabelle [25], lead to the possibility of automatizing the verification of such systems and provide certified code. They are computational tools that implement usually logic systems. Their design is tailored to automatize many (when possible all) steps of proofs. The theoretical background leads to see proofs as programs and programs as proofs. It is used to transform a proof that requirements are fulfilled in a model in a certified code.

Industrial examples of the usage of proof assistants are already present in companies as Mistubishi [7] and NASA.[2] Certified compilers developed using proof assistants are also in use [19].

Coq [4, 9] is one of the most prominent proof assistants. It deals with a high-level language to model, prove, and provide certified code automatically.

This work extends the usage of Coq to model instances of software interaction in Smart Cities domain by means of Cyber-Physical system as Finite State Machines [13], a standard approach [1, 10, 15, 28]. We focus on the strategy proposed by Privat et al. [28]. The usage of RGCoq library [13] (detailed on Sect. 18.3) reduces the effort of implementing the model constructions.

This chapter reviews and extends the discussion of Grilo and Lopes [13]. In the following sections, we review with more details the definitions and present the methodology of using Coq to provide certified code for Smart Cities. The automata model is extended to simplify the modelling process. It is well known that formal systems use to have high computational complexity (we remember that

[1]https://www.nytimes.com/2018/03/19/technology/uber-driverless-fatality.html.

[2]https://shemesh.larc.nasa.gov/fm/ftp/larc/PVS-library/.

even classical propositional logic SAT is NP-Complete). Trying to overcome this complexity, we introduce the certification of minimality of automata.

18.2 Smart Cities Domain Formalization

The adopted approach is to look at Smart Cities models as Cyber-Physical Systems [6, 8, 28]. The following methodology was presented by Privat et al. [28].

Instances of Smart Cities are similar to each other in a general sense (in the sense that every city has roads, streetlights, traffic lights, cameras, and other appliances that are intrinsic to Smart Cities). The difference between instances lies in the specific selection of the smart objects that compose an instance and on how these objects interact with each other.

To make the model more generic, a instance of a Smart City is described as the composition of those elements, where each element corresponds to a generic version of the real-life modelled objects, assuming that they belong to such a generic category. Figure 18.1 shows an example of a simple instance containing two roads, each traffic light in each road (denoted by the green and the red dots) and one of the roads containing a smart parking spot.

In Privat et al. [28], Smart Cities are modelled by categorizing its elements in two different groups.

1. City devices are devices that do not have a meaningful state on their own, where the only information relevant for modelling purposes are the states of the entities they keep track of, serving as controllers of the entities to which they are attached. Presence and heat sensors, USB WiFi dongles, photoelectric sensors, and electric switches are examples of such devices.
2. City entities are subsystems (entities) in which it is classified "useful" city equipment, such as traffic lights, street lights, toll, and train barriers. Such entities may be connected to each other (a set of traffic lights in adjacent streets that are simultaneously controlled) through sensors, since in the presented model

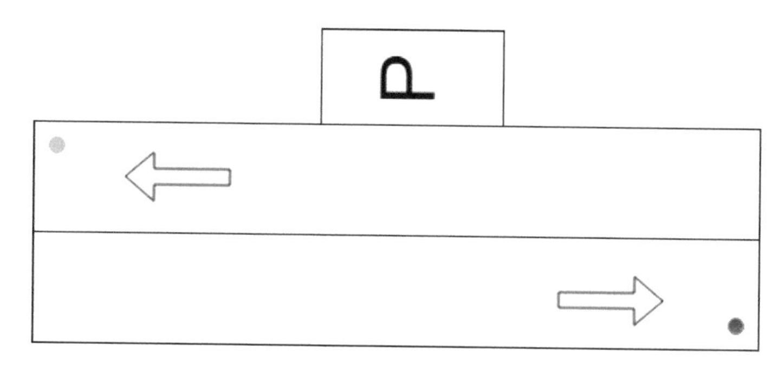

Fig. 18.1 Map of two roads in a Smart City

such entities does not have a network connection on their own and they may be monitored by means of available networked sensors.

Moving entities such as vehicles, people, or animals do not need to be accounted in the process of modelling. They may be treated as event triggers for modelled entities (e.g., a pedestrian passing by a streetlight with a presence sensor may trigger the event of turning on the light). If finer-grain modelling requires representing them individually, they can be modelled as entities identified by a unique proxy ID, not being necessary to use real data in order to ensure that they are not accounted twice.

The process of modelling Smart Cities in the context of Finite State Machines must capture real-time properties of the modelled entities without focusing into further details of the entity (such as internal functioning or the entity's manufacturer). Properties may be abstracted in terms of states and transitions between those states, removing uninteresting complex details about the modelled subsystems. A traffic light may be modelled with three states: "green," "yellow," and "red." Transitions between those states may be a temporizer which after a short time changes the state of the light.

Besides modelling city entities individually, the complexity of a Smart City with its interconnected entities can be unfolded. By using a graph, a city itself can be modelled as instances of related entities in different levels: a street lamp may contain a photoelectric sensor, while streets can be adjacent to each other. It is interesting to note that devices and non-modelled entities (such as flying birds or pedestrians) may be also represented in this graph because they may take a relevant role in the functioning of modelled entities.

Therefore, Privat et al. [28] present a framework which models instances of open Cyber-Physical Systems (also discussed by Felipe et al. [10]) where its main objective is to structure the middleware between ICT applications and instances of open Cyber-Physical Systems (Smart Cities, Smart Homes, or other environment similar to these).

The usage of Finite State Machines in the process of modelling is attractive because of the simplicity of the model, where at the same time it captures many properties one would like to model in Smart Cities domain, as Privat et al. [28] also point. In the end of this chapter we discuss how to extend to more robust automata formalism. This framework is defined as follows.

Definition 1 (Smart City Framework) The proposed framework by Privat et al. [28] is structured through three different "planes" as follows.

1. *physical plane* contains the physical entities and the sensors used by them, where the entities in this plane can be all sorts of material things, like legacy city appliances or relevant subspaces of a city;
2. *proxy instance plane* holds the instances of software representation for the entities in the physical plane, where external applications also interact with the system through this plane (a general view on the proxy instance plane is described in Fig. 18.2). The entities here modelled are instantiated with

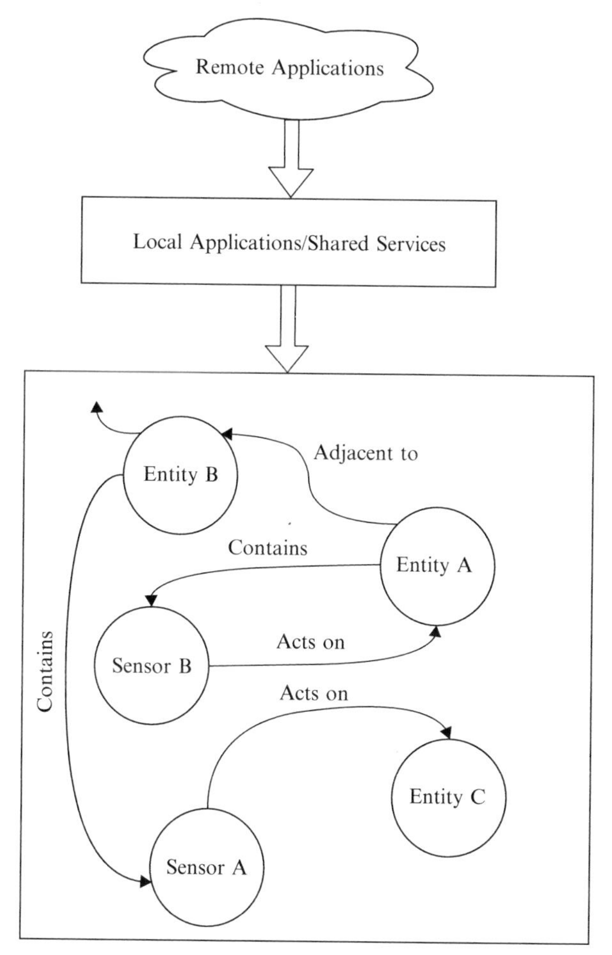

Fig. 18.2 An example of the structure of a proxy instance plane

information gathered from the model plane to provide a generic model of the entity modelled;
3. *model plane* has the generic reference models upon which a given representation relies, where external ontologies and/or knowledge bases would interface on this plane, gathering more generic information about the entities modelled. Here, entities are categorized as part of a generic taxonomy that can be organized as a directed acyclic graph that holds where all entities modelled in the proxy instance plane belong to.

The work proposed here focuses on the plane that captures the modelled entities and its relationship with their sensors (and with other entities), *proxy instance plane* that is organized in five different layers.

1. *devices proxy layer* is the layer that provides an interface to networked devices, independent on how they are networked. It contains the mapping of one to one

of the city devices that are in the physical plane, serving as actuators and sensors for city entities on that plane;
2. *entity proxy layer* is the layer that contains the entities themselves mapped from a given generic model that is in the *model plane*;
3. *entity groups proxy layer* is the layer that captures relationship between different entities that can be modelled as a group according to a given property;
4. *services and local applications layer* provides a layer that contains applications that operates on more than one entity mapped in the entity proxy layer. Also represented in this layer are applications that need to be close enough to the real environment because of specific constraints, e.g., physical constraints;
5. *instance graph* is the graph that contains all important relations between different city entities where it is possible to observe relationships such as entities that are adjacent to others, entities that belong to others, and where sensors observe/actuate.

In Sect. 18.3 we discuss its formalization in Coq.

18.3 Formalization in Coq

In this section we present how to formalize the concepts presented in Sect. 18.2 in Coq. We make use of Coq 8.7 (the by now up-to-date version) with the RGCoq[3] library to formalize them as Finite Automata. We take advantage of the simplicity of the Finite Automata model and its natural interpretation in Coq by means of RGCoq.

Hence, RGCoq lets their users to formalize and prove properties about models of software interaction for Smart Cities in Coq, having all apparatus that Coq offers at hand. Figure 18.3 shows a view of CoqIDE, Coq's official IDE.

RGCoq is a Coq library tailored to formalize and reason about Finite Automata. In what follows we present its background and how to model Cyber-Physical Systems using it. After that we show how to automatically transform it in certified code in Scheme, Haskell, and OCaml. The following methodology was proposed by Grilo and Lopes [13].

A Finite Automata is an abstract machine defined as below.

Definition 2 (Finite Automata) A Finite Automata (FA) is a tuple $A = (Q, \Sigma, \delta, q_0, Q_f)$ where

Q is a finite set of states,
Σ is the alphabet of the language,
δ : $Q \times \Sigma \rightarrow 2^Q$ is the transition relation,
q_0 is the initial state, and
Q_f is a set $Q_f \subseteq Q$ of final states.

[3] http://github.com/simasgrilo/RGCoq.

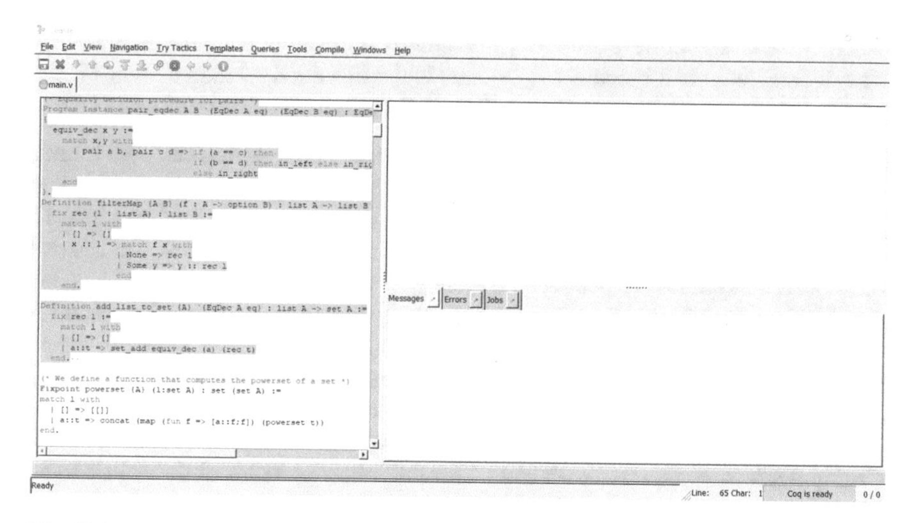

Fig. 18.3 CoqIDE, the official IDE for Coq

The *proxy instance plane* presented in Definition 1 as a Finite Automaton is achieved according to the following definition.

Definition 3 (Cyber-Physical Systems as FA)

- *Behavior* of a Cyber-Physical System can be modelled as states of a FA: a streetlight may have two states "on" and "off," representing the states where the light is on and the light is off, respectively. When modelling relation between entities, states of a FA can be seen as the related entities. As for the instance graph presented in Sect. 18.2, one can model the related entities as states of a FA.
- *Changing* of behavior of a Cyber-Physical System can be modelled as the transition relation of a FA. As for the instance graph, the relations of the entities can be formalized as transitions of the FA.

This definition is formalized in Coq as

Definition 4 (Cyber-Physical Systems as DFA in Coq)

```
Variable (S A : Type).
Context `{EqDec S eq} `{EqDec A eq}.
Record t := DFA {
  initial_state : S;
  is_final : S → bool;
  next : S → A → S;
  states : set S;
  alphabet : set A;
}.
```

where *initial_state* corresponds to q_0, *is_final* is a function that verifies if a given state is final, e.g., if it belongs to Q_f, *next* is the transition relation δ, *states* is the set Q, and *alphabet* is the set Σ according to Definition 2.

Because Definition 4 is defined inside a section within a module (both named *dfa*) in RGCoq, the usage of "`Variable` (S A : `Type`)" declares to Coq that inside this section it is expected two variables in runtime of type `Type` (`Type` being the most general type in Coq) which will define the states of the FA and the alphabet of the FA, respectively. By using "`Context` '{*EqDec S eq*} '{*EqDec A eq*}," Coq expects that (in runtime) both types denoted by the variables S and A to have an equality relation defined for its inhabitants (the usage of *EqDec* to define this equality relation is discussed in some detail in Sect. 18.4).

The `Record` defined in Definition 4 is parametrized by variables S and A. Hence, when instantiating this record, one may supply a variable of any kind to *initial_state*. Coq's inference system will then expect for the following fields of the record to be filled: whenever there is a variable S in the record's definition, a variable of the same type of the one supplied to *initial_state* may be supplied (the same happens in the case of A). The field *is_final* expects a function of type $S \rightarrow bool$ (a function that takes an argument of type S and returns a boolean value), to the transition relation *next* must be given a function of type $S \rightarrow A \rightarrow S$ (a function that takes two arguments, one of type S and another of type A and returns a value of type S). To both *states* and *alphabet* one has to give variables of type *set S* and *set A* (where *set* denotes sets implemented as lists in the library ListSet[4]).

For Deterministic Finite Automata, RGCoq defines functionalities such as *run* within the module *DFA* in which it is possible to check whether the given automaton recognizes the word given as input, *path* that shows all the states of a run with a given input on a DFA, and *get_all_reachable_states_from_a_state*, which with a DFA and a state returns all states that are reachable from the state given. This last definition is crucial to check the relationships between city entities in the instance graph.

The conversion from NFA to DFA may lead to an exponential grown of the model, sometimes with many redundant states. To reduce this problem, the minimization algorithm for DFA may be applied to identify and collapse equivalent states.

RGCoq also implements the verification of the presence of equivalent states. The function *is_minimal* certifies whether a DFA does not contain any equivalent states by implementing the algorithm described in [14]. Despite that in a worst-case analysis the model will be exponential, it is common that the model may be reduced. This leads to the possibility of using the smallest version of the modelled system before certifying it (i.e., reason about a smaller model).

RGCoq also defines the notion of Nondeterministic Finite Automata (NFA) as shown by the following record:

[4]https://coq.inria.fr/library/Coq.Lists.ListSet.html.

Definition 5 (NFA in RGCoq)

```
Variable (S A : Type).
Context `{EqDec S eq} `{EqDec A eq}.
Record t := NFA {
  initial_state : S;
  is_final : S → bool;
  next : A → S → set S;
  states : set S;
  alphabet : set A;
}.
```

where all definitions are the same as the ones presented in Definition 4, except for *next*, which now expects a function of type $A \to S \to set\ S$, representing the transition relation of a NFA, thus mapping a state and a terminal symbol to a set of states $Q_i \subseteq Q$.

RGCoq also defines for NFA the same functionalities described for DFA. The modelling using NFA is simpler and Coq supports nondeterminism. RGCoq will internally make it deterministic, in order to provide deterministic equivalent code. Therefore, functionalities defined for DFA (such as minimal verification and code extraction) can be applied to the instance modelled as NFA (after converting).

Nondeterministic Automata with ϵ-transitions are also formalized in RGCoq:

Definition 6 (NFA with ϵ-Transitions)

```
Variables ST A: Type.
Context `{EqDec ST eq} `{EqDec A eq}.
Record t := NFA_e {
  initial_state : ST;
  is_final : ST → bool;
  next : ST → set (nfa_epsilon_transitions.ep_trans ST A);
  states : set ST;
  alphabet : set A
}.
```

where the first two lines of code and all the fields of the record but *next* are defined as in Definition 5. Here, *next* (δ) is defined as a function that with a state of the NFA returns a set of possible transitions encapsulated by *ep_trans* (an inductive type defined in module *nfa_epsilon_transitions*), the ones that take a terminal as input and the ones that do not, called ϵ-transitions:

```
Inductive ep_trans S A :=
  | Epsilon : S → ep_trans S A
  | Goes : A → S → ep_trans S A.
```

with *Empty* as the constructor that encapsulates ϵ-transitions and *Goes*, the constructor that stands for the transitions defined with a symbol in the NFA's alphabet, therefore formalizing possible transitions for NFA with ϵ-transitions.

Using this formalization we may ensure properties inherent to each system and more general as termination, liveness, and absence of deadlocks. In Sect. 18.4 we show some examples where we assure some properties and provide certified code.

18.4 Some Examples

Grilo and Lopes [13] present the following two first examples from a discussion in Privat et al. [28]: the scenario of crossroads as an instance of open Cyber-Physical System. It is presented how to use Coq to model and reason about the aforementioned scenarios.

18.4.1 A Simple Parking Spot System

Figure 18.4 presents the modelling of an individual parking spot with a sensor that detects if the spot is taken by a car or not.

A single parking place can be either occupied or empty. The parking spot may be occupied only if there is a car on it and empty otherwise. This behavior is captured by the following inductive definition in Coq

`Inductive` *parkingPlace* :=
empty | *occupied*.

where *empty* denotes the state the spot is empty (there is no car parked there) and *occupied* stands for the case there is a car parked in the spot.

The sensor which acts in this parking place can either observe a car entering or leaving the parking place.

`Inductive` *sensorActivity* :=
carEntersSpot | *carLeavesSpot*.

with *carEntersSpot* meaning that the sensor in the parking place received the information that a car has entered the parking spot and *carLeavesSpot* denotes that the sensor has detected that a car inside the parking spot has left.

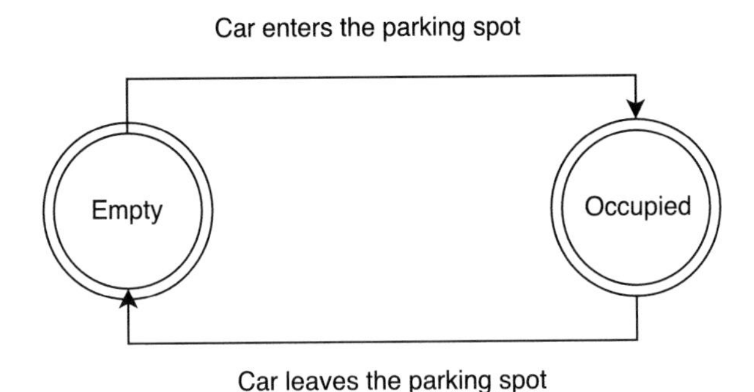

Fig. 18.4 A simplified model of a sensor acting in a parking place

Then, it is possible to define the behavior of the sensor which is monitoring the parking place: if the spot is "empty" and a car arrives and parks in it, the sensor receives information that a car is in the spot and it is now "occupied"; if the spot is "occupied," then the sensor may only detect information if the car leaves the spot, where it now knows that the parking place is "empty." Any other sequence of events should not trigger a change in the model's configuration. Therefore, by modelling this way the sensor must not be able to detect a car coming into the parking spot if there is already a car in there and also may not be able to detect the parking place being empty if it is already empty. The described behavior can be defined as follows.

```
Definition sensorChangesbehavior (s: option
parkingPlace) (a: sensorActivity) : option parkingPlace :=
match s with
  | Some empty ⇒ match a with
                  | carEntersSpot ⇒ Some occupied
                  | carLeavesSpot ⇒ None
                  end
  | Some occupied ⇒ match a with
                  | carEntersSpot ⇒ None
                  | carLeavesSpot ⇒ Some empty
                  end
  | None ⇒ None
end.
```

where it is used the *option* type available in Coq (a type that encapsulates an optional value; it is the same as *maybe* type in Haskell) to encapsulate the invalid transitions as discussed before.

The second line in Definition 4 tells Coq that for types S and A there will be an equality relation defined in runtime for the elements of both types. For this example, S is *parkingPlace* and A is *sensorActivity*. To define this relation, we use the tactic `Program`[5] which lets one write code as if in a regular programming language along with the proof that the written code meets its specification.

By using *Program*, the definition is saved in the proof context if all generated proof obligations are proven. Therefore, the equality relation for the elements in *parkingPlace* will be defined by using `Program` and can (but not limited to) be defined as follows:

```
Program Instance parkingEqDec : EqDec
  parkingPlace eq :=
     {equiv_dec x y :=
           match x, y with
           | empty,empty ⇒ in_left
           | occupied,occupied ⇒ in_left
```

[5] https://coq.inria.fr/refman/program.html.

```
            | empty,occupied ⇒ in_right
            | occupied,empty ⇒ in_right
            end
    }.
```

with *in_left* and *in_right* being notations for the constructors *left* and *right* of the inductive type *sumbool* (a type of booleans with a proof of their values). *EqDec*[6] is a class that gives a equality relation over a given type and *eq* is the equality relation between two elements of the same type.

The equality relation for *sensorActivity* can be defined by the same means:

```
Program Instance sensorEqDec : EqDec sensorActivity eq :=
    {equiv_dec x y :=
            match x, y with
            | carEntersSpot,carEntersSpot ⇒ in_left
            | carLeavesSpot, carLeavesSpot ⇒ in_left
            | carLeavesSpot,carEntersSpot ⇒ in_right
            | carEntersSpot,carLeavesSpot ⇒ in_right
            end
    }.
```

For the sake of simplicity, the model of the parking place lets both "empty" and "occupied" be final (valid, accepting) states for the model. The only non-valid state is denoted by *None*, one of the constructors of the *option* type in Coq:

```
Definition final (s: option parkingPlace) :=
  match s with
  | None ⇒ false
  | _ ⇒ true
  end.
```

Now it is possible to define a FA according to Definition 4:

```
Definition parkingSpotFA : dfa.t _ _ := {|
  dfa.initial_state := Some empty;
  dfa.is_final := final;
  dfa.next := sensorChangesbehavior;
  dfa.states := [Some empty; Some occupied];
  dfa.alphabet := [carEntersSpot;carLeavesSpot]
|}.
```

By defining the model as above, it is possible to prove certain properties about it. The following Lemma ensures that the model will not go to a invalid state only if the right conditions stated in the model's definition are met: it will go to the state "occupied" iff it is in the state "empty" and the sensor detects a car entering the parking place and it will only go to the state "empty" if it is in the state "occupied" and the sensor does not detect a car in the parking spot.

[6]https://coq.inria.fr/library/Coq.Classes.EquivDec.html.

```
Lemma sensorbehaviorSound : ∀ s: option parkingPlace, ∀ a: sensorActivity,
sensorChangesbehavior s a ≠ None ↔ (s = Some empty ∧ a = carEntersSpot)
∨ (s = Some occupied ∧ a = carLeavesSpot).
Proof.
intros;split.
- intros. destruct s. destruct p.
  + destruct a. left;split. reflexivity. reflexivity.
    simpl in H. destruct H. reflexivity.
  + right. destruct a. simpl in H. destruct H;reflexivity.
    split;reflexivity.
  + destruct H. simpl;reflexivity.
- intros. destruct H as [H2 | H3]. destruct H2. rewrite H. rewrite
H0. simpl. congruence.
  destruct H3. rewrite H. rewrite H0. simpl. congruence.
Qed.
```

By means of Extraction,[7] one can transform the verified model to code in languages as Scheme, Objective Caml and Haskell after verifying the desired properties in the formalized model. To enable extraction of certified code, one should use:

```
Require Extraction.
```

Therefore, commands such as `Extraction` *id* lets the user to have a preview of the extracted code in Coq's own console and `Extraction` "file" id_1 id_2 ... id_n that extracts recursively all terms from id_1 to id_n in the file named "file" become enabled (including all terms that id_I relies on, such as definitions used by id_i where $i = 1, 2, \ldots, n$).

For the shown example, the target extraction language used is Scheme. To set up this (the default target language is Objective Caml), one must use:

```
Require Extraction.
Extraction Language Scheme.
```

then

```
Extraction "parkingSpotCertified" parkingSpotFA
```

generates a .scm file (parkingSpotCertified.scm) containing the code of Listing 18.1.

Listing 18.1 Certified code for a parking spot generated by Coq

```
;; This extracted scheme code relies on some additional macros
;; available at http://www.pps.univ-paris-
;;              diderot.fr/~letouzey/scheme
(load "macros_extr.scm")
```

[7] https://coq.inria.fr/refman/extraction.html.

```
(define sensorChangesBehaviour (lambdas (s a)
  (match s
    ((Some p)
      (match p
        ((Empty)
          (match a
            ((CarEntersSpot) `(Some ,`(Occupied)))
            ((CarLeavesSpot) `(None))))
        ((Occupied)
          (match a
            ((CarEntersSpot) `(None))
            ((CarLeavesSpot) `(Some ,`(Empty)))))))
    ((None) `(None)))))

(define final (lambda (s)
  (match s
    ((Some _) `(True))
    ((None) `(False)))))

(define parkingSpotFA `(DFA ,`(Some ,`(Empty)) ,final
    ,sensorChangesBehaviour ,`(Cons ,`(Some ,`(Empty)) ,`(Cons
  ,`(Some ,`(Occupied)) ,`(Nil))) ,`(Cons ,`(CarEntersSpot)
      ,`(Cons ,`(CarLeavesSpot) ,`(Nil)))))
```

Listing 18.1 corresponds to the code written in Coq for Example 18.4.1, as stated by Letouzey [20]. To execute this code, one must follow the instructions given in the beginning of the extracted code.

18.4.2 Street Parking Spots by Distance

Taking advantage of the native compositional approach of Finite Automata, this proposal scales by composing states. Figure 18.5 illustrates another parking spot example where two roads cross each other and each road contains three different parking places, each of them has a sensor that presents the same behavior described in Sect. 18.4.1.

This example focuses on another formalism which is part of the framework proposed by Privat et al. [28] in which it uses the *instance graph* (as presented in Sect. 18.2) where one can capture the relationship between different entities of a city. The importance of this graph lies mainly on the property that it is possible to verify which entities may be affected by a given action, limiting the scope of such actions and being possible to avoid unwanted changes or potentially cascading side effects on entities interconnected by this graph.

Since it is not known (and not interesting to fix) in which road one can start verifying properties about the model, it can be modelled as a Nondeterministic Finite Automaton (NFA) with ϵ-transitions.

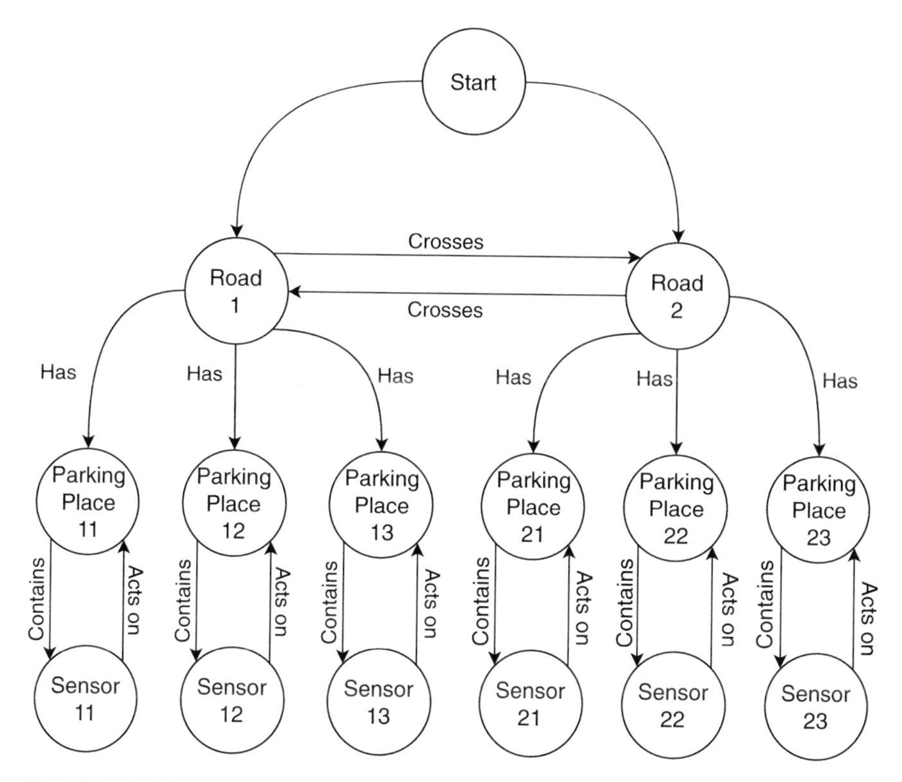

Fig. 18.5 Modelling of two roads crossing each other, each road having three parking spots

This scenario models a subset of a Smart City map that contains smart parking spots and its related entities. By modelling this situation, it is possible to validate how those entities interact with each other.

Since the non-labeled transitions can be triggered without any action taken (in other words, the verification can start in either road 1 or in road 2), it is possible to model it as ϵ-transitions. By making use of Definition 6, the system presented in Fig. 18.5 is formalized as:

```
Inductive systemStates :=
| start | road1 | road2 | parkingPlace11 | parkingPlace12 | parkingPlace13
| parkingPlace21 | parkingPlace22 | parkingPlace23 | sensor11 | sensor12 |
sensor13 | sensor21 | sensor22 | sensor23.

Inductive systemTransitions :=
| crosses | contains | actsOn | has.

Definition transitionNFA (state : systemStates) :=
  match state with
  | start ⇒ [Epsilon road1;Epsilon road2]
```

```
  | road1 ⇒ [Goes has parkingPlace11; Goes has parkingPlace12; Goes has
parkingPlace13]
  | road2 ⇒ [Goes has parkingPlace21; Goes has parkingPlace22; Goes has
parkingPlace23]
  | parkingPlace11 ⇒ [Goes contains sensor11]
  | parkingPlace12 ⇒ [Goes contains sensor12]
  | parkingPlace13 ⇒ [Goes contains sensor13]
  | parkingPlace21 ⇒ [Goes contains sensor21]
  | parkingPlace22 ⇒ [Goes contains sensor22]
  | parkingPlace23 ⇒ [Goes contains sensor23]
  | sensor11 ⇒ [Goes actsOn parkingPlace11]
  | sensor12 ⇒ [Goes actsOn parkingPlace12]
  | sensor13 ⇒ [Goes actsOn parkingPlace13]
  | sensor21 ⇒ [Goes actsOn parkingPlace21]
  | sensor22 ⇒ [Goes actsOn parkingPlace22]
  | sensor23 ⇒ [Goes actsOn parkingPlace23]
end.
Definition finalStates (s: systemStates) :=
  true.
```

with *systemStates* as the states of the FA, *systemTransitions* represents the transitions of the system, and *transitionNFA* is the transition relation of the modelled automaton. By simplicity, the final states are defined as a function that given a state, it always returns *true* (all states are valid, accepting states). Therefore, the model is defined in Coq as follows:

```
Definition roadsParkingNFA := {|
  nfa_epsilon.initial_state := start;
  nfa_epsilon.is_final := finalStates;
  nfa_epsilon.next := transitionNFA;
  nfa_epsilon.states := [start;road1;road2;parkingPlace11;
                         parkingPlace12;parkingPlace13;
                         parkingPlace21;parkingPlace22;
                         parkingPlace23;sensor11;sensor12;
                         sensor13;sensor21;sensor22;
                         sensor23];
  nfa_epsilon.alphabet := [crosses;contains;actsOn;has] |}.
```

To simulate such formalism, RGCoq provides a procedure (named *nfa_e_to_nfa*) that converts NFA with ϵ-transitions to NFA without those transitions according to [14], in order to use definitions formalized for NFA. Hence:

`Definition` *roadsParkingNFAnoEpsilon* := *nfa_epsilon.nfa_e_to_nfa roadsParkingNFA*.

creates an instance of a NFA from the ϵ-NFA modelled above. For this conversion to be done, one has to supply the equality relation as required by the second line in Definition 6 the same way it was required for Example 18.4.1.

RGCoq includes a C program in which given a inductive type of the states and its inhabitants returns the equality relation by using *EqDec* as done in Example 18.4.1. For this example, it generates 225 different "proof obligations" (here, each proof obligation carries on the proof that either two related elements of the inductive type are equal or not, corresponding to a line in the definition of the equality relation). After creating the NFA, it is possible to verify and prove properties about its specification.

Therefore if one needs to prove that in the specified model "Road 1" has exactly three parking spots, one can simply prove it by using:

```
Lemma road1_has_three_parking_spots : length (nfa.next roadsParkingNFAnoEpsilon has road1) = 3.
Proof. reflexivity. Qed.
```

It may be interesting to verify that an entity interacts directly with another: say, for example, that one wants to check that "road 1" contains the parking place named "parking place 13":

```
Lemma road1_has_parkingplace13 : In parkingPlace13 (nfa.next roadsParkingNFAnoEpsilon has road1).
Proof. cbv. right. right. left. reflexivity. Qed.
```

or that both roads must have more than two parking spots

```
Lemma both_roads_two_spots : length (nfa.next roadsParkingNFAnoEpsilon has road1) > 2 ∧ length (nfa.next roadsParkingNFAnoEpsilon has road1) > 2.
Proof. split.
- cbv. reflexivity.
- cbv. reflexivity.
Qed.
```

It is also possible to verify which entities are affected by others: RGCoq offers a procedure (named *nfa.get_all_related_states*) that given a NFA and a state of this NFA, it returns all states (entities) reachable from this state.

```
Lemma road2_sees_sensor23 : In sensor23 (nfa.get_all_related_states roadsParkingNFAnoEpsilon road2).
Proof. cbv. left. reflexivity. Qed.
```

After verifying the given model, certified code can be obtained the same way as described in Example 18.4.1. The target extraction language is also Scheme. Hence, by using

```
Extraction "roadsParkingNFACertified" roadsParkingNFA
```

the certified extracted code that represents the model introduced as an ϵ-NFA is shown in Listing 18.2.

Listing 18.2 Certified code for the model presented in Fig. 18.5 modelled in and genreated by Coq

```
;; This extracted scheme code relies on some additional macros
```

```
;; available at http://www.pps.univ-paris-
;;                diderot.fr/~letouzey/scheme
(load "macros_extr.scm")


(define transitionNFA (lambda (state)
  (match state
    ((Start) `(Cons ,`(Epsilon ,`(Road1)) ,`(Cons ,`(Epsilon
        ,`(Road2)) ,`(Nil))))
    ((Road1) `(Cons ,`(Goes ,`(Crosses) ,`(Road2)) ,`(Cons
        ,`(Goes ,`(Has) ,`(ParkingPlace11)) ,`(Cons ,`(Goes
      ,`(Has) ,`(ParkingPlace12)) ,`(Cons ,`(Goes ,`(Has)
        ,`(ParkingPlace13)) ,`(Nil))))))
    ((Road2) `(Cons ,`(Goes ,`(Crosses) ,`(Road1)) ,`(Cons
        ,`(Goes ,`(Has) ,`(ParkingPlace21)) ,`(Cons ,`(Goes
      ,`(Has) ,`(ParkingPlace22)) ,`(Cons ,`(Goes ,`(Has)
        ,`(ParkingPlace23)) ,`(Nil))))))
    ((ParkingPlace11) `(Cons ,`(Goes ,`(Contains) ,`(Sensor11))
        ,`(Nil)))
    ((ParkingPlace12) `(Cons ,`(Goes ,`(Contains) ,`(Sensor12))
        ,`(Nil)))
    ((ParkingPlace13) `(Cons ,`(Goes ,`(Contains) ,`(Sensor13))
        ,`(Nil)))
    ((ParkingPlace21) `(Cons ,`(Goes ,`(Contains) ,`(Sensor21))
        ,`(Nil)))
    ((ParkingPlace22) `(Cons ,`(Goes ,`(Contains) ,`(Sensor22))
        ,`(Nil)))
    ((ParkingPlace23) `(Cons ,`(Goes ,`(Contains) ,`(Sensor23))
        ,`(Nil)))
    ((Sensor11) `(Cons ,`(Goes ,`(ActsOn) ,`(ParkingPlace11))
        ,`(Nil)))
    ((Sensor12) `(Cons ,`(Goes ,`(ActsOn) ,`(ParkingPlace12))
        ,`(Nil)))
    ((Sensor13) `(Cons ,`(Goes ,`(ActsOn) ,`(ParkingPlace13))
        ,`(Nil)))
    ((Sensor21) `(Cons ,`(Goes ,`(ActsOn) ,`(ParkingPlace21))
        ,`(Nil)))
    ((Sensor22) `(Cons ,`(Goes ,`(ActsOn) ,`(ParkingPlace22))
        ,`(Nil)))
    ((Sensor23) `(Cons ,`(Goes ,`(ActsOn) ,`(ParkingPlace23))
        ,`(Nil))))))

(define finalStates (lambda (_) `(True)))

(define roadsParkingNFA `(NFA_e ,`(Start) ,finalStates
    ,transitionNFA ,`(Cons ,`(Start) ,`(Cons ,`(Road1) ,`(Cons
  ,`(Road2) ,`(Cons ,`(ParkingPlace11) ,`(Cons
      ,`(ParkingPlace12) ,`(Cons ,`(ParkingPlace13) ,`(Cons
  ,`(ParkingPlace21) ,`(Cons ,`(ParkingPlace22) ,`(Cons
      ,`(ParkingPlace23) ,`(Nil)))))))))) ,`(Cons ,`(Crosses)
  ,`(Cons ,`(Contains) ,`(Cons ,`(ActsOn) ,`(Cons ,`(Has)
      ,`(Nil)))))))
```

The code for the NFA without ϵ transitions generated and used in the example can be obtained the same way. For the present work, it is not annexed because of the size of the file obtained: by using

`Extraction` "roadsParkingNFACertified"

roadsParkingNFAnoEpsilon Coq will extract this definition recursively. In other words, it will also extract all definitions that *roadsParkingNFAnoEpsilon* relies on.

18.4.3 Smart Cars in Smart Cities

Another example following the idea presented in Sect. 18.4.2 is modelling a subarea of a Smart City containing a road and, say, three electric cars parked on recharging stations placed in this road. We also model some other roads containing these stations, being possible to verify the capability of these cars to go from the first road to the target road, considering if they have enough charge.

Figure 18.6 shows an overview of the aforementioned situation. Based on it, we model a scenario where there are four roads between them and there is a car parked in a recharging station in the first road. In this configuration, we can prove properties such as that a given car contains enough battery charge to reach the next road containing the target recharge station.

The modelling of the entities involved is similar to the other examples. We can define them as an inductive type as follows.

`Inductive` *cityEntities* :=
road1 | *road2* |*road3* | *road4* | *chargeTotem1* | *chargeTotem2* | *chargeTotem3* | *chargeTotem4* | *chargeTotem5* | *chargeTotem6* | *car*: *nat* → *cityEntities*.

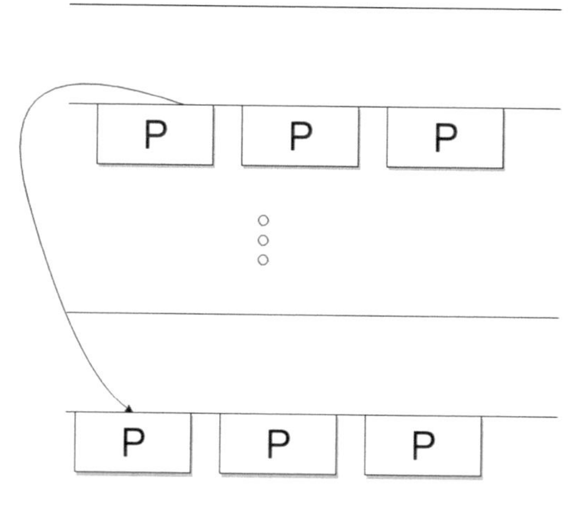

Fig. 18.6 Graphic description of the situation presented

It is interesting to point that the entity *car* denotes the car as a function that takes as argument a natural number, which is the amount of battery charge the car has. The relation between these entities is formalized as

`Inductive` *cityRelations* := *contains*| *isParked* | *adjacent*.

We consider that the model has a nondeterministic behavior in the sense that a given car can head to any of the charging stations that are positioned in the target road. Also, since the roads are interconnected, the car has the possibility to go forth and back between those roads. Therefore, we can model the situation as follows.

```
Definition nfaex := {|
  nfa.initial_state := car 5;
  nfa.is_final := final;
  nfa.next := cityInteractions;
  nfa.states := setEntities;
  nfa.alphabet := setRelations |}.
```

where the car we want to reason about is the initial state, final is a function that verifies the car reached the target road (in this example, it is the last road, *road4* that contains the tokens), *cityInteractions* is the definition that denotes how the entities interact with each other, *setEntities* is a set that contains all entities modelled, and *setRelations* is a set containing all relations. Figure 18.7 shows the instance graph of the modelled situation. For a matter of simplicity, it does not show *chargeTotem5* and *chargeTotem6* (although they are present).

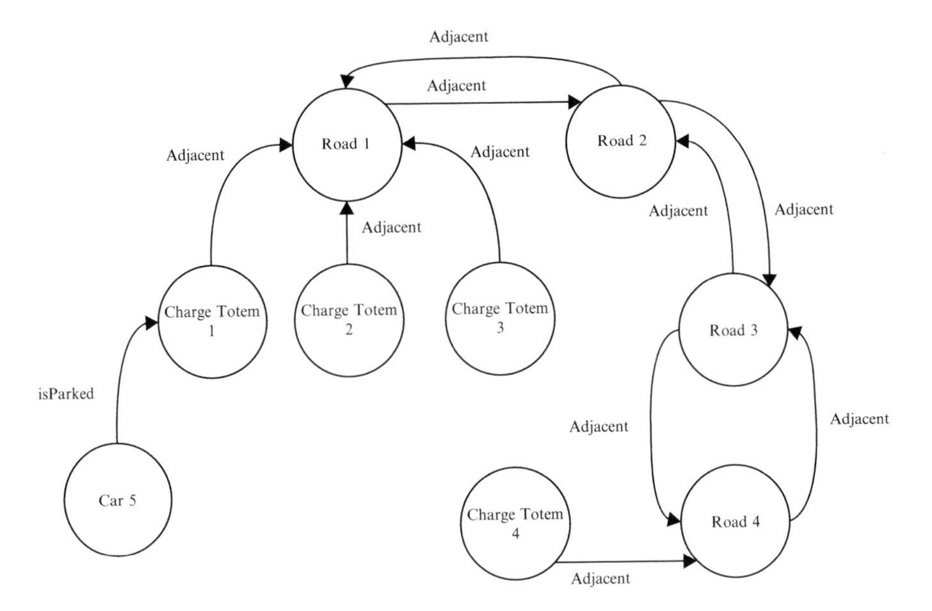

Fig. 18.7 Instance graph for the presented example

In order to certify that the car will have enough fuel to reach the desired recharge station, we consider that for each unit of fuel the vehicle has, it can pass through a road in the modelled scenario.[8]

By ensuring that the roads are interconnected by the relation *adjacent* with

`Lemma` *roadsInterconnected* : *cityInteractions adjacent road1* = [*road2*] ∧ *cityInteractions adjacent road2* = [*road1*;*road3*] ∧ *cityInteractions adjacent road3* = [*road2*;*road4*] ∧ *cityInteractions adjacent road4* = [*road3*], we can run the automaton to certify that the car has enough fuel to head to its destiny.

In the instance hereby modelled, the car we want to model contains five units of battery charge. By considering a step that consumes the battery, we only take in account the roads that the car goes by during the run. Therefore, since the trace of the automaton considers the car itself (since it is the initial state) and the charging station it is parked on is *road1*, we consider the two first entities (*car1* and *road1*) not to be battery-consuming situations. Therefore

`Lemma` *carCanReachRoad4* : (*length* (*nfa.path nfaex* [*isParked*;*adjacent*;*adjacent*;*adjacent*;*adjacent*])) - 2 < 5,

is an example of a property that can be formalized and proved, where *carCanReachRoad4* is a lemma that asserts that the modelled car in *nfaex* has enough battery to reach the last road (Road 4 in Fig. 18.7).

18.5 An Overview of the Approach

Critical systems need their safety ensured in order to guarantee that they will not provoke physical or financial damage caused by failure. This can be obtained by mathematically ensuring that the model is sound following its specification and other desired properties.

The approach hereby presented models Smart Cities software interaction by modelling them as Finite State Machines by means of Finite Automata based on the framework proposed by Privat et. al. [28]. It formalizes such theory in a proof assistant, enabling the possibility of model, and proves properties about the instances modelled in a computational environment.

These are only the first steps on automatizing this approach, consisting of a simple yet effective framework to model Smart Cities software interaction [13]. This approach can also be extended in order to ease its usage and to be able to model richer scenarios in specific domains. The scenario shown in Sect. 18.4.3 can be extended to reason about more cars, automatically returning the optimal solution of which car to use.

In order to increase the expressibility of the framework, implementations of more expressive automata models, such as Constraint Automata [2, 3, 17] are

[8]More realistic fuel decreasing functions may be modelled, such as an exponential decreasing based on time, topography, etc.

in development and will constitute valuable tools. Such extensions may enable the model to capture more properties, such as the ones based on how entities exchange data, interacting with each other. They extensions will lead to a richer framework, capable of modelling and certifying more properties in the Smart Cities domain.

Acknowledgements This work was partially supported by CNPq, CAPES, and FAPERJ.

References

1. Ai, Y., Peng, M., Zhang, K.: Edge computing technologies for internet of things: a primer. Digital Commun. Netw. **4**(2), 77–86 (2018)
2. Arbab, F.: Reo: a channel-based coordination model for component composition. Math. Struct. Comput. Sci. **14**(3), 329–366 (2004)
3. Arbab, F.: Coordination for component composition. Electron. Notes Theor. Comput. Sci. **160**, 15–40 (2006). Proceedings of the International Workshop on Formal Aspects of Component Software (FACS 2005)
4. Bertot, Y., Castéran, P.: Interactive Theorem Proving and Program Development: Coq'Art: The Calculus of Inductive Constructions. Springer, Berlin (2013)
5. Bochot, T., Virelizier, P., Waeselynck, H., Wiels, V.: Model checking flight control systems: the airbus experience. In: 31st International Conference on Software Engineering-Companion Volume, 2009. ICSE-Companion 2009, pp. 18–27. IEEE, Piscataway (2009)
6. Cassandras, C.G.: Smart cities as cyber-physical social systems. Engineering **2**(2), 156–158 (2016)
7. de Souza Silva, N.: Verificação formal de sistemas embarcados em carro elétrico. Master's thesis, Universidade Federal de Goiás (2015)
8. Derler, P., Lee, E.A., Vincentelli, A.S.: Modeling cyber–physical systems. Proc. IEEE **100**(1), 13–28 (2012)
9. Dowek, G., Felty, A., Herbelin, H., Huet, G., Murthy, C., Parent, C., Paulin-Mohring, C., Werner, B.: The COQ Proof Assistant: User's Guide: Version 5.6. INRIA (1992)
10. Felipe, E., Santana, Z., Chaves, A.P., Gerosa, M.A., Kon, F., Milojicic, D.S.: Software platforms for smart cities: concepts, requirements, challenges, and a unified reference architecture. ACM Comput. Surv. **50**(6), 1–78 (2017)
11. Gerhart, S., Craigen, D., Ralston, T.: Case study: Paris metro signaling system. IEEE Softw. **11**(1), 28–32 (1994)
12. Ghaemi, A.A.: A cyber-physical system approach to smart city development. In: IEEE International Conference on Smart Grid and Smart Cities (2017)
13. Grilo, E., Lopes, B.: Formalization and certification of software for smart cities. In: International Joint Conference on Neural Networks (IJCNN), pp. 662–669. IEEE, Piscataway (2018)
14. Hopcroft, J.E., Motwani, R., Ullman, J.D.: Automata Theory, Languages, and Computation, vol. 24, International edn. Addison-Wesley, Boston (2006)
15. Klein, C., Kaefer, G.: From smart homes to smart cities: opportunities and challenges from an industrial perspective. In: Balandin, S., Moltchanov, D., Koucheryavy, Y. (eds.) Next Generation Teletraffic and Wired/Wireless Advanced Networking, pp. 260–260. Springer, Berlin (2008)
16. Knight, J.C.: Safety critical systems: challenges and directions. In: Proceedings of the 24th International Conference on Software Engineering, pp. 547–550. ACM, New York (2002)
17. Kokash, N., Arbab, F.: Formal Behavioral Modeling and Compliance Analysis for Service-Oriented Systems, pp. 21–41. Springer, Berlin (2009)

18. Lee, E.A.: Cyber physical systems: design challenges. In: 11th IEEE International Symposium on Object Oriented Real-Time Distributed Computing (ISORC), Orlando, 5–7 May 2008 (2008)
19. Leroy, X.: Formal certification of a compiler back-end or: programming a compiler with a proof assistant. In: ACM SIGPLAN Notices, vol. 41, pp. 42–54. ACM, New York (2006)
20. Letouzey, P.: A new extraction for coq. In: International Workshop on Types for Proofs and Programs, pp. 200–219. Springer, Berlin (2002)
21. Leveson, N., et al.: Medical devices: The therac-25. Appendix of: Safeware: System Safety and Computers. Addison-Wesley, Boston (1995)
22. Leveson, N.G., Turner, C.S.: An investigation of the therac-25 accidents. IEEE Comput. **26**(7), 18–41 (1993)
23. Loveland, D.W.: Automated Theorem Proving: a Logical Basis. Elsevier, New York (2014)
24. Milner, R.: Some directions in concurrency theory. Futur. Gener. Comput. Syst. **88**, 163–164 (1988)
25. Nipkow, T., Paulson, L.C., Wenzel, M.: Isabelle/HOL: A Proof Assistant for Higher-Order Logic, vol. 2283. Springer, Berlin (2002)
26. Ostro, J.S.: Formal methods for the specification and design of real-time safety critical systems. J. Syst. Softw. **18**(1), 33–60 (1992)
27. Peleska, J.: Formal methods for test automation-hard real-time testing of controllers for the airbus aircraft family. In: IDPT'02, vol. 1 (2002)
28. Privat, G., Zhao, M., Lemke, L.: Towards a shared software infrastructure for smart homes, smart buildings and smart cities. In: International Workshop on Emerging Ideas and Trends in Engineering of Cyber-Physical Systems (2014)

Printed in the United States
By Bookmasters